MAMMALS

of

CANADA

Tamara Eder • Gregory Kennedy

Lone Pine Publishing

Lone Pine Publishing
10145 – 81 Avenue
Edmonton, Alberta T6E 1W9

Website: www.lonepinepublishing.com

Library and Archives Canada Cataloguing in Publication

Eder, Tamara, 1974–
    Mammals of Canada / Tamara Eder, Gregory Kennedy.

Includes index.
ISBN 978-1-55105-857-3 (bound).
ISBN 978-1-55105-856-6 (pbk.)

    1. Mammals—Canada—Identification. I. Kennedy, Gregory, 1956– II. Title.

QL721.E34 2011   599.0971   C2010-905474-1

Editorial Director: Nancy Foulds
Project Editor: Nicholle Carrière
Technical Reviewer: David Kirk, PhD
Editorial: Genevieve Boyer, Nicholle Carrière, Kelsey Everton, Wendy Pirk
Production Coordinator: Gene Longson
Layout & Production: Volker Bodegom, Janina Kuerschner
Maps: Volker Bodegom, Janina Kuerschner
Cover Design: Gerry Dotto
Cover Photo: Mountain Lion, © 2010 John Giustina / Getty Images
Photo Management: Randy Kennedy
Photo Editing: Kamila Kwiatkowska, Gary Whyte
Photo & Illustration Credits: All reasonable efforts have been made to obtain permission
from respective copyright holders for the images used in this book. A complete list of
credits can be found on page 447, which constitutes an extension of this copyright page.

We acknowledge the financial support of the Government of Canada through the
Canada Book Fund (CBF) for our publishing activities.

PC: 20

Contents

Acknowledgements

Over the years, Lone Pine Publishing's field guides have developed into standard references for plants and animals, and have become an underlying force for the appreciation and conservation of wildlife in North America. Thanks are extended to the growing family of professional and backyard naturalists who have offered their inspiration and expertise to help build Lone Pine's expanding library. Thanks go to Chris Fisher and Don Pattie for their contributions to previous books in this series. In addition, thanks go to Gary Ross, Ian Sheldon, Kindrie Grove and George Penetrante, whose skilled illustrations have brought each page to life. Thank you as well to the photographers (listed on page 447) for graciously allowing us to include so many of their images on these pages.

We are very grateful to Nicholle Carrière, Volker Bodegom and the Lone Pine staff for transforming a wide-ranging collection of words, maps, drawings and photographs into another exceptional Lone Pine nature guide that will allow Canadians to learn about and enjoy wild mammals in their backyards and afield.

American Bison
p. 38

Mountain Goat
p. 42

Muskox
p. 46

Bighorn Sheep
p. 50

Dall's Sheep
p. 54

Pronghorn
p. 58

Fallow Deer
p. 62

Wapiti
p. 64

Mule Deer
p. 68

White-tailed Deer
p. 72

Moose
p. 76

Caribou
p. 80

Feral Pig
p. 84

Feral Horse
p. 86

HOOFED MAMMALS

WHALES, DOLPHINS & PORPOISES

Harbour Porpoise
p. 92

Dall's Porpoise
p. 96

Atlantic White-sided
Dolphin
p. 100

Pacific White-sided Dolphin
p. 102

Orca
p. 106

Beluga
p. 110

Narwhal
p. 114

North Atlantic
Right Whale
p. 116

Bowhead Whale
p. 120

Humpback Whale
p. 124

Grey Whale
p. 128

CARNIVORES

Mountain Lion
p. 134

Canada Lynx
p. 138

Bobcat
p. 142

CARNIVORES

Western Spotted Skunk
p. 146

Striped Skunk
p. 148

American Marten
p. 150

Fisher
p. 152

Short-tailed Weasel
p. 156

Long-tailed Weasel
p. 158

Black-footed Ferret
p. 160

Least Weasel
p. 162

American Mink
p. 164

Wolverine
p. 166

American Badger
p. 170

Northern River Otter
p. 174

Sea Otter
p. 178

Raccoon
p. 182

CARNIVORES

Harbour Seal
p. 186

Ringed Seal
p. 190

Grey Seal
p. 192

Harp Seal
p. 194

Bearded Seal
p. 196

Hooded Seal
p. 198

Northern Elephant Seal
p. 200

Walrus
p. 204

Northern Fur Seal
p. 208

Northern Sea Lion
p. 210

California Sea Lion
p. 214

Black Bear
p. 216

Grizzly Bear
p. 220

Polar Bear
p. 224

CARNIVORES

Coyote
p. 228

Grey Wolf
p. 232

Arctic Fox
p. 236

Swift Fox
p. 240

Red Fox
p. 244

Grey Fox
p. 248

RODENTS

Nutria
p. 254

North American Porcupine
p. 256

Meadow Jumping Mouse
p. 260

Western Jumping Mouse
p. 261

Pacific Jumping Mouse
p. 262

Woodland Jumping Mouse
p. 263

Western Harvest Mouse
p. 264

Keen's Mouse
p. 265

RODENTS

White-footed Mouse
p. 266

Deer Mouse
p. 267

Northern
Grasshopper Mouse
p. 269

Bushy-tailed Woodrat
p. 270

Norway Rat
p. 272

Black Rat
p. 274

House Mouse
p. 275

Southern Red-backed Vole
p. 277

Northern Red-backed Vole
p. 278

Western Heather Vole
p. 279

Eastern Heather Vole
p. 280

Rock Vole
p. 281

Long-tailed Vole
p. 282

Singing Vole
p. 283

Montane Vole
p. 284

Prairie Vole
p. 285

Tundra Vole
p. 286

Creeping Vole
p. 287

Meadow Vole
p. 288

Woodland Vole
p. 289

Water Vole
p. 290

Townsend's Vole
p. 292

Yellow-cheeked Vole
p. 293

Sagebrush Vole
p. 294

Muskrat
p. 295

Brown Lemming
p. 297

Northern Bog Lemming
p. 298

Southern Bog Lemming
p. 299

Northern
Collared Lemming
p. 300

Ungava
Collared Lemming
p. 301

RODENTS

Victoria
Collared Lemming
p. 302

Ogilvie Mountain
Collared Lemming
p. 303

Richardson's
Collared Lemming
p. 304

American Beaver
p. 305

Olive-backed Pocket Mouse
p. 309

Great Basin Pocket Mouse
p. 310

Ord's Kangaroo Rat
p. 311

Northern Pocket Gopher
p. 313

Plains Pocket Gopher
p. 315

Yellow-pine Chipmunk
p. 316

Least Chipmunk
p. 318

Red-tailed Chipmunk
p. 320

Townsend's Chipmunk
p. 321

Eastern Chipmunk
p. 322

Hoary Marmot
p. 324

Yellow-bellied Marmot
p. 326

Woodchuck
p. 328

Vancouver Island Marmot
p. 330

Columbian
Ground Squirrel
p. 332

Franklin's
Ground Squirrel
p. 334

Golden-mantled
Ground Squirrel
p. 336

Arctic
Ground Squirrel
p. 338

Richardson's
Ground Squirrel
p. 339

Cascade Golden-mantled
Ground Squirrel
p. 341

Thirteen-lined
Ground Squirrel
p. 342

Black-tailed Prairie Dog
p. 344

Eastern Grey Squirrel
p. 346

Eastern Fox Squirrel
p. 348

Douglas's Squirrel
p. 349

Red Squirrel
p. 350

RODENTS

Northern Flying Squirrel
p. 352

Southern Flying Squirrel
p. 354

Mountain Beaver
p. 355

HARES & PIKAS

Eastern Cottontail
p. 358

Mountain Cottontail
p. 360

European Rabbit
p. 362

Snowshoe Hare
p. 364

Arctic Hare
p. 366

European Hare
p. 368

White-tailed Jackrabbit
p. 370

Collared Pika
p. 372

American Pika
p. 373

BATS

California Myotis
p. 377

Western Small-footed Myotis
p. 378

Long-eared Myotis
p. 379

Keen's Myotis
p. 380

Eastern Small-footed Myotis
p. 381

Little Brown Myotis
p. 382

Northern Myotis
p. 384

Fringed Myotis
p. 385

Long-legged Myotis
p. 386

Yuma Myotis
p. 387

Western Red Bat
p. 388

Eastern Red Bat
p. 389

Hoary Bat
p. 390

Silver-haired Bat
p. 392

Eastern Pipistrelle
p. 393

Big Brown Bat
p. 395

Spotted Bat
p. 396

BATS

Townsend's
Big-eared Bat
p. 397

Pallid Bat
p. 399

INSECTIVORES & OPOSSUMS

American Shrew Mole
p. 402

Coast Mole
p. 403

Townsend's Mole
p. 404

Hairy-tailed Mole
p. 405

Eastern Mole
p. 406

Star-nosed Mole
p. 407

Arctic Shrew
p. 409

Pacific Water Shrew
p. 410

Masked Shrew
p. 411

Long-tailed Shrew
p. 412

Smoky Shrew
p. 413

Gaspé Shrew
p. 414

Hayden's Shrew
p. 415

Pygmy Shrew
p. 416

Maritime Shrew
p. 417

Merriam's Shrew
p. 418

Dusky Shrew
p. 419

American Water Shrew
p. 420

Preble's Shrew
p. 422

Olympic Shrew
p. 423

Trowbridge's Shrew
p. 424

Tundra Shrew
p. 425

Barren-ground Shrew
p. 426

Vagrant Shrew
p. 427

Northern
Short-tailed Shrew
p. 428

Least Shrew
p. 429

Virginia Opossum
p. 430

Terrestrial and Marine Ecoregions of Canada

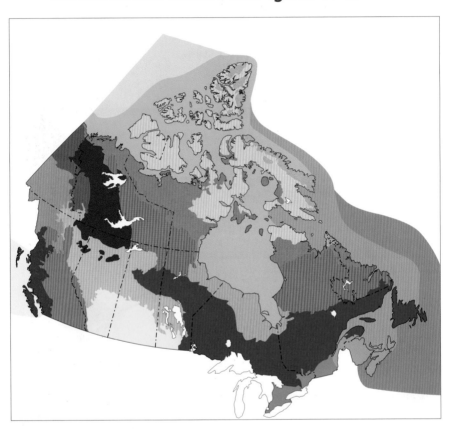

Arctic and Taiga

 Arctic Cordillera

Northern Arctic

Southern Arctic

Taiga Shield

Taiga Plain

Hudson Plain

Pacific and Western Mountains

Taiga Cordillera

Boreal Cordillera

Pacific Maritime

Montane Cordillera

Central Plains

 Boreal Plain

Prairie

Boreal Shield

Boreal Shield

Mixedwood Plain

Mixedwood Plain

Atlantic

Atlantic Maritime

Marine

Pacific Maritime

Arctic Basin

Arctic Archipelago

Northwest Atlantic Marine

Atlantic Marine

Introduction

Few things characterize wilderness as well as wild animals, and few animals are more recognizable than our fellow mammals. In fact, many people use the term "animal" when they really mean "mammal"—they forget that birds, reptiles, amphibians, fish and all the many kinds of invertebrates are animals, too.

Mammals come in a wide variety of colours, shapes and sizes, but they all share two characteristics that distinguish them from the other vertebrates—only mammals have real hair, and only mammals nurse their young from mammary glands (the feature that gives this group its name). Other, less well-known features that are unique to mammals include a muscular diaphragm, which separates the lower abdominal cavity from the cavity that contains the heart and lungs, and a lower jaw that is composed of a single bone on each side. Additionally, a mammal's skull joins the first vertebra at two points of contact—a bird's or reptile's skull has only one point of contact, which is what allows birds to turn their heads so far around. As well as setting mammals apart from all other kinds of life, these characteristics also identify humans as part of the mammalian group.

Whether you want to watch a Sea Otter playing in a kelp bed, catch a glimpse of a Muskox on the tundra or listen to the haunting sound of a Grey Wolf's howl, the varied habitats of Canada provide many spectacular mammal-watching opportunities. Despite the pressures of human development, much of Canada's forests, grasslands and mountain regions are internationally recognized destinations for visitors who are interested in rewarding natural experiences.

To honour this treasure is to celebrate North America's intrinsic virtues, and this book is intended to provide readers with the knowledge they need to appreciate the rich variety of mammals in this country. Whether you are a naturalist, a photographer, a wildlife enthusiast or all three, you will find terrific opportunities everywhere in Canada to satisfy your greatest wilderness expectations.

The Natural Regions

The natural regions of Canada are extremely diverse. This country, the second largest in the world, represents a little more than 40 percent of North America's land area and encompasses a dramatic variety of landscapes. Extensive boreal forests, clear blue lakes, pristine rivers, rolling grasslands, diverse deciduous forests, expansive tundra and icy Arctic waters all contribute to the scenic beauty and ecological uniqueness of this country. The most easterly point of Canada is Cape Spear in Newfoundland, and the most westerly point is Mount St. Elias in the Yukon Territory, giving Canada a total width of 9306 km. Canada's most northerly point (only 725 km south of the North Pole) is Cape Columbia on Ellesmere Island, and its most southerly point is Middle Island in Lake Erie. The total north-south distance is 4634 km. Canada's total area is 9,984,670 km², and with 202,080 km of the country bordered by water, it boasts the longest coastline of any nation in the world.

The wildlife and wildlife associations in Canada are linked to the geological, climatic and biological influences of the country's varied biogeography. For simplification, this book divides Canada into six natural regions: Pacific and Western Mountains, Central Plains, Arctic and Taiga, Boreal Shield, Mixedwood Plain and Atlantic. These natural regions contain 20 different terrestrial or marine "ecozones," each listed accordingly. Looking at these regions in detail can lead to a better understanding of the mammals of Canada and how they interact with each other.

Pacific and Western Mountains

This natural region contains marine, coastal and interior ecozones. Parts of this lush region receive the highest annual rainfall in Canada, and the result is a moist temperate rainforest of Douglas-fir, western hemlock and western red-cedar. Scenic mountain peaks rise above the forest over much of the region, and some of this mountainous area is protected in national and provincial parks. The interior parts of the region are colder than the coastal areas, and mammals that may be encountered include Caribou, Grizzly Bears and Wolverines. Black Bears and Mountain Lions also occur in high numbers. The Pacific and Western Mountains region also boasts more than 27,000 km of coastline and has a wealth of marine life. Many places along the coast offer excellent opportunities to see Sea Otters and migrating Grey Whales.

Terrestrial Ecozones: Taiga Cordillera, Boreal Cordillera, Montane Cordillera, Pacific Maritime

Marine Ecozone: Pacific Marine

Central Plains

The Central Plains is made up of two seemingly very different ecozones, the Boreal Plain and the Prairies. The similarity of these two ecozones is more geological than biological. As the name suggests, the Central Plains vary from nearly flat to gently rolling, the trademark of glacier-swept land. The Boreal Plain covers northern Alberta and central Saskatchewan, as well as some of British Columbia, the Northwest Territories and Manitoba. Forest vegetation is dominated by spruce, pine and tamarack. Many mammals are at home in the Boreal Plain, including the Woodland Caribou, Black Bear, American Marten, Fisher and Canada Lynx.

The Prairies are the most altered of any of Canada's ecozones, owing to the conversion of its fertile natural grasslands to agricultural croplands. Little of the original prairie remains, though growing public interest has resulted in newly protected areas of mainly native grassland and many private endeavours to restore native grassland habitat. One of Canada's newest national parks is Grasslands National Park in southwestern Saskatchewan, which protects large tracts of untouched prairie. Mammals of the Prairies ecozone include the American Bison, White-tailed Deer, Coyote, American Badger, Ord's Kangaroo Rat, Richardson's Ground Squirrel, Black-tailed Prairie Dog and Plains Pocket Gopher.

Terrestrial Ecozones: Boreal Plain, Prairies

Arctic and Taiga

About half of Canada lies within the Arctic and Taiga natural region, an area much more diverse than its name suggests. Within this region are six terrestrial and two marine ecozones, more than are found in any other of Canada's natural regions. Dominated by taiga and tundra, this region is fragile and slow to recover from disturbance. Taiga is a vast, poorly drained landscape that supports extensive forests of spruce, fir and tamarack mixed with wetlands, bogs and patterned fens. Mammals of the taiga include Moose, Black Bears, Grey Wolves, American Martens, Short-tailed Weasels, Northern Flying Squirrels, Northern Bog Lemmings and Hoary Bats. Where permafrost sets in, the taiga gives way to the treeless tundra. Mosses, lichens, flowering plants and low shrubs make up the bulk of the vegetation, and mammal life is surprisingly abundant. The iconic mammal of the tundra is the Muskox, living alongside others such as the Caribou, Arctic Fox, Grey Wolf, Wolverine, Arctic Hare, Arctic Shrew and a variety of lemming species. The marine ecozones are home to Polar Bears, Bowhead Whales, Belugas, Narwhals and many seal species.

Terrestrial Ecozones: Northern Arctic, Arctic Cordillera, Taiga Plain, Taiga Shield, Southern Arctic, Hudson Plain

Marine Ecozones: Arctic Archipelago, Arctic Basin

Boreal Shield

The famous boreal forest of the Boreal Shield region is the largest forested region in Canada, covering 1.8 million km² and representing almost 20 percent of the country's landmass. Geologically, it is also part of the Canadian Shield. The boreal forest is very similar to the more northerly taiga forest, except that its climate is somewhat milder and it is slightly more diverse biologically. Still, snow is present for eight or nine months of the year, and summer temperatures rarely rise above 25°C. The main tree species in the boreal forest are black spruce, white spruce, tamarack, jack pine (in well-drained areas) and balsam fir, but some broad-leaved trees such as paper birch, trembling aspen and balsam poplar also occur. Pristine lakes and rivers add to the scenic beauty of this forested wilderness, and tourism is a year-round interest. A variety of mammals inhabit this region, including Moose, Black Bears, Fishers, Canada Lynx, Snowshoe Hares, Red Squirrels and American Beavers.

Terrestrial Ecozone: Boreal Shield

Mixedwood Plain

The Mixedwood Plain is Canada's smallest natural region and includes the Carolinian Forest in southern Ontario and the long, narrow plains along the St. Lawrence River in Québec. The winters here are mild and relatively short, creating a much longer growing season than elsewhere in the country. Geologically, this region is part of the Great Lakes and St. Lawrence Lowlands, and sedimentary deposits underlie the rich, deep soils typical of this area. The dominant forest trees here are broad-leaved species such as sugar maple, American beech, American basswood, red ash, white oak and butternut. Unfortunately, most of the forested areas have been cleared to make way for extensive human settlement, industry and agriculture. The Mixedwood Plain is home to many mammals, including White-tailed Deer, Grey Foxes, American Badgers, Northern River Otters, Raccoons, Eastern Grey Squirrels, Southern Flying Squirrels, Eastern Cottontails, Woodland Voles and

Virginia Opossums. While a few of these species, such as the Raccoon and the White-tailed Deer, have adapted to human habitation, some are now found only where pockets of the original habitat remain.

Terrestrial Ecozone: Mixedwood Plain

Atlantic

The Atlantic natural region encompasses one terrestrial and two marine ecozones. The terrestrial ecozone includes all of New Brunswick, Nova Scotia and Prince Edward Island, as well as some of Québec. The Atlantic Maritime ecozone has an irregular and lengthy coastline that measures about 11,200 km. Acadian forest dominates more than 70 percent of the region. The climate is cool and moist owing to the moderating effects of the ocean. Common mammals of the region include the Moose, Coyote, Black Bear, Raccoon, Striped Skunk, Bobcat and Eastern Chipmunk.

The two marine ecozones are moderated by the Gulf Stream and are significantly warmer than nearby Arctic waters. Marine mammals in the region include Blue, Humpback, Fin and Minke whales, as well as Belugas, dolphins and six different seal species.

Terrestrial Ecozone: Atlantic Maritime

Marine Ecozones: Atlantic Marine, Northwest Atlantic Marine

Human-Altered Landscapes

The impact of human activity on natural environments is visible throughout Canada, but it is most noticeable in the southern parts of the country. Cities, roadways, agriculture, forestry and mining are just a few of the impacts we have had on the land. Many of the most common plants and animals that are found in these altered landscapes did not occur there before modern human habitation and transportation. The House Mouse, Norway Rat and European Hare are some of the highly successful exotics that were introduced to North America from Europe and Asia. As well, the distributions of many native mammals have changed as a result of habitat degradation and fragmentation. Mammals such as wolves and bears are extirpated in many parts of their former ranges because of dramatic changes in the landscape, but others such as Raccoons, White-tailed Deer and Coyotes have actually expanded their ranges.

Seasonality

The seasons in Canada greatly influence the lives of mammals. Aside from bats and marine mammals, most species are confined to relatively slow forms of terrestrial travel. As a result, they have limited geographic ranges and must cope in various ways with the changing seasons.

The rising temperatures, melting snow and warm rains of spring bring renewal. Many mammals bear their young at this time of year, and an abundance of food cycles through the food chain—lush new growth provides ample food for herbivores, and the numerous herbivore young become easy prey for the carnivorous mammals. While some small mammals, particularly the shrews and rodents, mature within weeks, the offspring of larger mammals depend on their parents for much longer periods.

During the warmest time of the year, the animals' bodies have recovered from the strain of the previous winter's food scarcity and spring's reproductive efforts, but summer is not a time of relaxation. To prepare for the upcoming winter, some animals must eat large amounts of energy-rich foods to build up fat reserves, while others work furiously to stockpile food caches in safe places.

Autumn is the time when some mammals begin hibernation. For others, such as the large ungulates, fall is the time for mating. Some small mammals such as voles and mice mate every few months or even year-round.

Winter varies in intensity and duration among the different regions of the country. In southern areas, winters are mild and generally less stressful. In northern regions, however, winter can be an arduous, life-threatening challenge for many mammals. For herbivores, high-energy foods are difficult to find, often requiring more energy to locate than they provide in return. The animals' negative energy budget gradually

weakens most herbivores through winter, and those not sufficiently fit at the onset of winter end up feeding the equally needy carnivores, which ironically find an ally in winter's severity. Voles and mice also find advantages in the season—an insulating layer of snow buffers their elaborate trails from the worst of winter's cold. Food, shelter and warmth are all found in this thin, subnivean layer, and the months devoted to food storage now pay off. Winter eventually wanes, and the warmth and life of spring prevails.

An important aspect of seasonality is its effect on the composition of an area's mammal population. You will typically see a different group of mammals in an area in winter than you will in summer. Many mammals, such as ground squirrels and bears, are dormant in winter. Conversely, many ungulates may be more visible in winter because of the lack of foliage and their tendency to enter open meadows in their search for edible vegetation.

Watching Mammals

Many mammals are most active at night, so the best times for viewing are during the "wildlife hours" at dawn and dusk. At these times of day, mammals emerge from their hideouts and move through areas where they can be more easily encountered. In winter, hunger may force certain mammals to be more active during daylight hours. When conditions are more favourable in spring and summer, some mammals may become less active during the day.

The protected areas and parks of Canada offer excellent opportunities for mammal watching, but large areas of this country are still wilderness and wildlife is never far away. Many of the larger mammals, in particular, can be viewed easily from a vehicle along the many roadways that cut through the countryside. If you walk backcountry trails or hike through forests, however, you can find yourself right in the homes of certain mammals.

Although people have become more conscious of the need to protect wildlife, the pressures of increased human visitation have nevertheless damaged critical habitats, and some mammals experience frequent harassment. Modern wildlife viewing demands sensitivity and common sense. While some of the mammals that are encountered in Canada appear easy to approach, it is important to respect your own safety as well as the welfare of the animal being viewed. This advice seems obvious for the larger species (though it is ignorantly dismissed in some instances), but it applies equally to small mammals. Honour both the encounter and the animal by demonstrating a respect appropriate to the occasion. Here are some points to remember for ethical wildlife watching:

- Confine your movements to designated trails and roads, wherever provided. Doing so allows animals to adapt to human use in the area and also minimizes your impact on the habitat.

- Avoid dens and resting sites, and never touch or feed wild animals. Baby animals are seldom orphaned or abandoned, and it is against the law to take them away.

- Stress is harmful to wildlife, so never chase or flush animals from cover. Use binoculars and keep a respectful distance, for the animal's sake and often your own.

- Leave the environment, including both flora and fauna, unchanged by your visits. Take home only pictures and memories.

- Pets are a hindrance to wildlife viewing. They may chase, injure or kill other animals, so control your pets or leave them at home.

- Take the time to learn about wildlife and the behaviour and sensitivity of each species.

Canada's Top Mammal-Watching Sites

1. Kluane National Park and Reserve
2. Ovayok Territorial Park
3. Pond Inlet and Sirmilik National Park
4. Gwaii Haanas National Park Reserve
5. Pacific Rim National Park
6. Victoria and the Gulf Islands
7. Tweedsmuir North Provincial Park
8. Jasper National Park
9. Yoho National Park
10. Banff National Park
11. Kootenay National Park
12. Wood Buffalo National Park
13. Elk Island National Park
14. Suffield National Wildlife Area
15. Prince Albert National Park
16. Grasslands National Park
17. Riding Mountain National Park
18. Churchill and Wapusk National Park
19. Woodland Caribou Provincial Park
20. Polar Bear Provincial Park
21. Pukaskwa National Park
22. Algonquin Provincial Park
23. La Mauricie National Park
24. Baie-Sainte-Catherine
25. Torngat Mountains National Park
26. Forillon National Park
27. Digby Neck
28. Cape Breton Highlands National Park
29. Gros Morne National Park
30. Witless Bay Ecological Reserve

Canada's Top Mammal-Watching Sites

Mountain Parks

The Rocky Mountains of Canada offer some of the best opportunities for mammal watching in North America. Abundant hiking trails make it easy to access prime habitat where mammals can be observed. The following list of mountain parks is by no means exhaustive, but it does describe some of the most accessible parks that give visitors the highest reward in terms of wildlife encounters.

Banff National Park

Canada's first national park is magnificent and well set up for wildlife viewing. A wide range of enjoyable excursions provide world-renowned scenery and excellent opportunities to encounter stunning flora and fauna. Lake Louise, Peyto Lake and Moraine Lake are three of the jewels of the park. The looping alpine meadow trails at Bow Summit are accessible to everyone and provide visitors with good chances to see Golden-mantled Ground Squirrels and Bighorn Sheep. Warm evenings around the Vermilion Lakes are an excellent time to see Moose.

Jasper National Park

Grizzly Bears, Black Bears, Gray Wolves, Caribou, Wapiti, Mule Deer, Moose, Mountain Lions and Bighorn Sheep all roam the wilds of this large park. Look for sheep and wolves at the north end of Jasper Lake, Caribou, American Mink and Moose along the Medicine River and bears along the Icefields Parkway. Visitors can hike on the toe of the Athabasca Glacier and learn about geological processes firsthand.

Yoho National Park

The melting Daly Glacier, nestled among the towering peaks of the Continental Divide, gives rise to spectacular Takakkaw Falls, whose water plummets 380 m into the wild Yoho River. Grizzly Bears, Mountain Lions, Hoary Marmots and Mountain Goats are regular inhabitants. Moose, deer and Wapiti may be found at the salt lick near the amazing natural bridge. Yoho's Burgess Shale World Heritage Site contains the fascinating fossil remains of marine animals estimated to be 530 million years old.

Kootenay National Park

This park's ochre-tinted paint pots and myriad of wild plants were once used for ceremonies and survival by members of the Ktunaxa Nation, who also hunted Mule Deer, Snowshoe Hares and American Beavers. In more recent times, tourists have been attracted to this special place by the hot springs and the many wildlife-enticing natural salt licks.

Kluane National Park and Reserve

Encompassing Canada's highest mountain, Mount Logan (5959 m), Kluane National Park is a wild landscape easily accessible from Whitehorse, Yukon. Although much of the park is high mountains and glaciers, there is wonderful habitat for wildlife. About 105 species of birds are found in the park, as well as a variety of mammals. Grizzly Bears, Black Bears, Grey Wolves, Wolverines, Canada Lynx, Dall's Sheep, Mountian Goats, Caribou, Snowshoe Hares and Arctic Ground Squirrels are all abundant.

Parks in the Central Grasslands

The prairie grasslands of Canada have only recently been considered a habitat in decline. Spurred by the bounty of fertile soils, farmers have long been converting the grasslands into croplands. Although lacking the obvious charisma of the scenic mountains and their ever-present Bighorn Sheep, the thrill of visiting unfenced native grasslands touches something primordial within us and lingers long after we have returned to our urban lifestyles. Too few areas of natural prairie remain—it is Canada's most endangered habitat.

Grasslands National Park

The undisturbed mixed- and short-grass prairie in Grasslands National Park is some of the last natural grassland left in Canada. Three of our rarest mammals occur in this park, and for two of them, it is their only Canadian range. The Swift Fox, Black-tailed Prairie Dog and Black-footed Ferret help make the endangered prairie habitat here a functioning ecosystem. Thanks to Swift Fox reintroduction programs, the population in the park numbers about 100 individuals. Black-footed Ferrets were recently reintroduced and appear to be surviving. The grasslands are their primary habitat, and their main prey, Black-tailed Prairie Dogs, are numerous.

Suffield National Wildlife Area

Within CFB Suffield, 50 km north of Medicine Hat, Alberta, is a region of relatively intact mixed-grass prairie that is now designated the Suffield National Wildlife Area (SNWA). The region owes its uncultivated state to its unique military history. Burrowing owls are common among the grasslands here, and it is one of the few known ranges for the Swift Fox. The Western Harvest Mouse and Ord's Kangaroo Rat both live in SNWA as well.

Arctic and Tundra Region Parks

Much of Canada's landmass is Arctic tundra, and although this region is often thought of as "barren," there is an abundance of wildlife present, including a variety of mammals. There are no fences in the far north, few roads and even fewer cities, so wildlife is not necessarily confined to protected parks as is often the case in the southern parts of Canada. Nevertheless, protected parks tend to offer good access, which helps confine the disturbance caused by human visitors to localized areas.

Pond Inlet and Sirmilik National Park (Baffin Island)

One of the most scenic hamlets in the Canadian Arctic, Pond Inlet is located near the northern tip of Baffin Island. It is nearly surrounded by mountains that glisten white from their year-round snow cover. In the sea, there is a nearby ice floe edge, a good place for finding marine mammals. Belugas, Narwhals, Bowhead Whales, Orcas, Polar Bears and Harp, Ringed and Bearded seals can all be seen in this dramatic landscape.

Ovayok Territorial Park (Victoria Island)

Located very near to the hamlet of Cambridge Bay, this territorial park includes Pelly Mountain (called Ovayuk in Inuinnaqtun), a large esker that is a traditional Inuit hunting ground. There are over 120 archaeological sites in the park. An abundance of birds and mammals inhabit the area, as well as several billion mosquitoes. Birds such as sandhill cranes, Arctic loons, tundra swans, and parasitic jaegers are easily seen around the many lakes and ponds. A good number of Muskox live in the park, as do Barren-ground Caribou, Arctic Hares, Arctic Foxes and Brown Lemmings.

Polar Bear Provincial Park

This park, the most northerly one in Ontario, is accessible only by aircraft. The effort required to get here, however, is worth the reward, because this park protects over 2 million ha of low-lying, unspoiled tundra. Peat soils and muskeg are found throughout the region, and an obvious treeline occurs in the park. North of treeline, the plant life includes caribou lichen and reindeer and sphagnum mosses; south of treeline, there are stunted willows, spruces and tamarack. Large numbers of Caribou live here, as do Moose, American Martens, Arctic Foxes, Walruses, Belugas, Ringed Seals and Bearded Seals. The high density of Polar Bears in this park in early winter attracts many wildlife enthusiasts, researchers and photographers.

Churchill and Wapusk National Park

Nicknamed the "Polar Bear Capital of the World," Churchill is a small town in northern Manitoba on the shores of the Hudson Bay. About 45 km south of Churchill is Wapusk National Park. Access is limited in this park because it protects maternal denning sites for Polar Bears, but it remains a world-class location for viewing and photographing these magnificent animals. In addition, thousands of Belugas congregate in the warmer estuary of the Churchill River in July and August, where females will give birth.

Western Canada

From British Columbia to Manitoba, the western provinces and territories are less populated than the eastern provinces, and there are plenty of opportunities for wilderness exploration and wildlife encounters.

Tweedsmuir North Provincial Park

A destination for true wilderness enthusiasts, Tweedsmuir park has no facilities or supplies. Located in west-central British Columbia, this park protects an almost-pristine mountain wilderness. Grizzly Bears, Black Bears, Grey Wolves, Mountain Lions, Wolverines, Caribou, Moose, Mountain Goats, Hoary Marmots and many kinds of small mammals can be encountered in this park.

Wood Buffalo National Park

Canada's largest national park, Wood Buffalo National Park is located on the border between Alberta and the Northwest Territories. The park is a little more than 44,000 km² in area and represents the Boreal Plain ecozone. Within its boundaries are the only known nesting sites for the critically endangered Whooping Crane and the world's largest free-roaming herd of Wood Bison. A total of 46 different mammal species inhabit this large park, including Grizzly Bears, Black Bears, Grey Wolves, Canada Lynx, Wolverines and Fishers. There are also large caves in the park that are important hibernation sites for Northern Myotis.

Elk Island National Park

Elk Island National Park is a little less than an hour's drive east of Edmonton, Alberta. This is a fairly small national park, but it protects an important habitat of aspen parkland and grassland. North America's largest land mammal, the American Bison, and the smallest, the Least Shrew, are both found here. Famous for its bison, this park is one of the most accessible places to meet herds of both Plains Bison and Wood Bison. Many other mammals find refuge here from Alberta's intensive agricultural and mining activity, including Coyotes, Wapiti, deer, Moose, American Mink, weasels and American Beavers. There are even occasional records of Mountain Lions and Grey Wolves within the park.

Prince Albert National Park

Located north of Saskatoon, Saskatchewan, Prince Albert National Park includes the cabin of Grey Owl, a famous conservationist and naturalist of the early 1900s. This park protects vast coniferous and aspen forests, as well as patches of fescue grasslands. The mix of forest and grassland in the southern part of the park supports Plains Bison and American Badgers. Other mammals in the park include Black Bears, Grey Wolves, Red Foxes, Moose, Wapiti, Northern River Otters, Red Squirrels and Northern Flying Squirrels.

Riding Mountain National Park

Located on the highest point of the Manitoba Escarpment, Riding Mountain National Park was designated in 1986 as a Biosphere Reserve by UNESCO. This park is dominated by forested parkland with intermixed grassland areas. Within the park boundaries is a large population of Black Bears—one of the highest densities in North America. Other mammals include Mountain Lions, Canada Lynx, Grey Wolves, Northern River Otters, deer, Moose, North American Porcupines and several species of squirrels. There is also a small herd of bison in the centre of the park.

Eastern Canada

From Ontario to the Maritimes, there are many fabulous locations for hiking, canoeing and kayaking. Each site offers unique wildlife assemblages and amazing landscapes, perfect for wildlife enthusiasts and photographers.

Woodland Caribou Provincial Park

As its name suggests, Woodland Caribou Provincial Park supports a large number of Caribou. This park is situated along the Manitoba-Ontario border, and access is restricted to air, water or rough forest roads. Canoeing is a major attraction in this park, and the connected lakes and rivers offer more than 1600 km of water routes. The mammals found here are those common to the boreal forest. As well, there are a few species present that are typical of more southwestern habitats. In addition to Caribou, there are Moose, Black Bears, American Beavers, Northern River Otters, Muskrats, American Mink, Fisher, Wolverines, weasels, Canada Lynx, Red Foxes and Grey Wolves. This park is also home to one of Ontario's only well-documented colonies of Franklin's Ground Squirrels.

Pukaskwa National Park

The Pukaskwa wilderness of Lake Superior's north shore—part of the ancient Canadian Shield landscape—is a region of forested hills, rough ridges and rocky-shored lakes. The majority of the park is forested, and the dominant trees include spruce, fir, cedar, birch and aspen. Human activity within the park is a concern because protecting this unique part of Ontario's natural heritage is essential. In this wilderness is a small, isolated population of Caribou, as well as good numbers of Grey Wolves, Black Bears, Moose, American Mink, Canada Lynx and White-tailed Deer. Several important research projects here have studied the interactions between Grey Wolves, Moose and Caribou. Other mammals in the park include Snowshoe Hares, North American Porcupines, chipmunks, Northern Flying Squirrels and Woodland Jumping Mice.

Algonquin Provincial Park

This famous park owes much of its scenic beauty to the unique transition environment within its boundaries. Southern areas of the park are mainly deciduous forest, whereas northern areas mark the beginning of the coniferous boreal forest. Algonquin's accessibility and enormous biological diversity make it one of the finest wildlife-viewing parks in Ontario. At least 45 mammal species have been recorded here, and it is the best place in the province to see Moose or listen to the nocturnal howls of Grey Wolves. Other mammals such as Red Foxes, Fishers, American Mink, Northern River Otters and American Beavers also live here and are frequently seen by visitors to the park.

La Mauricie National Park

La Mauricie National Park protects 536 km² of the Québec Laurentians between Montréal and Québec City. Although much of the forest in this park was logged during the early 1900s, the mixed forests of this southern Canadian Shield region have now regrown. Some of the wildlife species in this park are boreal and at their southern limits, whereas others inhabit warmer regions to the south and are at their northern limits here. Wood turtles, though rare in Canada, can be found along the banks of slow-moving streams. At least 40 species of mammals inhabit this park, including Black Bears, Red Foxes, Grey Wolves, Coyotes, American Mink, Northern River Otters, Fishers, Raccoons, North American Porcupines and Woodchucks.

Forillon National Park

Encompassing the northern end of the Appalachian Mountains, as well as salt marshes, forests, sand dunes and rugged coastline, Forillon National Park protects 244 km² on the tip of the Gaspé Peninsula in Québec. Seaside cliffs provide important nesting sites for a variety of seabirds, and this park also gives visitors ample opportunity to encounter both forest and marine mammals, including Black Bears, Bobcats, Red Foxes, Moose, Grey Seals and Harbour Seals. There are seven species of whales that have been seen from shore here, from the small Harbour Porpoise to the enormous Blue Whale. There is even an isolated herd of Caribou in this park.

Torngat Mountains National Park

The northernmost region of Labrador is protected in a national park that comprises some 9700 km². The Torngat Mountains of the Arctic Cordillera lie within this wildness park, and the confluence of mountain, boreal and Arctic ecozones results in a wide variety of mammals. Polar Bears, Black Bears, Red Foxes, Arctic Foxes, Caribou, Humpback, Fin and Minke whales and Ringed, Harp and Hooded seals are all part of the dramatic diversity of mammals found here.

Gros Morne National Park

Gros Morne National Park, on the west coast of Newfoundland, is a UNESCO World Heritage Site comprising some 1800 km² of mountain and forest landscape. Geologically, the mountains here are 20 times older than the Rockies. The forest includes black spruce, white spruce, balsam fir and some red maple and yellow birch. There are also marshes, bogs and sand dunes. Black Bears, Canada Lynx, Woodland Caribou, Red Foxes, Snowshoe Hares, Little Brown Bats and Northern Myotis are among the mammals found here. Moose were introduced to Newfoundland in the early 1900s, and with up to 8000 individuals, this park may well have the highest Moose density in the world. Sometimes Harbour Seals and a few whale species are seen along the coastline.

Cape Breton Highlands National Park

Nova Scotia's Cape Breton Island Highlands National Park encompasses a beautiful landscape of mountains and coastline. Covering 950 km², this park protects both Acadian mixed forest and boreal forest. A variety of seabirds inhabit the coastal areas, and pinnipeds and whales are sometimes seen as well. Mammals here are diverse, from Black Bears to the small and rare Gaspé Shrew. This park has a long and varied history of human settlement as well.

West Coast Marine Sites

Canada's West Coast boasts an extensive coastline and numerous islands, amounting to prime habitat for many kinds of marine mammals. In addition to providing opportunities for encountering terrestrial mammals, the following sites are singled out for their diversity of marine mammals.

Gwaii Haanas National Park Reserve

Located on the southern end of Haida Gwaii (Queen Charlotte Islands), this protected area is a popular destination for hikers and kayakers, as well as tourists interested in the history of the Haida people. The relatively intact coastal old-growth forests here are an example of some of the finest temperate rainforests on the Pacific Coast. The park protects a complex island ecosystem where many varieties of animals thrive. Northern Sea Lions, Harbour Seals, dolphins, porpoises, Grey Whales and Humpback Whales are all common here. Kayaking is an excellent way to experience the biodiversity of this park.

Pacific Rim National Park

Perhaps British Columbia's most famous park, Pacific Rim National Park offers an outstanding array of wildlife encounters and adventures for thrill-seekers, including the renowned West Coast Trail. Long stretches of beach, rocky islets and rugged, forested terrain characterize much of this park. It encompasses luxuriant old-growth rainforests and is one of the best places in the province for viewing marine mammals. Migrating Humpback Whales and Grey Whales can be observed from shore, and Northern Sea Lions can be seen on the rocks in summer. As well, this open coastline is prime habitat for Sea Otters.

Victoria and the Gulf Islands

The waters surrounding Vancouver Island are ideal for spotting Orcas, Grey, Humpback and Minke whales, Dall's Porpoises, Harbour Seals, Northern Elephant Seals, Northern Sea Lions and California Sea Lions. The ideal way to encounter these marine mammals is by boat. Even a simple ferry ride can provide excellent sightings of many species. In Victoria and nearby coastal towns, there are numerous whale-watching companies that specialize in locating and providing opportunities for visitors to quietly view any and all of these species.

East Coast Marine Sites

Since the decline of the fishing industry on the East Coast, whale watching has become a new source of revenue for many locals. A great many species of seals and whales are encountered in the sheltered waters of the region, and many more can be seen off the open coastline. The following locations are in prime spots for observing many marine mammal species, and a wealth of trip operators can be found in the best seasons.

Digby Neck

The long, narrow peninsula on the southwestern part of Nova Scotia separates St. Mary's Bay from the Bay of Fundy. Small towns dot this peninsula, and whale-watching trips are ubiquitous. Boat trips in this region offer excellent chances to see many species of whales, including Fin, Minke, Humpback and North Atlantic Right whales and Harbour Porpoises. Other species recorded in this region include Right, Sperm, Blue, Sei and Pilot whales, along with Orcas, Belugas and Atlantic White-sided Dolphins.

Baie-Sainte-Catherine

Baie-Sainte-Catherine is located in the upper Charlevoix region of Québec alongside the Saguenay River estuary. In nearby waters during summer, hundreds of resident Minke Whales and Belugas are joined by Blue, Fin and Humpback whales. From mid-June to early October is the best time to encounter these marine mammals, and in some cases, they can easily be spotted from land. Local tour operators will know the best place, and a short boat trip can offer once-in-a-lifetime whale sightings.

Witless Bay Ecological Reserve

The Witless Bay Ecological Reserve on the east coast of Newfoundland protects four islands that are of critical important to nesting seabirds. North America's largest colony of Atlantic puffins is found in this reserve. In the surrounding waters, there are 22 species of whales that may be encountered. Fin, Minke, Humpback and North Atlantic Right whales are abundant.

About This Book

This guide describes 188 species of wild and feral mammals that have been reported in Canada. Domestic farm animals such as cattle, sheep and llamas are not described. Although many whale species are known to occur in Canadian waters, only some of those that can be found in nearshore waters are included. Humans have lived in this part of the world at least since the end of the last Pleistocene glaciation, but the relationship between our species and the natural world is well beyond the scope of this book.

Organization

Biologists divide mammals (class Mammalia) into a number of subgroups, called orders, which form the basis for the organization of this book. Nine mammalian orders have wild or feral representatives in Canada: even-toed hoofed mammals (Artiodactyla), odd-toed hoofed mammals (Perissodactyla), whales, dolphins and porpoises (Cetacea), carnivores (Carnivora), rodents (Rodentia), hares and rabbits (Lagomorpha), bats (Chiroptera), insectivores (Insectivora) and opossums (Didelphimorphia). In turn, each order is subdivided into families, which group together the more closely related species. For example, within the carnivores, the Wolverine and the American Mink, which are both in the weasel family (Mustelidae), are more closely related to each other than either is to the Striped Skunk, which is in its own family (Mephitidae).

Mammal Names

Although the international zoological community closely monitors the use of scientific names for animals, common names, which change with time, local language and usage, are more difficult to standardize. In the case of birds, the American Ornithologists' Union has been very effective in standardizing the common names used by professionals and recreational naturalists in North America. There is, as yet, no similar organization to oversee and approve the common names of mammals, which can lead to some confusion.

For example, many people apply the name "mole" to pocket gophers, which are burrowing mammals that leave loose piles of dirt in fields and reminded early settlers of the moles they knew in Europe. To add to the confusion, most people use the name "gopher" to refer not to pocket gophers, but to ground squirrels. If you consider nonmammalian species, it gets even worse. The name "gopher" is used in many parts of North America to denote a species of snake and even a tortoise!

You may think that such confusion is limited to the less-charismatic animal species, but even some of the best-known mammals are victims of human inconsistency. Most Canadians know the identity of the Moose, but this name can cause great confusion for European visitors. The species that we know as the Moose, *Alces alces*, is called Elk in Europe ("elk" and *alces* come from the same root), whereas in North America, Elk is one of the common names for *Cervus canadensis*, a species similar to the Red Deer in Europe. The blame for this confusion falls on the early European settlers, who misapplied the name Elk to populations of *Cervus canadensis*. In order to resolve the confusion, many naturalists prefer to use the name Wapiti for *Cervus canadensis*, an effort supported here in this guide.

Despite the lack of an official list of mammal common names, there are some widely accepted standards, such as the "Revised Checklist of North American Mammals North of Mexico, 2003" (Jones et al, 2003, Occasional Papers, Museum of Texas Tech University, No. 229). *Mammals of Canada* follows that checklist for the scientific names and for most of the common names. Readers should also know that other sources have attempted to standardize mammal common names on a worldwide basis.

Range Maps

Mapping the range of a species is a problematic endeavour because mammal populations fluctuate, distributions expand and shrink, and dispersing or wandering individuals are occasionally encountered in unexpected areas. The range maps included in this book are intended to show the distribution of breeding or self-sustaining populations in the region and not the extent of individual specimen records. Land ranges are shown in red, and marine ranges are shown in dark blue. For certain species accounts, the range of a mammal is shown by question marks, which indicate uncertainty over the species' presence in that area.

Similar Species

Before you finalize your decision on the identity of a mammal, check the **Similar Species** section of the account, which briefly describes other mammals that could be mistakenly identified as the species you are considering. By concentrating on the most relevant field marks, the subtle differences between species can often be reduced to easily identifiable characteristics. As you become more experienced at identifying mammals, you might find you can immediately shortlist an animal to a few possible species. The **Similar Species** section lets you quickly glean the most relevant field marks to distinguish between those species, thereby expediting the identification process.

Bat and Shrew Keys

For the bats and shrews, a basic identification key is included. These keys will be helpful in identifying a species, especially if you have a specimen in hand. The keys are dichotomous—that is, they present you with pairs of descriptions, only one of which will apply to your specimen. To identify a bat or shrew, simply work your way through the key, choosing one of the paired alternatives and then moving to the next set of choices, as indicated by the number at the end of the line. When the line ends in a name rather than a number, you have identified your specimen.

HOOFED MAMMALS

The hoofed mammals include the "megaherbivores" of Canada; they fall into the largest size class of terrestrial mammals, and they all eat plants almost exclusively. All native hoofed mammals in the country, as well as the introduced Feral Pig, belong to the order Artiodactyla (even-toed hoofed mammals). Even-toed hoofed mammals have either two or four toes on each limb. If there are four toes, the outer two, which are called dewclaws, are always smaller and higher on the leg, touching the ground only in soft mud or snow. Horses, which are not native to North America, belong to the order Perissodactyla (odd-toed hoofed mammals) and have just a single toe on each foot.

Another difference between the two orders of hoofed mammals is in the structure of their ankle bones. The ankle bones of all even-toed hoofed mammals are grooved on both their upper and lower surfaces, which enables these animals to rise from a reclined position with their hindquarters first. This ability means that the large hindleg muscles are available for fight or flight more quickly than in odd-toed hoofed mammals such as horses, which must rise front first. As well, the even-toed hoofed mammals have incisors only on the lower jaw. They have a cartilaginous pad at the front of the upper jaw instead of teeth.

Cattle Family (Bovidae)

Bison, sheep and goats are distinguished from our other hoofed mammals by the presence of true horns in both the male and female. These horns are never shed, and they grow throughout an animal's life. They consist of a keratin sheath (keratin is the main type of protein in our fingernails and hair) over a bony core that grows from the frontal bones of the skull. Like deer, all bovids are cud chewers, and they have complex, four-chambered stomachs to digest their meals.

Pronghorn Family (Antilocapridae)

This exclusively North American family contains just the one species. The Pronghorn has only two toes (no dewclaws), and, like the other native artiodactyls, it lacks upper canine teeth as well as upper incisors. Both sexes have true horns, but, unlike bovids, the Pronghorn sheds and regrows the keratinous sheath each year (the bony core is not shed). The dark-coloured sheath—but not the bony core—is branched, hence the name "pronghorn." Females may or may not have horns.

Deer Family (Cervidae)

All adult male cervids (and female Caribou) have antlers, which are bony out-growths of the frontal skull bones and are shed and regrown annually. In males with an adequate diet, subsequent sets of antlers are larger each year, and some species develop more tines. New antlers are soft and tender, and they are covered with "velvet," a layer of skin with short, fine hairs and a network of blood vessels to nourish the growing antlers. The ant-lers stop growing in late summer, and as the velvet dries up, the deer rubs it off. Cervids are also distinguished by the presence of scent glands in pits just in front of the eyes. Their lower canine teeth look like incisors, so there appear to be four pairs of lower incisors.

Swine Family (Suidae)

The Swine family includes eight species that originated in Eurasia and Africa. Most of them are restricted to their native ranges, but the Domestic Pig (*Sus scrofa*) has been widely introduced in most areas of human habitation. Despite its name, the Domestic Pig now ranges free in many parts of North America, and these feral popu-lations can become quite large. These free-ranging pigs are now called Feral Pigs. The European Wild Boar is another wild form of *Sus scrofa* that was also intro-duced to North America, and these two varieties of swine sometimes hybridize. Feral Pigs have upper incisors and upper canines. The canines are modified into tusks.

Horse Family (Equidae)

All members of this family, which also includes zebras and donkeys, have a sin-gle toe on each limb, a bushy dorsal mane and a long, well-haired tail. Although horselike animals were once native to North America, they disappeared from our continent more than 10,000 years ago. The herds of Feral Horses that are now found in several places through the Rockies and the American West are descended from domestic horses.

American Bison
Bison bison

"Plains Bison"

Historically, few areas of central Canada were left untouched by the American Bison, the largest terrestrial mammal in North America. From the northern boreal forest to the western foothills to the central grasslands, bison roamed much of this country and left their impressive marks on the landscape. Evidence of their once-great presence can still be found, particularly on the Prairies. Large boulders isolated on the grasslands are often smoothly polished and set in shallow pits—stark evidence of thousands of years of itchy bison rubbing their hides for relief. More eerie still are the bones that spill out of prairie riverbanks, a testament to an epic past.

Although frequently referred to as "buffalo," the bison is in a different genus from the buffalo species of Asia and Africa. The

RANGE: The historical range extended from the southeastern Yukon south to northern Mexico and east to the Appalachian Mountains. Free-ranging herds are now almost exclusively restricted to protected areas. Many small herds are raised on fenced game ranches.

Total Length: 2.4–3.9 m

Shoulder Height: 1.3–1.8 m

Tail Length: 28–39 cm

Weight: 360–1000 kg

American Bison is divided into two sub-species, the Wood Bison (*B. b. athabasacae*) and the Plains Bison (*B. b. bison*). Wood Bison are the larger and more northerly of the two. One of the best distinguishing field marks is that the highest point of the hump of a Wood Bison is forward of the forelegs, whereas the highest point of the hump of a Plains Bison is directly above its forelegs. Another good field mark is that the horns of a Plains Bison rarely extend past the top of its head, whereas the horns of a Wood Bison extend well past the top of the head.

Little more than a century ago, the American Bison was near extinction and most of the remaining few were found on one man's ranch in northern Montana. From 1907 to 1924, Don Michel Pablo sold 709 of his bison to the Canadian government. These were Plains Bison, and they were transferred to Elk Island National Park, Alberta. Shortly after, all but 48 of them were moved to the newly established Buffalo National Park near Wainwright. The herd soon increased to about 10,000 animals. To prevent over-crowding, several thousand bison were shipped north by rail and barge to Wood Buffalo National Park, which had been established in 1922 to protect one of the few remaining wild bison populations. This wild population, however, was Wood Bison, and the two subspecies inevitably interbred in the park. Additionally, the Wainwright bison brought with them tuberculosis and brucellosis infections, which spread to the healthy wild bison and continues to hinder conservation efforts. The Wainwright park was closed as a bison range in 1939, and its residents went to either Elk Island National Park or Wood Buffalo National Park. Virtually all of the Plains Bison today are descend-ants of those that were returned plus the original 48 that were left behind in Elk Island. Most of the bison in Wood Buffalo National Park are hybrids, but a few pure Wood Bison were found in a remote area of the park in the 1960s. Most of these were sent to the southern part of Elk Island National Park, where they have increased in number and remain separate

"Wood Bison"

from the Plains Bison in the north sec-tion of the park. Plains Bison have been successfully reintroduced to their native prairie habitat in the relatively new Grasslands National Park in southwest-ern Saskatchewan, and Wood Bison have been successfully reintroduced to parks in Alaska, British Columbia and the Yukon.

DESCRIPTION: The front end is covered with long, shaggy, woolly, dark brown hair that abruptly becomes shorter and lighter brown behind the shoulders. The head is massive and appears to be carried low because of the high shoulder hump and massive forequarters. Both sexes have short, round, curved, black horns that grow upward. The legs are short and clothed in shaggy hair. The tail is long and has a tuft of hair at the tip. Males are slightly larger and heavier than females. Calves are reddish at birth but become darker by their first autumn.

HABITAT: Although the American Bison was originally most abundant on the Prairies, it also historically inhabited alpine tundra and areas of boreal forest and aspen parkland with abundant short vegetation.

FOOD: Most of the diet is made up of grasses, sedges and forbs. In winter, the American Bison sometimes browses on shrubs, cattails and lichens, but grasses remain the primary food. To get at vegetation under the snow, a bison will paw away the snow or push it to the side with its head if the snow is not too crusted.

DEN: Historically, American Bison were nomadic so did not have permanent dens. Bison typically bed down at night

walking trail

and during the hottest part of the day. After a herd has been in an area for a while, the animals will leave behind wallows—dusty, saucerlike depressions where the bison rolled and rubbed repeatedly.

REPRODUCTION: After a gestation of 9 to 10 months, a cow bison typically gives birth to a single calf in May. The calf weighs about 20 kg and is able to follow its mother within hours of birth. It begins to graze at about one week of age, but it is not weaned until it is about seven months old. A cow typically mates for the first time at two or three years of age. Bulls are sexually mature then, too, but competition from older males customarily prevent younger males from breeding until they are seven to eight years old.

hoofprint

SIMILAR SPECIES: The **Moose** (p. 76) has a similarly coloured coat, but it is taller, with long, thin legs and a much longer and leaner body overall. The smaller **Muskox** (p. 46) has longer hair, horns that "wrap" downward, and it lives on the tundra.

Moose

Mountain Goat
Oreamnos americanus

male

Acrophobia—the fear of heights—is unknown to the Mountain Goat. This nimble bovid is at home on rocky cliffs, so the very heights that instill fear in many people are comfortable and easily navigable for this animal. Although it is called a goat, it is not a member of the genus *Capra*, the true goats. All true goats and the Mountain Goat, however, belong to the same subfamily Caprinae, along with Muskox and sheep. This subfamily has respresentatives around the world, including the gorals, serows, tahr and chamois of Asia and Europe.

The Mountain Goat has several physical characteristics that help it live in such precarious surroundings. The hard outer ring of each hoof surrounds a softer, spongy

RANGE: The natural range extends from southeastern Alaska south through the Coast Mountains into the Washington Cascades and southeast through the Rockies into Idaho and Montana.

Total Length: 1.2–1.6 m

Shoulder Height: 0.9–1.2 m

Tail Length: 9–14 cm

Weight: 45–136 kg

central area that provides a good grip on rocky surfaces. The dewclaws are long enough to touch the ground on soft surfaces, and they provide greater "flotation" on weak snow crusts. To keep the Mountain Goat relatively comfortable in subzero temperatures and the strong winter winds that sweep along mountain faces, its winter coat consists of a thick, fleecy undercoat topped by guard hairs more than 15 cm long. When spring arrives, the goat begins to shed "blankets" of thick hair, which fall in pieces, often in the animal's dusting pits dug high on the sides of mountains. At this time, the Mountain Goat is not in its picturesque prime. Its short, neat, white summer coat comes in by June and continues to grow to form the thick winter coat.

DESCRIPTION: The coat of this stocky, hump-shouldered animal is white and usually shaggy, with longer guard hairs surmounting the fleecy undercoat. When the Mountain Goat is viewed from the side, the chest appears deep, but when viewed head-on, the chest looks remarkably thin. The lips, nose, eyes and hooves are black. Both sexes may sport a noticeable beard, which is longer in winter. The short legs often look like they are clothed in breeches in winter because the hair of the lower legs is much shorter than that of the upper leg. The tail is short, and the ears are relatively long. Both sexes have narrow, black horns. The billy's horns are

thicker and curve backward along a constant arc. The nanny's horns are narrower and tend to rise straight from the skull and then bend sharply to the rear near their tips. The kid has a grey-brown stripe along its back.

HABITAT: Mountain Goats generally occupy steep slopes and rocky cliffs in alpine and subalpine areas, where low temperatures and deep snow are common, a habitat choice that helps protect the animals from predation. Although they typically inhabit treeless areas, Mountain Goats may travel through dense subalpine or montane forests going to and from salt licks in April and May and in early autumn. In summer, they tend to be seen more frequently at lower elevations but move to the highest windswept ledges in winter to find vegetation that is free of snow cover.

FOOD: This adaptable herbivore varies its diet according to its environment. In one study, Mountain Goats ate shrubs almost exclusively, with the balance of the diet coming from mosses, lichens and forbs;

in another study, only one-quarter of the diet was shrubs and three-quarters was grasses, sedges and rushes. The Mountain Goat's winter feeding areas are generally separate from the summer areas.

DEN: Mountain Goats bed down in shallow depressions scraped out in shale or dirt at the base of a cliff. Clumps of white hair are often scattered in the vicinity of a scrape.

REPRODUCTION: To avoid predators, a nanny moves to an isolated rock edge to give birth. In May, after a gestation of five to six months, a nanny bears a single kid (75 percent of the time) or twins, weighing 2.9 to 3.8 kg. The kid is precocial and can stand on its own within hours. After a few days, the kid starts eating grasses and forbs, but it is not weaned until it is about six weeks old. Both sexes become sexually mature after about 2.5 years. Nannies and kids form loose groups of up to about 50 members. A nanny mates every other year. During the mating season, billies and nannies come together in groups, and billies dig rutting pits and engage in dominance scuffles. After mating, billies disperse in small groups of two or three individuals.

hoofprint

walking trail

SIMILAR SPECIES: The **Bighorn Sheep** (p. 50) has brown upperparts and a whitish rump patch. Its brown horns are either massive and thick at the base (rams) or flattened (ewes) but never round, thin, stilettolike or black like a Mountain Goat's horns. **Dall's Sheep** (p. 54) in the white form can appear similar but lacks the long hair and distinctive black horns.

Bighorn Sheep

Muskox
Ovibus moschatus

male

The sight of Muskox roaming the open tundra conjures up sensations of epic Pleistocene wildness. Like the Pronghorn, the Muskox is the sole surviving member of its genus; both species were contemporaries of the Woolly Mammoth, Mastodon and Woolly Rhinoceros. Muskox survived the last Ice Age in refugia, ice-free areas where many plants and animals maintained a foothold. Their adaptations to cold that served them so well during the late Pleistocene now confine them to Arctic regions.

Like the Arctic Wolf, the Muskox is adapted not just to the cold of the Arctic, but to the darkness as well. Able to tolerate

RANGE: Muskox are found on the Arctic mainland and on several islands of the archipelago. They have been successfully reintroduced to parts of their historic range in Canada, as well as to Greenland, Alaska, Russia, Norway and Sweden.

Total Length: *Male:* 2.1–2.6 m; *Female:* 1.5–2.2 m

Shoulder Height: 1.0–1.2 m

Tail Length: 9–10 cm

Weight: *Male:* 180–410 kg; *Female:* 160–200 kg

five months of darkness a year, the Muskox is adept at finding the food it needs to survive. Its hard, sharp hooves are used to dig into the crusted snow to uncover the sedges and woody plants that sustain it through winter. Rather than seeking shelter in low areas, herds tend to roam on high, exposed areas in winter, where the high winds help scour away the snow, exposing the vegetation. The Muskox's body also has a fairly high amount of fat, a characteristic shared among many Arctic mammals.

The cumulative effect of these adaptations to cold means that the Muskox cannot tolerate warm temperatures and risks overheating with the exertion of running. Because it cannot run for more than just a short sprint, its best method of protection against a predator is to stand and face it. The Muskox's main predator is the Arctic Wolf, although Polar Bears and even Grizzly Bears may hunt it during the summer months. When a predator approaches a herd, the strongest Muskox form a defensive circle around the young and weak members. As the intruder draws nearer, individual Muskox dart out of the circle and charge the intruder.

The wool of the Muskox's thick winter undercoat is called *qiviut*. Qiviut is a highly valued wool because it is very warm and not itchy. As with sheep's wool, the qiviut can be collected without harming the animal. Because Muskox domesticate easily, a growing industry in the North specializes in Muskox qiviut and meat.

Muskox once had a circumpolar distribution, inhabiting the Arctic tundra, but overhunting decimated their numbers everywhere except the central Canadian Arctic. In the last few decades, Muskox have been reintroduced to much of their former range, including Norway, Sweden, Russia, Alaska and northern Greenland. The current population of Muskox is only about 100,000 to 125,000 individuals worldwide, of which about 75,000 are in Canada. They are protected from commercial hunting, but subsistence hunters still take a small number.

DID YOU KNOW?

With the long, windbreaking guard hairs and insulating qiviut, the winter coat of a Muskox is about eight times warmer than sheep's wool.

Warming Arctic temperatures pose a serious threat to the Muskox. When winter temperatures rise above 0°C and then freeze again, an impenetrable layer of ice can form over the ground. This ice prevents Muskox from getting the food they need, and starvation—a looming threat during any winter—can claim many.

ALSO CALLED: Oomingmak.

DESCRIPTION: Well adapted to the cold, the Muskox is short and stout with long, ground-length hair. The long guard hairs are dark brown or grizzled, and in winter, a thick, lighter brown or greyish undercoat develops. This undercoat is shed in spring. Both males and females have horns that curve downward over the head and then turn up at the tip. A male's horns meet in the centre, creating a heavily ridged and thickened area called a "boss." A female's horns do not touch in the centre and have pale hairs between them. The shoulder hump makes the head appear to hang low, like that of the American Bison. Its stocky build gives the impression of a larger animal, but the Muskox is really quite small. Males have a strong musky odour, hence the name.

HABITAT: Muskox live on the tundra, usually on well-vegetated sedge slopes and valleys. They are often found near wetlands and watercourses with an abundance of sedges, grasses and forbs. Muskox are nonmigratory, and their movements between summer and winter ranges is not more than 80 km.

FOOD: In summer, Muskox feed mainly sedges, grasses, willows and some aquatic vegetation. Their winter food consists of sedges and woody plants, and they will dig through snow to reach the plants.

DEN: Muskox live in groups of five to 50 individuals (the average is 15), and though bulls may summer alone, they rejoin the herd in winter. There is no den, and their home territory is small. As cud-chewers, Muskox rest many times throughout the day, wherever they are. Shelter, such as ridges or slopes,

walking trail

is not sought, but the animals do huddle together for warmth, and during severe storms, they lie down with their backs to the wind.

REPRODUCTION: Some bulls live alone during summer, but they rejoin their herds in late August. Breeding takes place from August to September, and males compete by clashing horns and their heavily shielded heads. A male will mate with as many females as he can win access to. Females give birth to a single calf (rarely twins) in mid-April to early June after a gestation of eight to nine months. The calf nurses for 15 to 18 months. If a calf makes it through its first winter, it has a good chance of living a long life. Females are sexually mature at three years of age, and a female usually produces one calf every two years. Muskox can live for about 20 years.

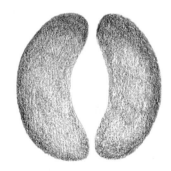

hoofprint

SIMILAR SPECIES: The **American Bison** (p. 38) is larger, with short, brown hair, and it has a different range.

American Bison

Bighorn Sheep
Ovis canadensis

male

No matter where you travel in North America, the mountains of Canada are unsurpassed for their diversity of hoofed mammals. It seems fitting, therefore, that one of the most recognizable and revered ungulates, the Bighorn Sheep, is a favourite symbol of the mountain wilderness. Although the Bighorn Sheep has a well-developed sense of balance and is at home on steep slopes and rocky ledges, it is also common along roadsides in mountain parks and preserves. Two subspecies are found in Canada: the California Bighorn Sheep (ssp. *californiana*) and the Rocky Mountain Bighorn Sheep (ssp. *canadensis*).

Now that the days of hunting Bighorn Sheep in protected areas have long passed,

RANGE: From the Rocky Mountains of Alberta and west-central British Columbia, the Bighorn Sheep's range extends southeast through Montana and south through California and New Mexico into northern Mexico.

Total Length: 1.2–1.9 m

Shoulder Height: 75–105 cm

Tail Length: 8–15 cm

Weight: 53–156 kg

many animals wander comfortably in areas of human activity. Observers who are unobtrusive and nonaggressive may be rewarded with glimpses of the sheep's natural behaviour amid the beautiful mountain scenery. As friendly and quiet as a Bighorn Sheep may appear, however, always remember that if it feels threatened or uncomfortable, it can easily demonstrate its strength and agility in an intimidating charge.

A Bighorn lamb that is too young or too small to have mastered the sanctuary of cliffs is particularly vulnerable to Coyotes and Grey Wolves. A newborn lamb may become prey for an eagle, Mountain Lion or Bobcat, as well. If it survives its first year, however, the Bighorn will most likely live a long life—few of its natural predators can match the Bighorn Sheep's surefootedness and agility.

The magnificent courtship battles between Bighorn rams have made them favourites of TV wildlife specials and corporate advertising. During October and November, adult rams establish a breeding hierarchy based on the relative sizes of their horns and the outcomes of their impressive head-to-head combats. In battle, opposing rams rise on their hindlegs, run a few steps toward one another and smash their horns together with testosterone-fuelled fervour. Once the breeding hierarchy has been established, mating takes place, after which the rams and ewes tend to split into separate herds. For the most part, the rams abandon their head-butting encounters until the next autumn, but broken horns and ribs are reminders of their hormone-induced clashes.

ALSO CALLED: Mountain Sheep.

DESCRIPTION: This robust, brownish sheep has a bobbed tail and a large, white rump patch. The belly, the insides of the legs and the end of the muzzle are also white. The brown coat is darkest in autumn, gradually fading with winter wear. In June and July, the sheep looks motley while the new coat grows in. "Bighorn" is a well-deserved name because

the circumference of a ram's horn can be as much as 45 cm at the base. The curled horns can be 110 cm long and spread 65 cm from tip to tip. Heavy ridges, the pattern of which is unique to each individual, run transversely across the horn. A deep groove forms each winter, which makes it possible to determine a sheep's age from its horns. The ewe's horns are shorter and noticeably more flattened from side to side than the ram's horns. Also, the ewe's horns are only up to 30 cm long and never curl around to form even a half circle, whereas an older ram's horns sometimes form a full curl or more.

HABITAT: The Bighorn Sheep is most common in unforested, mountainous areas where cliffs provide easy escape routes, but it can thrive outside the mountains as long as precipitous slopes are present near appropriate food and water. Some populations live along steep riverbanks and even in the gullied badlands of desert environments.

FOOD: The diet consists primarily of broad-leaved, nonwoody plants and grasses. Exposed, dry grasses on windswept slopes provide much of the winter food. Bighorn Sheep exhibit an incredible appetite for salt, and to fulfill this need, herds may travel many kilometres, even through dense forests, to reach natural salt licks. They often eat soil alongside highways to get the road salt that is applied during winter, which unfortunately increases the number of collision fatalities.

DEN: The Bighorn Sheep typically beds down for the night in a depression that is about 1.2 m wide and up to 30 cm deep. The depression usually smells of urine and is almost always edged with the sheep's droppings.

REPRODUCTION: Typically, after a gestation period of about six months, a ewe gives birth to a single lamb in seclusion on a remote rocky ledge in late May or early June. The ewe and her lamb rejoin the herd within a few days. Initially, the lamb nurses every half hour; as it matures, it nurses less frequently, until it is weaned at about six months old. Lambs are extremely agile and playful, jumping and running about, scaling rock outcroppings, engaging in mock fights and even jumping completely over one another—all activities that prepare them for escaping predators later in life.

hoofprint

walking trail

SIMILAR SPECIES: Dall's Sheep (p. 54) has thinner horns and a more northerly range. The **Mule Deer** (p. 68) also has a large, whitish rump patch and an overall brown colour, but bucks typically have branched antlers (rather than unbranched, curled horns), and have neither antlers nor horns. The **Mountain Goat** (p. 42), which sometimes shares habitat with the Bighorn Sheep, is white, not brown, and its shorter horns are black and cylindrical.

Dall's Sheep

Dall's Sheep
Ovis dalli

male

"Stone Sheep"

In the northern regions of the Rockies, Dall's Sheep appear as tiny, greyish spots on the wilderness palette. Closely related to the Bighorn Sheep, the Dall's Sheep is easy to identify because of its long, wide-spreading, spiralled horns. It is sometimes called the "Thinhorn Sheep" because the ram's horns are relatively thin at the base compared to the massive horns of Bighorn rams. Dall's ewes also have horns, but they are much reduced in size and shape.

There are two distinct races of the Dall's Sheep, and some authorities have

RANGE: The Dall's Sheep occurs in all but the extreme northern and western parts of Alaska, across the Yukon, in the western mountains of the Northwest Territories and in the mountains of northern British Columbia.

Total Length: 1.4–1.8 m

Shoulder Height: 75–105 cm

Tail Length: 7–11 cm

Weight: 45–100 kg

male

white form

even considered them separate species. Both subspecies occur in Canada. The northern one (ssp. *dalli*) is found in the Yukon, the Northwest Territories and British Columbia. It has a predominantly white coat. The southern subspecies (ssp. *stonei*), which is sometimes called the "Stone Sheep," has a darker grey coat and only occurs in British Columbia. In areas where the two forms interbreed, their offspring often have grey backs and white heads, legs and rumps.

The composition of a herd of Dall's Sheep changes with the seasons. Ewes and lambs form nursery groups from early summer until autumn, during which time the rams journey high into the mountains. Some of the rams may group together, but often the oldest and dominant rams remain solitary. The separation of the two groups means that each has less competition for food. The ewes with lambs remain in areas of better cover and less exposure. Even if the grazing is poorer, these kinds of areas provide better protection from predation.

In autumn, the two groups come together for the mating season, which is a busy and dramatic time for Dall's Sheep. The rams engage in vigorous courtship battles to determine their status. Competing rams rise up on their hindlegs and lunge toward their opponent in the same manner as the more famous Bighorn Sheep. After a few head-on blows, the rams push and shove each other until one of them turns away. The dominant ram wins the chance to mate with the most ewes.

Bands of Dall's Sheep must often cross extensive lowlands as they migrate from their summer ranges to their winter ranges. During this time, they are away from the safety of the cliffs, and in open terrain, they are vulnerable to predation by Grey Wolves, Mountain Lions, Canada Lynx, Wolverines and bears. On occasion, a golden eagle may swoop down and take a young lamb.

ALSO CALLED: Thinhorn Sheep, Stone Sheep.

DESCRIPTION: The Dall's Sheep found in the mountains of north-central BC are slate brown to almost blackish overall, except for white on the muzzle, forehead, rump patch and the inside of the hindlegs. The pure white form of the Dall's Sheep can be found farther north. The horns and hooves are light amber, and the iris of the eye is golden brown. A ram's horns are thicker than a ewe's and spiral widely. The horns of a ewe are short and curved backward, never achieving the complete spirals sometimes exhibited by a ram's horns.

HABITAT: In summer, Dall's Sheep occupy alpine tundra slopes to 2000 m elevation. They descend to drier south- or southwestern-facing slopes in winter. Bands of Dall's Sheep may travel long distances outside their typical habitats to find mineral licks.

FOOD: Broad-leaved herbs are favoured foods in spring and summer, with grasses and seeds making up most of the winter diet. The branch tips of willows, as well as pasture sage, cranberry, crowberry and mountain avens, are also consumed in winter.

DEN: The Dall's Sheep does not keep a den, but it is seldom far from steep, rocky cliffs,

> **DID YOU KNOW?**
>
> When Dall's Sheep rams engage in their autumn head-butting contests, the sound of their horns clashing together can be heard more than 2 km away.

which serve as escape cover from eagles and carnivores. At night, a Dall's Sheep beds down wherever it is, choosing an elevated site with good visibility. In rocky areas, it paws the ground to remove the larger stones and create a gravelly bed.

REPRODUCTION: Usually a single lamb (occasionally twins) is born in the second or third week of May, following a gestation of slightly less than six months. The lamb lies close to its mother at first, but within a few days, it is clambering about the cliffs. By the time they are a month old, the lambs form groups and begin to feed on plants, but they are not weaned until they are three to five months old. A ewe first breeds in her second autumn, and she may mate with several rams during the day or two when she is receptive. A ram is typically seven to eight years old before he gets a chance to mate.

SIMILAR SPECIES: The **Bighorn Sheep** (p. 50) is brownish overall, the ram has more massive horns, and the species occurs to the south of the Dall's Sheep's range. The **Mountain Goat** (p. 42) is all white and has black, stilettolike horns, a longer coat and often a beard.

Bighorn Sheep

Pronghorn

Antilocapra americana

male

Through the blurred, heat-shimmered light of a grassland afternoon, the shape of a Pronghorn emerges from the brown landscape to stand and stare. Just as suddenly, it turns and retreats to the open plains. The Pronghorn superficially resembles a deer, and it is often called an antelope, but it has no close living relatives—it is the sole member of an ancient family of hoofed mammals that dates back 20 million years. This animal's unique, pronged horns are neither antlers nor true horns because only the outer keratin sheath is shed each year, not the bony core.

In open landscapes, the Pronghorn's phenomenal eyesight serves it well in detecting predators. The large eyes protrude so far out from the sides of its head that it

RANGE: The Pronghorn is found through much of western North America, from southern Alberta and Saskatchewan southwest into Oregon and south through California and western Texas into northern Mexico.

Total Length: 1.3–1.4 m

Shoulder Height: 83–100 cm

Tail Length: 6–17 cm

Weight: 35–60 kg

has stereoscopic vision to the rear as well as in front. A Pronghorn is rarely seen first.

Should danger press, a Pronghorn will erect the hairs of its white rump patch to produce a mirrorlike flash that is visible at a great distance. Speed, which comes easily and quickly to the Pronghorn, is this animal's chief defence, and even three-day-old fawns can outrun a human. The Pronghorn is the swiftest of North America's land mammals, and it is among the fastest in the world, second only to the cheetah. With its efficient metabolism, powered by an extremely large heart and lungs for its body size, the Pronghorn can run at about 90 km/h for several minutes at a time. Its lack of dewclaws is also thought to be a result of its adaptation for speed.

Until the 1920s, when the Canadian government established reserves on the Prairies to protect the dwindling populations, Pronghorn numbers declined rapidly as a result of overhunting. By the late 1940s, Pronghorn numbers seemed to have stabilized, and the government permanently closed the reserves. The Pronghorn is still considered a game animal by some, and hunting still occurs. Fortunately, the efforts of government and private programs have helped increase its numbers.

For all its speed, the Pronghorn is a poor jumper, and the fencing of rangelands throughout the Prairie provinces contributed to its decline. Although the fences still remain, many are now constructed without the bottom strand of wire, leaving a gap of about 46 cm. A herd of running Pronghorns surprises many passing motorists when, one after another, the animals hardly break stride to deftly dip beneath the lowest strand of wire in a fence.

ALSO CALLED: American Antelope.

DESCRIPTION: The upperparts, legs and tail are generally tan. The belly, lower sides and lower jaw are white, and there are two broad, white bands across the throat and a large, white rump patch. Both sexes may have horns, but those of the doe are never as long as her ears and

DID YOU KNOW?

Although Pronghorns typically give humans a wide berth, they can display extreme curiosity, and a piece of plastic or a rag caught on a fence and waving in the breeze will often entice individuals to approach.

do not have the ivory-coloured tips seen on the buck's horns. The buck's horns are up to about 40 cm long. They are straight near the base, and then bear a short branch, or "prong," before curving backward or inward to sharp tips. A doe's horns are about 12 cm long and are rarely pronged. The muzzle is black, and on the buck, the black extends over the face to the horn bases. The buck also has a broad, black stripe running from the base of the ear to behind the lower jaw. There is a short, black-tipped mane on the nape of the neck. There are no dewclaws on the legs.

HABITAT: The Pronghorn is a staunch resident of treeless areas. It inhabits open, often arid grasslands, grassy brushlands and semi-deserts and avoids woodlands. It is adapted to exploit the patchy mosaic of grasses and forbs created by large, grazing herbivores such as the American Bison.

FOOD: The winter diet is composed almost exclusively of sagebrush and other woody shrubs. In spring, the diet switches to rabbitbrush, snowbrush, snowberry and sagebrush for 67 percent of the intake, forbs make up 17 percent, alfalfa and crops about 15 percent and grasses only one percent.

DEN: Because it is a roaming animal that remains active day and night—alternating short naps with watchful feeding—the Pronghorn does not maintain a home bed.

REPRODUCTION: Forty percent of does bear a single fawn with their first pregnancies, but 60 percent of first pregnancies and nearly all subsequent pregnancies result in the birth of twins. A doe finds a secluded spot on the prairie to give birth

in June, following a gestation period of 7.5 to 8 months. Fawns lie hidden in the grass at first, and their mothers return to nurse them about every 1.5 hours. The doe gradually reduces the frequency of nursing, and when a fawn is about three weeks old and capable of outrunning potential predators, mother and young rejoin the herd. Some does may breed during the short, mid- to late September breeding season of their first year, but most do not breed until their second year.

hoofprint

walking trail

gallop group

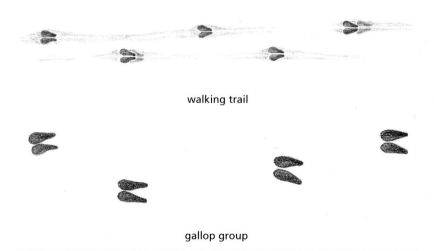

SIMILAR SPECIES: The **Mule Deer** (p. 68) is larger, with a white rump, and the buck has antlers, not black horns. The **White-tailed Deer** (p. 72) does not have a white rump. Neither deer has the white throat bands or white lower sides of the Pronghorn.

Mule Deer

Fallow Deer
Dama dama

With populations in at least 38 countries, the Fallow Deer is one of the most widely introduced ungulates in the world. It is native to the Mediterranean countries, Asia Minor and possibly parts of northern Africa. The first introduction of this deer into the British Isles and Europe occurred because of Europeans' interest in new game species. The beauty of this deer's antlers made it a prize fit for princes and kings.

The intentional introduction of foreign species is often ill-conceived and ill-fated; many nonnative populations die out either from harsh environmental conditions or from an inability to defend against local predators. On the other hand, populations of introduced species may succeed too well and outcompete native animals to the point of extirpation or extinction. The introduction of

RANGE: This species has been widely introduced throughout North America, especially in parks. In British Columbia, it is only found on James and Sidney islands, with a few sighting on the Saanich Peninsula.

Total Length: 1.4–1.8 m

Shoulder Height: about 1 m

Tail Length: 16–19 cm

Weight: 40–80 kg

the hardy Fallow Deer has resulted in a decline of some native deer species.

Although the original introductions of the Fallow Deer were for hunting purposes, it has recently been introduced because of its adaptability and beauty. Most deer species remain shy and wary of humans, but the Fallow Deer semi-domesticates easily, making it a popular addition to public and private parks. In British Columbia, the species was introduced to several of the Gulf Islands, where its populations are stable. Attempts to introduce it to Vancouver Island and Saltspring Island were unsuccessful.

DESCRIPTION: This small deer is usually light brown with white spots, but individuals can be white, cream, yellowish, silver, greyish or even black. The undersides are white. A black stripe runs along the spine from the nape of the neck onto the long tail, and there is a conspicuous white line along the flanks. The hindlegs are slightly longer than the forelegs, which elevates the rump. The male's antlers are distinctly palmate and flattened on the terminal tines, giving the antlers a "top-heavy" appearance.

HABITAT: Worldwide, these deer inhabit a variety of habitats, such as open areas within forests, grasslands, brushy hills, savannah and rolling parkland. Most populations are found in warm, humid climates, but some herds inhabit cool, humid areas or warm, dry areas.

DID YOU KNOW?

The Latin word *dama* is a general term for deer or deerlike animals. Thus, the scientific name *Dama dama* loosely translates as "just a deer, just a deer."

FOOD: The Fallow Deer's diet changes through the year. It eats grasses and other green vegetation when they are abundant. In autumn and winter, it consumes many nuts from trees and shrubs.

DEN: Fallow Deer live in herds that roam through good foraging areas. At night, they bed down in the grass and leave unmistakable imprints in the vegetation. When a female is ready to give birth, she becomes secretive and finds a hiding place in bushes or other cover, where she forms a bed in the vegetation. The female and her fawn continue to use this hiding place for about one week after she gives birth, and then they rejoin the rest of the herd.

REPRODUCTION: Rut peaks in October, when dominant males control a group of females. Subordinate males are chased away by the dominant male. After a gestation of 33 to 35 weeks, the female gives birth to one fawn, which is weaned in five to nine months. Females are sexually mature at six months of age but do not breed until they are at least 16 months old. Males do not mate until they are four years old.

SIMILAR SPECIES: The **Mule Deer** (p. 68) and the **White-tailed Deer** (p. 72) have smaller, nonpalmated antlers, and only their fawns have spots. The **Wapiti** (p. 64) is larger and has larger, nonpalmated antlers and a distinctive yellowish rump.

Mule Deer

Wapiti
Cervus canadensis

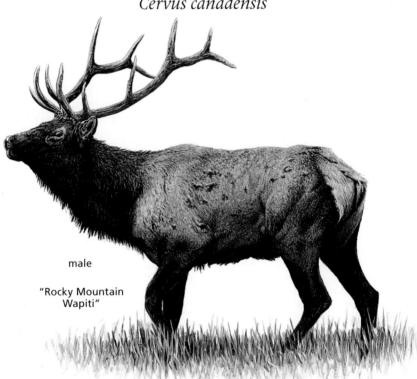

male

"Rocky Mountain
Wapiti"

The pitched bugle of the bull Wapiti is, in some parts of Canada, as much a symbol of autumn as the first frost, golden leaves and migrating geese. The Wapiti has likely always held some form of fascination for humans, as evidenced by Native hunting and lore, but it is another of North America's large mammals that suffered widespread extirpation during the time of settlement and agricultural expansion across the continent.

The dramatic decline of the Wapiti in North America during the 19th century prompted widescale conservation efforts

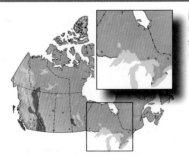

RANGE: The Wapiti occurs from northeastern British Columbia to southern Manitoba and south to California, Arizona and New Mexico. It has been reintroduced to Ontario.

Total Length: 2.0–2.6 m

Shoulder Height: 1.2–1.5 m

Tail Length: 12–18 cm

Weight: 175–495 kg

for remnant populations and far-reaching reintroduction programs to form new herds. Even the great numbers of Wapiti currently seen in mountain parks owe their presence to mitigating human efforts.

Fortunately for Wapiti, much of their range has become more accessible to grazing, even during winter. Artificially lush golf courses and agricultural fields supply high-quality forage throughout the year, whereas roads, townsites and other human activities have eliminated most major predators—except, of course, humans. In wilder areas, Wapiti are typically most active during the daytime, particularly near dawn and dusk, but they often become nocturnal in areas of high human activity where hunting occurs.

In Canada, there are three subspecies of Wapiti: Rocky Mountain Wapiti (ssp. *nelsoni*), Roosevelt Wapiti (ssp. *roosevelti*) and Manitoban Wapiti (ssp. *manitobensis*). The Rocky Mountain Wapiti have pale-coloured sides and flanks, whereas the Roosevelt subspecies is usually darker in colour, and the males tend to develop a "cup" on the royal tine of their antlers. This cup gives the base of the antlers a slightly palmate appearance. Manitoban Wapiti have slightly smaller antlers relative to their body size than the other subspecies. New DNA studies may contradict this subspeciation, as preliminary studies have indicated the possibility that all extant North American Wapiti are, in fact, the same subspecies.

Wapiti form breeding harems to a greater degree than other deer, and the bugle of the male is associated with mating in autumn. A bull Wapiti that is a harem master expends a considerable amount of energy during the autumn rut—his fierce battles with rival bulls and the upkeep of cows in his harem demand more work than time permits—and, if snows come early, he starts winter in a weakened state. Once the rut is over, however, bulls can put on as much as half a kilogram per day if conditions

are good. Cows and young, on the other hand, usually see the first frost while they are fat and healthy. This disparity makes sense in evolutionary terms: cows enter winter pregnant with the next generation, whereas, once winter arrives, the older bulls' major contributions are past.

ALSO CALLED: Elk.

DESCRIPTION: The summer coat is generally golden brown. The winter coat is longer and greyish brown. Year-round, the head, neck and legs are darker brown, and there is a large, whitish to orangey rump patch bordered by black or dark brown fur. The oval metatarsal glands on the outside of the hocks are outlined by stiff, yellowish hairs. A bull Wapiti has a dark brown throat mane, and he starts growing antlers in his second year. By his fourth year, the bull's antlers typically bear six points to a side, but there is considerable variation both in the number of points a bull will have and the age at which he acquires the full complement of six. A bull rarely has seven or eight points. The antlers are shed by March, and new ones begin to grow in late April, becoming mature in August.

hoofprint

gallop print

HABITAT: Although Wapiti prefer open forests and grasslands, they sometimes range into coniferous forests or brushlands.

FOOD: The Wapiti is one of the most adaptable grazers. Woody plants and fallen leaves frequently form much of the winter and autumn diet. Sedges and grasses make up 80 to 90 percent of the diet in spring and summer. Because the diet is deficient in minerals, salt is a necessary dietary component for all animals that chew their cud. Wapiti may travel great distances to find salt-rich soil.

DEN: The Wapiti does not keep a permanent den but often leaves flattened areas of grass or snow where it has bedded down to sleep.

REPRODUCTION: A cow Wapiti isolates herself from the herd to give birth to a single calf between late May and early June, following an 8.5-month gestation. The calf can stand and nurse within an hour, but the female keeps the calf hidden for 10 days. Within two to four weeks, the cow and calf rejoin the herd. The calf is weaned in autumn.

walking trail

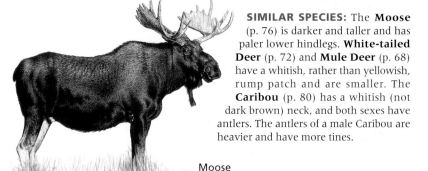

SIMILAR SPECIES: The **Moose** (p. 76) is darker and taller and has paler lower hindlegs. **White-tailed Deer** (p. 72) and **Mule Deer** (p. 68) have a whitish, rather than yellowish, rump patch and are smaller. The **Caribou** (p. 80) has a whitish (not dark brown) neck, and both sexes have antlers. The antlers of a male Caribou are heavier and have more tines.

Moose

Mule Deer
Odocoileus hemionus

male

"Rocky Mountain
Mule Deer"

The Mule Deer, an inhabitant of much of Canada since prehistoric times, continues to thrive in mountains, grasslands and even in fragmented landscapes. It tends to frequent open areas in parks and other protected areas and can be bold, conspicuous and quite tolerant of human visitors.

One of the Mule Deer's best-known characteristics is its bouncing gait, which is known as "stotting" or "pronking." When it stots, a Mule Deer bounds and lands with all four legs simultaneously, so that it looks like it's using a pogostick. This fascinating gait allows the deer to move safely and rapidly across and over the

RANGE: Widely distributed throughout western North America, this deer ranges from the southern Yukon southeast to Minnesota and south through California and western Texas into northern Mexico.

Total Length: 1.3–1.7 m

Shoulder Height: 90–105 cm

Tail Length: 11–22 cm

Weight: 31–215 kg

many obstructions it encounters in the complex brush and hillside areas it typically inhabits. Although stotting is characteristic of the Mule Deer, this animal also walks, trots and gallops perfectly well. When disturbed, a retreating Mule Deer will often stop for a last look at whatever disturbed it before disappearing completely from view.

Mule Deer feed at dawn, at dusk and well into the night. They have great difficulty travelling through snow that is more than 50 cm deep, so they are unable to occupy high, mountainous areas in winter. To avoid the snow, they migrate to lower elevations at the onset of winter, often into residential and suburban areas, which have buried grasses and dormant ornamentals that are much to their liking.

During the mating season, Mule Deer bucks compete for the does that are in estrus. Two bucks will battle using their antlers, each trying to force the other's head lower than his own. The weaker of the two eventually surrenders and usually leaves the area. Rarely, the antlers of two bucks become locked during these competitions, and if they are unable to free themselves, both bucks inevitably perish from starvation, predation or battle wounds.

In regions where both the Mule Deer and the White-tailed Deer occur, they do hybridize on occasion. Hybrid male offspring are sterile, and though hybrid females are fertile, all hybrids seem to have higher mortality rates than the pure species, which may be why hybrids are rarely seen.

There are three subspecies of Mule Deer found in Canada: the Sitka Deer (ssp. *sitkensis*) is found in coastal British Columbia and on the islands, the Black-tailed or Columbian Deer (ssp. *columbianus*) is found in the rest of British Columbia, and the Rocky Mountain Mule Deer (ssp. *hemionus*) is found east of the Coast Range crest, through the Rockies and most of the plains.

DESCRIPTION: As its name suggests, the Mule Deer has large ears. The dark forehead contrasts with both the face and

DID YOU KNOW?

Although the Mule Deer is usually silent, it can snort, grunt, cough, roar or whistle, and a fawn will sometimes bleat. Even people who have observed deer extensively may be surprised to encounter one vocalizing.

upperparts, which are tan in summer and dark grey in winter. There is a dark spot on either side of the nose. The large, whitish rump patch is divided by the short, black-tipped tail. The throat and underparts are white year-round. The buck has fairly heavy, upswept antlers that are equally branched (bifurcated) into forked tines. The metatarsal glands on the outside of the lower hindlegs are 10 to 15 cm long.

HABITAT: Summer habitats vary from lowland coulees and dry brushland to alpine tundra. The bucks tend to move to higher altitudes and form small bands, whereas the does and fawns remain at lower altitudes. In drier regions, both sexes are often found in streamside habitats. Mule Deer thrive in the early successional stages of forests, so they are often found where fire or logging removed the canopy a few years before.

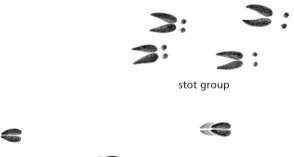

stot group

walking trail

FOOD: Grasses and forbs form most of the summer diet. In autumn, Mule Deer consume the foliage and twigs of shrubs. In winter, they make increasing use of twigs and woody vegetation, and they graze in hayfields adjacent to cover.

DEN: The Mule Deer leaves oval-shaped depressions in grass, moss, leaves or snow where it lays down to rest or chew its cud. It typically urinates upon rising; a doe steps to one side first, but a buck will urinate in the middle of the bed.

REPRODUCTION: Following a gestation period of 6.5 to 7 months, a doe gives birth to one to three (usually two) fawns in May or June, each typically weighing from 3.5 to 3.8 kg at birth. Fawns are born with light dorsal spots, which they carry until the autumn moult in August. They are weaned at four to five months of age and become sexually mature at 1.5 years.

hoofprint

SIMILAR SPECIES: The **White-tailed Deer** (p. 72) has much shorter metatarsal glands and a much smaller rump patch that is usually hidden by the reddish to greyish brown upper surface of the tail. A buck's antlers consist of a main beam with typically unbranched tines. The **Wapiti** (p. 64) is larger and has a yellowish or orangey rump patch. The male has a dark mane on the throat. The **Pronghorn** (p. 58) has black horns and white throat bands, and the lower half of its sides are white.

White-tailed Deer

White-tailed Deer
Odocoileus virginianus

male

Given the current status of the White-tailed Deer in Canada, it is hard to imagine that before the arrival of Europeans, this graceful animal was found only in small, isolated populations. Historically, this deer was uncommon, but with the spread of agricultural development and forest fragmentation, the White-tailed Deer has become quite widespread. In many regions, it is regularly seen in croplands and open fields.

The White-tailed Deer is a master at avoiding detection, so in wilderness areas, it can be frustratingly difficult to observe. It is very secretive during daylight hours, when it tends to remain concealed in thick shrubs or forest patches. Once the sun begins to set, however, the White-tailed Deer leaves its daytime resting spot, moving gracefully and weaving an intricate path through dense shrubs and over fallen trees, to travel to a foraging site. Despite its mastery of the terrain, this animal is still vulnerable to predators—the deer is aware that danger could be lurking in any shadow, and its nose

RANGE: From the southern third of Canada, the White-tailed Deer ranges south into the northern quarter of South America. It is largely absent from Nevada, Utah and California. It has been introduced to New Zealand, Finland, Prince Edward Island and Anticosti Island.

Total Length: 1.4–2.1 m

Shoulder Height: 70–115 cm

Tail Length: 21–36 cm

Weight: 50–200 kg

and ears constantly twitch. The major threats are Grey Wolves and humans, though fawns and old or sick individuals may also be easy prey for Coyotes.

Speed and agility are good defences against most predators, but all deer are vulnerable during winter in the northern parts of their range. Snow and a scarcity of high-energy food leave the deer with a negative energy budget from the first deep snowfalls in autumn until green vegetation emerges in spring. In spite of their slowed metabolic rates during winter, some deer may starve before spring arrives; these victims of winter weather provide food for scavengers.

In national parks and protected areas, White-tailed Deer may become habituated to the presence of humans and can sometimes be closely approached. Doing so can be perilous, however, especially around does protecting their young. Also, if a person approaches or touches a fawn, the doe may abandon it.

Although real danger exists in approaching any wild animal too closely, reports that White-tailed Deer are responsible for far more human fatalities annually than all North American bears misrepresent this animal. While true, these statistics include human fatalities resulting from vehicle collisions with deer. Each year, several hundred thousand deer are involved in accidents on North American roads.

ALSO CALLED: Flag-tailed Deer, Whitetail.

DESCRIPTION: The upperparts are generally reddish brown in summer and greyish brown in winter. The belly, throat, chin and underside of the tail are white. There is a narrow, white ring around the eye and a band around the muzzle. A buck starts growing antlers in his second year. The antlers first appear as unbranched "spikehorns"; in later years, generally unbranched tines grow off the main beam. The main beams, when viewed from above, are usually heart-shaped. The metatarsal gland on the outside of the lower hindleg is about 2.5 cm long.

DID YOU KNOW?

The White-tailed Deer is named for the bright white underside of its tail. A deer raises, or "flags," its tail when it is alarmed. The white flash of the tail communicates danger to nearby deer and provides a guiding signal for following individuals.

HABITAT: The optimum habitat for a White-tailed Deer is a mixture of open areas and young forests with suitable cover. This deer frequents valleys, stream courses, woodlands, meadows and abandoned farmsteads with tangled shelterbelts. Areas cleared for roads, parking lots, summer homes, logging and mines support much of the vegetation on which the White-tailed Deer thrives.

FOOD: During winter, the leaves and twigs of evergreens, deciduous trees and brush make up most of the diet. In early spring and summer, the diet shifts to forbs, some grasses and even mushrooms. On average, a White-tailed Deer eats 2 to 5 kg of food per day.

DEN: A deer's bed is simply a shallow, oval, body-sized depression in leaves or snow.

Favoured bedding areas—often in secluded spots with good all-around visibility where deer can remain safe while they are inactive—will have an accumulation of new and old beds.

REPRODUCTION: A White-tailed doe gives birth to one or two fawns (rarely three) in late May or June, after a gestation of 6.5 to 7 months. At birth, a fawn weighs about 2.9 kg, and its coat is reddish with white spots. The fawn can stand and suckle shortly after birth, but it spends most of the first month lying quietly under the cover of vegetation. It is weaned at about four months. A few well-nourished females may mate as autumn fawns, but most do not mate until their second year.

hoofprint

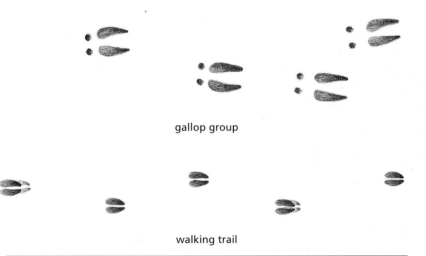

gallop group

walking trail

Mule Deer

SIMILAR SPECIES: The **Mule Deer** (p. 68) has longer metatarsal glands and a larger rump patch. A buck Mule Deer's antlers are bifurcated. The **Caribou** (p. 80) is larger and is brownish grey to white rather than reddish brown. The male Caribou also has much larger antlers. The **Wapiti** (p. 64) is larger, and its rump patch is yellowish.

Moose
Alces alces

male

The monarch of Canadian forests and lush wetlands, the Moose is a handsome animal that provides a thrilling sight for tourists and wildlife enthusiasts alike. People who know it only from TV cartoon characterizations may not have such feelings for the Moose, but those who have followed its trails through snow and mosquito-ridden bogs respect its abilities.

The Moose is the largest living deer in the world, and in much of Canada, there are excellent opportunities to see this majestic ungulate. Recent studies have shown the Moose population in some areas to be quite dynamic. In general,

RANGE: In North America, this Holarctic species ranges through most of Canada and Alaska. Its range has southward extensions through the Rocky and Selkirk mountains, into the northern Midwest states and into New England and the northern Appalachians. The Moose is expanding into the farmlands of Alberta, Saskatchewan, North Dakota and South Dakota, from which it was absent for many decades.

Total Length: 2.5–3 m

Shoulder Height: 1.7–2.1 m

Tail Length: 9–19 cm

Weight: 230–540 kg

this species appears to be increasing in number and range, but some regions—especially in southern parts of the country—report declining numbers. A factor contributing to low Moose densities may be an increase in parasites resulting from warmer temperatures.

The Moose's long legs, short neck, humped shoulders and big, bulbous nose may lend the animal an awkward appearance, but these features all serve it well in its environment. With its long legs, the Moose can easily cross streams and step over downed logs and forest debris. Snow, which seriously impedes the progress of other deer and predators, is no obstacle for the Moose, which lifts its legs straight up and down, creating very little snow drag.

The Moose has a huge battery of upper and lower cheek teeth that are perfectly suited for chewing the twigs that make up most of its winter diet. The big, bulbous nose and lips hold the twigs in place so the lower incisors can nip them off.

Winter ticks are often a problem for the Moose. A single animal can carry more than 200,000 ticks, and their irritation causes the Moose to rub against trees for relief. With excessive rubbing, a Moose will lose much of its guard hair, resulting in the pale grey "ghost" Moose that are sometimes seen in late winter. Winter deaths are usually the result of blood loss to the ticks rather than starvation—the twigs, buds and bark of deciduous trees and shrubs that form the bulk of the animal's winter diet are rarely in short supply.

The Moose's common name can also be traced to the animal's diet—the Algonquian called it *moz*, which means "twig eater." Its summer diet of aquatic vegetation and other greenery seems quite palatable and varied, but even then, more than half the intake is woody material.

DESCRIPTION: The dark, rich brown to black upperparts fade to lighter, often greyish tones on

> **DID YOU KNOW?**
>
> The Moose is an impressive athlete. Individuals have been known to run as fast as 55 km/h, swim continuously for several hours, dive to depths of 6 m and remain submerged for up to a minute.

the lower legs. The head is long and almost horselike. The nose is humped, and the upper lip markedly overhangs the lower lip. In winter, a mane of hair as long as 15 cm develops along the spine over the humped shoulders and along the nape of the neck. In summer, the mane is much shorter. Both sexes usually have a large dewlap, or "bell," hanging from the throat. Only the bull Moose has antlers. Unlike the antlers of other deer, the Moose's antlers emerge laterally, and many of the tines are palmate, meaning they are merged throughout much of their length, giving the antler a shovel-like appearance. Wapiti-like antlers are common in young bulls, and they are the only type seen in Eurasian individuals today.

A cow Moose has a distinct light patch around the vulva. A calf is brownish to greyish red during its first summer.

HABITAT: Typically associated with the northern coniferous forest, the Moose is most numerous in the early successional stages of willows and poplars. In less-forested foothills and lowlands, it frequents streamside or brushy areas with abundant deciduous, woody plants. In summer, it may range well up into tundra areas.

FOOD: About 80 percent of the Moose's diet is woody matter, mostly twigs and branches. It prefers deciduous trees and shrubs over conifers. In summer, it also feeds on aquatic vegetation. Sometimes a Moose sinks completely below the surface of the water to acquire succulent aquatics, but these plants never make up a large part of the diet.

DEN: The Moose makes its daytime bed in a sheltered area, much like other members of the deer family, and it leaves ovals of flattened grass from its weight. Other signs around the bed include large, oval droppings, browsed vegetation and tracks.

REPRODUCTION: In May or June, after a gestation of about eight months, a cow bears one to three (usually two) unspotted calves, each weighing 10 to 16 kg. To avoid wolves, cows often give birth on islands. The calves begin to follow their mother on her daily routine when they are about two weeks old. A few cows breed in their second year, but most will wait until their third year.

hoofprint

walking trail

Feral Horse

SIMILAR SPECIES: With its large size and long head, the Moose resembles a bay or black **Feral Horse** (p. 86) more than any native mammal. The **Caribou** (p. 80) is smaller and lighter in colour, and the bull does not have the lateral, palmate antlers of a bull Moose. The **Wapiti** (p. 64) is shorter and lighter brown in colour, with a yellowish rump patch.

Caribou
Rangifer tarandus

The Caribou carves out a living in the deep snows and blackfly-infested fens where most other deer species do not venture. It appears to do best in areas of expansive wilderness that allow it to undertake seasonal migrations between summer and winter feeding grounds. This specialist is better adapted to cold climates than other deer—even the Caribou's nose is almost completely furred.

The winter coat has hollow guard hairs up to 10 cm long that top a fine, fleecy, insulating undercoat. These guard hairs provide excellent flotation (as well as insulation) when the animal swims across rivers and lakes during its migrations. The Caribou's broad hooves serve it well over rough terrain and allow it to dig through snow to expose edible lichens. The bristle-like hairs that cover a Caribou's feet in winter may help prevent the snow from abrading its skin when the animal digs. This feeding strategy has been one of the Caribou's best-known characteristics for centuries—the name "Caribou" comes from the Mi'kmaq word *halibu*, which means "pawer" or "scratcher."

Unlike all other North American cervids, both sexes of Caribou grow antlers, but not all shed their antlers at the same time: mature bulls shed their large, sweeping racks in December; younger bulls retain their antlers until February; and cows keep theirs until April and begin growing new ones within a month. After losing their antlers, the bulls become subordinate to

RANGE: The North American range of this Holarctic animal extends across most of Alaska and northern Canada, from the Arctic Islands into the boreal forest and south through the Canadian Rockies and the Columbia and Selkirk mountains.

Total Length: 1.7–2.4 m

Shoulder Height: 0.9–1.7 m

Tail Length: 13–23 cm

Weight: 90–110 kg

the still-antlered cows, which are better equipped to defend desirable feeding sites.

The fragmentation of Caribou populations is of serious concern to resource managers, biologists and naturalists. There are still a few places where you can be assured of seeing these threatened animals, but they tend to move throughout their large home range. In winter, Caribou are more frequently seen in the southern parts of their range. In the provinces, movement is largely in response to food availability, and the animals tend to be in smaller bands than the Caribou of the territories. The seasonal movements of these small groups of Caribou hardly compare to the incredible migrations of their more northerly counterparts. In the North, some Caribou populations make the longest migration of any terrestrial mammal, up to 5000 km per year. In early spring, tens of thousands of these highly migratory animals work their way to northern Yukon and Alaska to calve and raise their young.

There was a time when the Caribou of North America were considered four separate species, but now all the North American Caribou and the Reindeer of Eurasia are classified as one species. The best-known subspecies in Canada include the Woodland Caribou (*R. t. caribou*), Barren-ground Caribou (*R. t. groenlandicus*) and Peary Caribou (*R. t. pearyi*).

DID YOU KNOW?

Lichens, the Caribou's favourite winter food, grow very slowly and are frequently restricted to tundra and older spruce and fir forests, but a herd's erratic movements typically prevent it from overgrazing one particular area.

ALSO CALLED: Forest Caribou, Mountain Caribou.

DESCRIPTION: In summer, a Caribou's coat is brown or greyish brown above and lighter below, with white along the lower side of the tail and hoof edges. The winter coat is much lighter, with dark brown or greyish brown areas on the upper part of the head, the back and the front of the limbs. Both sexes have antlers, but the bull's are much larger. Two tines come off the front of each main antler beam, and one lower "brow" tine is palmate near the tip. All other tines come off the back of the main beam, an arrangement that is unique to the Caribou.

HABITAT: Caribou are found in two habitats: forests of spruce, fir and pine or tundra areas in northern parts of the country.

FOOD: The most important food item for Caribou is fruticose lichens, but other

lichens, grasses, sedges, mosses, forbs and mushrooms also contribute to the summer diet. In winter, Caribou paw at the ground for lichens or eat arboreal lichens within easy reach. As well, winter foods include the buds, leaves and bark of both deciduous and evergreen shrubs. This restless feeder takes only a few mouthfuls before walking ahead, pausing for a few more bites and then walking on again.

hoofprint

walking trail (in snow)

walking trail (hard surface)

DEN: As with other cervids, the Caribou's bed is a simple, shallow, body-sized depression, usually in snow in winter and in leaves or grass in summer. In winter, the Caribou usually lies with its body at right angles to the sun on exposed frozen lakes—an arrangement that probably helps it absorb more solar energy. Entire herds will sometimes lie in the same orientation.

REPRODUCTION: Calving occurs in late May or June after a gestation of about 7.5 months. The female bears one unspotted calf (rarely twins) weighing about 5 kg. The calf is able to follow its mother within hours of birth and begins grazing when it is two weeks old. A calf is weaned after four or five months but probably stays with its mother for its first year. A cow usually first mates when she is 1.5 years old; most males do not get a chance to mate until they are at least three to four years old.

White-tailed Deer

SIMILAR SPECIES: The rectangular head and heavy body of the Caribou distinguish it from the other members of the deer family. The **White-tailed Deer** (p. 72) and **Mule Deer** (p. 68) are redder. The **Moose** (p. 76) is larger and darker brown. The **Wapiti** (p. 64) has a dark brown neck. Both sexes of Caribou bear antlers and even calves may bear spikes, a feature that distinguishes them from the females or young of other deer species.

Feral Pig
Sus scrofa

When we enter wildlife areas and national parks and think of the potentially dangerous animals therein, the first animals that conjure fear in us are bears and Mountain Lions. Oddly enough, we should really be thinking of the Feral Pig, which is a ferocious animal when wounded, cornered or with young. If you see distinctive patches of torn-up earth where a pig has been rooting around, be cautious because the animal may still be in the vicinity. The Feral Pig does not have good eyesight but its hearing is acute, and it will know of your presence long before you are aware it is nearby.

The Feral Pigs that are found in North America are descended from escaped domestic pigs, from European Wild Boars introduced for hunting or from hybrids of the two. The wild, pure-blood Feral Pig is different in appearance than the farm-raised variety, but the two varieties are able to interbreed. When Wild Boars were introduced to North America from Europe, they were first contained in preserves for hunting purposes. Many individuals escaped and crossbred with existing domestic varieties. In some provinces, the Feral Pig population has exploded, and these ravenous newcomers can destroy cropland.

ALSO CALLED: Wild Boar, Wild Pig, Wild Hog, Feral Hog.

RANGE: Feral Pigs occur in many parts of North America, especially the southern United States. Feral Pig populations in Canada are found in British Columbia, Alberta, Saskatchewan, Manitoba and Québec.

Total Length: 1.2–1.8 m

Shoulder Height: 55–110 cm

Tail Length: up to 30 cm

Weight: *Male:* 75–200 kg; *Female:* 35–150 kg

DESCRIPTION: The Feral Pig is a medium-sized hoofed mammal, just slightly shorter in height than the Mule Deer. Like a domestic pig, it has a sensitive, flexible disc at the end of the snout. The fur is coarse, dense and usually grey, brown or black, but some individuals may be mottled with white. In winter, the Feral Pig has a thick undercoat. Along the ridge of the back are long, dark, bristly hairs. The tail is sparsely furred and hangs straight down. The Feral Pig has tusks, or modified canines, that continue growing throughout its life. The upper tusks curl out and up from the mouth and may be up to 10 cm long. The larger lower tusks curl slightly outward from the mouth and may be up to 23 cm long. Unlike native artiodactyls, the Feral Pig has both upper and lower incisors.

HABITAT: The Feral Pig inhabits a variety of regions such as forested mountain areas, brushy areas, marshes, swamps, ravines and ridges.

FOOD: Feral Pigs are omnivorous and will eat almost anything. In nut-bearing forests in autumn, they dine heavily on acorns, walnuts and pecans. At other times of the

> **DID YOU KNOW?**
>
> Feral Pigs have an extremely well-developed sense of smell. For this reason, pigs are famous for their ability to "sniff out" truffles, a fungal delicacy that grows underground. The pigs that are used to find truffles are muzzled, so they can locate the truffle but not eat it.

year—or in other habitats—they eat green vegetation, roots, tubers, fruit, crayfish, frogs, salamanders, eggs, fledgling birds, rabbits, newborn fawns and carrion.

DEN: At night, Feral Pigs sleep in hollowed-out depressions or places of trampled vegetation. Pregnant sows hollow out a shallow "farrowing nest" in the ground and line it with grasses or other vegetation.

REPRODUCTION: Mating occurs throughout the year, but there are two seasonal peaks, from November to March and in July. Dominant males mate first, followed by young and subordinate males. Gestation is 16 weeks, whereupon the female gives birth to a litter of 1 to 12 young. The young are born in a grass-lined depression made by the sow. The piglets have several longitudinal stripes along each side, but they lose these by the time they are six weeks old. The young are weaned when they are three months old. A sow and her young often feed together as a family group, and, in some places, families join and form herds of up to 50 individuals.

SIMILAR SPECIES: Feral Pigs from domestic stocks tend to have finer fur, rounder bodies and shorter legs, and they lack tusks. Hybrids of Feral Pigs and domestic pigs have intermediate characteristics.

hoofprint

walking trail

Feral Horse
Equus caballus

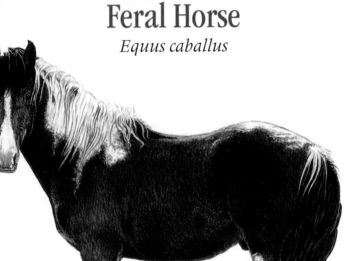

Feral Horses in North America are descended from domesticated populations and have lived in western Canada for centuries. They can usually be distinguished from their domestic kin by their much longer manes and tails and their patterns of behaviour. Most of Canada's Feral Horses live in British Columbia, Alberta and the Yukon. However, the most famous population in Canada is the one on Sable Island, Nova Scotia, which numbers about 300 horses.

Horse herds can have a different assortment of males and females, depending on the herd type. An accumulation of young bachelors is one type of herd. Males, usually over the age of two, leave their parent herd and may band together for a while because no herd stallion—the dominant male—will permit them near

RANGE: Feral Horses occur in pockets from the Yukon, British Columbia and Alberta through Montana, Wyoming, Utah and Colorado. Isolated herds live on Sable Island, Nova Scotia, and Assateague Island in the United States.

Total Length: up to 2 m

Shoulder Height: 1.0–1.7 m

Tail Length: up to 0.9 m

Weight: 265–390 kg

his mares. These young males stay together until they either find mares of their own or are strong enough to steal mares from an older stallion.

A second type of herd is the mixed herd, in which a number of mares, a single adult stallion, foals and a few young males live and forage together. The mares in this herd type are closely guarded by the stallion and are not free to come and go, though they are the ones that decide the herd's daily activities.

Another type of herd has two stallions, a number of subordinate mares and perhaps a few foals. In this grouping, the subdominant, usually younger, stallion exhibits "champing" behaviour, in which he approaches the dominant stallion nose to nose with his ears forward in a gesture of friendly respect. The subdominant stallion is usually the offspring of one of the mares in the harem. As he matures and becomes the dominant stallion's equal, the subdominant male's "champing" behaviour may become more threatening and, ultimately, a duel occurs. The two stallions face each other with their ears back, necks arched and tails high, then fight standing side by side, biting, kicking and pushing each other off balance. Eventually, one stallion is defeated and runs off, possibly to find other mares to make a new harem. The mares of the original harem stay together and accept the control of the victorious stallion. Only rarely does a mare leave to join a different herd.

Feral Horses uses their teeth to groom the mane, neck and withers of other horses, which helps develop and maintain the bonds between herd members. Biting flies seem to be a serious irritant to the Feral Horse, and after too many bites, a horse may be in a state of extreme distress. To rid itself of the flies, the Feral Horse may walk into thick foliage to scrape the flies off, roll in mud to cover and soothe the skin or submerge itself in water.

ALSO CALLED: Wild Horse, Mustang.

DESCRIPTION: Feral Horses are extremely variable in size and colour because of their domestic roots. They may be a solid colour, ranging from black to white, or they may be spotted, bay or have various other colour patterns. White markings, such as a star or blaze, are often seen on the face. Horses have both upper and lower incisors and small (if any) canines. They generally have a long mane and tail. Their hooves are semicircular and uncloven, and they lack dewclaws.

HABITAT: Feral Horses prefer areas of abundant vegetation close to watercourses,

but they are so adaptable that they may be found from deserts to alpine tundra. In Canada, they are found in woodland areas, foothills, dry ridges, brushlands and even on marshy plateaus.

FOOD: As grazers, Feral Horses spend as much as 80 percent of daytime hours grazing. Even at night, they sleep only about 50 percent of the time—the rest of the night they continue to graze. These horses are herbivores and consume mainly grasses and forbs during summer.

hoofprint

walking trail

In winter, they eat woodier vegetation such as the twigs or bark of shrubs.

DEN: The Feral Horse makes no den for sleeping, but if it lies down, a "bed" is visible where the grass was flattened. Although the Feral Horse can lie down, it usually sleeps standing up. As soon as the horse slips into sleep, a highly specialized tendon in each leg locks the knee and prevents the leg from collapsing. When the horse wakes, the tendon is released and the animal can move. This mechanism evolved as a defensive strategy. If the animal is awakened and detects danger, it can immediately run rather than having to take the time to rise from a prone position.

REPRODUCTION: A mare gives birth to one foal per year after a gestation of about 11 months. Mating may occur during spring, summer or autumn, often just a few days after a mare delivers her foal. The foal is precocious, and within a few hours of birth, it is able to run with its mother and the rest of the herd. The foal is weaned shortly before the next foal is born. If a mare is weak or if food is scarce, she may have a foal only every second year.

Moose

SIMILAR SPECIES: Although a large, dark-coloured Feral Horse may be mistaken for a **Moose** (p. 76) from a distance, the resemblance is restricted to the general size and colour. No other animal has the same combination of a long-haired mane and tail and a single, uncloven hoof on each foot. Unlike the native ungulates, the Feral Horse has neither antlers nor horns.

WHALES, DOLPHINS & PORPOISES

All whales belong to the order Cetacea and are distinguished from other mammals by their nearly hairless bodies, paddlelike forelimbs, lack of hindlimbs, fusiform bodies and powerful tail flukes. There are at least 80 species worldwide, classified into two suborders according to whether they have teeth (suborder Odontoceti) or baleen (suborder Mysticeti). The toothed whales are far more numerous and diverse, with some 70 species worldwide. There are only 11 species of baleen whales worldwide, but this group comprises the largest cetaceans.

At least 27 species of cetaceans are known to occur in Canadian waters; the best known and most commonly seen are the Grey Whale, Humpback Whale, North Atlantic Right Whale, Beluga, Orca, Atlantic White-sided Dolphin, Pacific White-sided Dolphin and Dall's Porpoise. With luck, and in the right locations, these species can be seen while they feed in bays and inlets or pass by on their annual migrations. Less-common species can be encountered unexpectedly at any time, especially on boat trips into open waters.

While whale watching, you may be lucky enough to see any of a number of whale displays. In a "breach," some or all of the whale's body rises out of the water and splashes back in. "Lobtailing" refers to a whale forcefully slapping its tail flukes on the surface of the water—not to be confused with "fluking," in which the flukes are raised clear above the water before a dive. A whale is "spyhopping" when it rises almost vertically out of the water, just far enough to have a look around. "Logging" is a form of rest; individuals float at the surface alone or in a close group, all facing the same direction.

Porpoise Family (Phocoenidae)

The porpoises, which number only six species worldwide, are often mistakenly referred to as dolphins, which they superficially resemble. The largest porpoise rarely reaches more than 2 m in length, and the smallest, the critically endangered Vaquita (*Phocoena sinus*), is no more than 1.5 m long. Unlike dolphins, porpoises do not have a distinct beak, and their heads are quite rounded. Their body shape is a bit more robust than the streamlined dolphins, and their flippers are typically small and stubby. Porpoises are generally timid—when they surface for air, they rise only long enough for a quick breath, and then roll rapidly back under the water.

Ocean Dolphin Family (Delphinidae)

This family includes some of the best-loved cetaceans, Bottlenose Dolphins (*Tursiops* spp.) and Orcas. Although many people call the Orca a whale, it is actually the largest dolphin in the world. All delphinids have a sleek, fusiform shape and are generally free of callosities and barnacles. Many of them exhibit a high brain-to-body weight ratio (known as the "encephalization quotient" or EQ) and are considered the most intelligent of the cetaceans. The Bottlenose Dolphin has the highest EQ, with a ratio similar to that of a chimpanzee.

Beluga & Narwhal Family (Monodontidae)

This small family of whales includes only the Beluga and the Narwhal, polar species that follow the seasonal formation and retreat of ice in the Arctic. These two whale species have overlapping ranges, and temporary nursery colonies have even formed with members from both species. These toothed whales do not have fused neck vertebrae—an unusual character-istic for cetaceans—and they are able to turn their heads, nod and look around in a manner unlike other whales. Neither species has a dorsal fin, an adaptation that allows for better manoeuvring under and around the ice.

Bowhead & Right Whale Family (Balaenidae)

This family includes two kinds of whales, the Bowhead Whale and the right whales. The Bowhead Whale is found only in Arctic and Subarctic waters, whereas right whales have a much larger distribution. Once found throughout the world's cold and temperate oceans, right whales are endangered and their total population now is probably only a few thousand individuals. Among their most distinctive features are the growths over the face and head. These callosities are masses of keratin topped with barnacles and teeming with whale lice. The whale lice may be pink, white, orange or yellow, and they give the cal-losities their unique colour. These callosities grow in all the places on the whale's head where a human would have hair. Specifically, they are found on the chin, the upper "lip" and above the eye.

Rorqual Family (Balaenopteridae)

Rorquals, or baleen whales, which number only a few species worldwide, represent some of the largest whales on earth, including the Blue Whale (*Balaenoptera musculus*), the largest animal on our planet. The name "rorqual" is derived from the Norwegian word *rorhval,* meaning "furrow," and refers to the pleats, or folds, in the skin of the throat. When the whale gulps a massive volume of food-rich water into its mouth, these pleats unfold, allowing the throat to distend into an enor-mous, balloonlike shape. The whale does not swallow the water; instead, the pleats contract and the tongue moves forward to push the water out through the baleen. Any crustaceans or fish are trapped inside the mouth to be swallowed whole.

Grey Whale Family (Eschrichtiidae)

The unique Grey Whale is the sole member of this family. It shares some characteris-tics with the rorqual whales, but it is dissimilar enough to be classified on its own. Like the rorquals, the Grey Whale has throat pleats that expand when water is drawn into the mouth, but the Grey's throat pleats are fewer and much less effective. The Grey Whale also has a heavier appear-ance, an arched mouth and yellow-ish brown baleen. It is believed to carry more parasites, such as barnacles and whale lice, than any other whale.

Harbour Porpoise
Phocoena phocoena

The little Harbour Porpoise is the most widespread member of its family. It favours coastal waters such as estuaries, shallow bays and tidal channels, and in Canada, it has even been seen quite far inland up major rivers. Harbour Porpoises are common on both coasts, especially in bays of the Inside Passage on the West Coast and in Trinity Bay and the Bay of Fundy on the East Coast.

In the past, this porpoise was a familiar sight to boaters and sailors of coastal waters, but its population is declining. In some regions, alarming numbers of Harbour Porpoises have washed up either dead or dying. Although the reasons are not clear, many scientists suspect that high levels of toxins and pollutants impair the immune systems of these and other cetaceans, which makes them more susceptible to life-threatening diseases.

Drowning deaths are another main reason why the Harbour Porpoise is declining. Because of this porpoise's feeding habits, it is frequently caught in bottom-set gill nets, a kind of net that hangs deep in the water like a curtain. Almost every kind of deep net can trap and drown cetaceans, but the most sinister are nets that have been thrown away or lost by fishermen. An untended net catches numerous cetaceans, sea turtles, fish and other sea creatures until it is so heavy that it sinks to the bottom.

The solution to this problem is not easy, but researchers have been working on ways to warn animals, particularly porpoises, where the nets are. A specially made device called a "pinger" can be fitted to the nets. The pinger emits a sound that

porpoises do not like, so they steer clear of the net. In the future, devices such as this may help save the lives of thousands of porpoises and other cetaceans.

On the West Coast, particularly around southern Vancouver Island, Harbour Porpoises can hybridize with Dall's Porpoises. The hybrid offspring usually have the overall shape of a Dall's Porpoise but the colouration of a Harbour Porpoise. In all cases, the hybrids are the offspring of a female Dall's Porpoise and a male Harbour Porpoise. These hybrid porpoises may represent as much as two percent of the population in the southern

RANGE: The Harbour Porpoise is found in cold-temperate and Subarctic waters of the Northern Hemisphere. It occurs in coastal waters up to 300 m deep but prefers waters no deeper than 200 m.

Total Length: up to 2 m; average 1.5 m

Total Weight: up to 90 kg; average 45–60 kg

Birth Length: 66–89 cm

Birth Weight: 5 kg

DID YOU KNOW?

The deepest dive ever recorded by a Harbour Porpoise was 224 m.

Gulf Islands. Unlike many hybrid mammals, the hybrid porpoises appear to be fertile; two known female hybrids have been seen with calves travelling near them. Hybridization has been documented in other cetaceans as well, most notably between Blue Whales and Fin Whales.

DESCRIPTION: This porpoise is the smallest cetacean in Canadian waters.

Its markings are not distinctive, so the best way to identify it is by general body shape and size, behaviour and the sound it makes to breathe. The Harbour Porpoise is black on the back and flippers, slate grey on the sides and head, and white on the belly. In some individuals, the colour change is gradual, whereas in others, the change is well defined, especially on the tailstock. Even on one individual, the colour pattern may not be the same

from one side of the body to the other. The face is small, the head tapers gently and the mouth angles slightly upward. This porpoise has black "lips," and one or two black streaks lead back to its all-black flippers. The dorsal fin and flukes are black as well, and the flukes are pointed slightly backward and are notched in the middle. Upon close examination, you can see little bumps, or tubercles, on the leading edge of the dorsal fin and flippers.

BLOW: The Harbour Porpoise makes a remarkable sound when it breathes. When this porpoise breaks the surface, it makes a quick sneezing sound, which has earned it the nickname "Puffing Porpoise."

OTHER DISPLAYS: Like other porpoises, Harbour Porpoises show little of themselves above water. They rarely perform acrobatics like dolphins, and they are not

as fast or as active as Dall's Porpoises. Only their comical sneezing sound attracts the attention of whale watchers. Harbour Porpoises do not like intruders, however, and if they are uncomfortable and want to hide, they will lie silently and motionlessly just below the surface of the water. This behaviour also occurs at night and probably helps them remain undetected by predators. On rare occasions, Harbour Porpoises have been seen bow riding. When they swim quickly, they may "porpoise" in and out of the water.

GROUP SIZE: Harbour Porpoises usually live in small groups of two to 10 individuals. Some groups have up to 20 members, and good feeding waters can attract from 50 to several hundred porpoises.

FOOD: This porpoise feeds in midwater or on the bottom, usually within 60 m of the suface, and it only rarely ventures into water more than 200 m deep. Main foods include small schooling fish such as anchovies or herring.

REPRODUCTION: Both male and female Harbour Porpoises mature sexually between three and four years of age. Mating usually occurs in late summer, and calves are born 10 to 11 months later. The newborns are dull brown, and for the first few hours, they have birth lines or creases circumscribing their bodies. They are weaned at five to nine months of age. This species is not long-lived and has a lifespan of about eight years.

SIMILAR SPECIES: The **Dall's Porpoise** (p. 96) is black and white. The **Atlantic White-sided Dolphin** (p. 100) and **Pacific White-sided Dolphin** (p. 102) tend to have larger fins and flippers and more pronounced beaks.

Dall's Porpoise

Dall's Porpoise
Phocoenoides dalli

Dall's Porpoises are a welcome sight for boaters and whale watchers. These high-speed swimmers can provide hours of delight for human spectators. They seem tolerant of human company, and the approach of boats rarely startles them.

Despite their name, these animals do not actually "porpoise" through the water the way dolphins and other small cetaceans do. Instead, they surface only long enough for a quick breath. In doing so, they create the distinctive conical splashes of water that are typical of the species.

Dall's Porpoises appear to undertake short migrations along the West Coast. In summer, they tend to move northward, and in winter, they move farther south. They may also move between inshore and offshore waters, perhaps in response to food availability. In some years, for unexplained reasons, mass assemblies of a few thousand Dall's Porpoises have been recorded in passages near Alaska and northern British Columbia.

Many cetacean species are believed to navigate long distances using magnetism. Researchers have found magnetic material in the brain and other tissues of Dall's Porpoises. These findings suggest that these porpoises may be able to use the Earth's magnetic field for navigation.

Worldwide efforts to protect whales have had many admirable results. Unfortunately, Dall's Porpoises are still being hunted on a massive scale for human consumption. Several countries, especially Japan, take a total of at least 14,000 per year—some take as many as 45,000. Several thousand more are accidentally killed in fishing nets. It is not known how long the species can sustain such losses; the Dall's Porpoise is not currently classed as endangered, but few reliable population estimates are available.

RANGE: This species is found in the North Pacific between 30° N and 62° N latitude, both in the open ocean and close to land.

Total Length: up to 2.4 m; average 1.8 m

Total Weight: up to 220 kg; average 140 kg

Birth Length: 76–91 cm

Birth Weight: about 20 kg

In addition to hunting, Dall's Porpoises appear to be particularly vulnerable to fishing nets. Some of their preferred prey are deepwater fish species that come nearer to the surface at night. This may account for some of the unintentional nighttime netting of Dall's Porpoises. Also, some studies suggest that their sonar may not be able to detect nets with a mesh finer than 5 cm.

Studies have shown that toxic contamination may also be causing population decline. Organochlorides and other toxins in the tissues of Dall's Porpoises can cause reduced testosterone in males and affect the viability of calves. Contamination that simultaneously affects both reproduction and survivability can have long-term negative consequences for some populations.

ALSO CALLED: Spray Porpoise, True's Porpoise, White-flanked Porpoise.

DESCRIPTION: Often mistaken by inexperienced observers for a baby Orca, the Dall's Porpoise is distinctly black and white. Its head is black and tapers to a narrow mouth. The "lips" of the small mouth are usually black but may be white on some individuals. The black body is extremely robust for its length, and there is a large, white patch on the belly and sides. In some Dall's Porpoises, the white patch extends from in front of the flippers to the tailstock; in others, it begins about one-third of the way down the body. The black, triangular dorsal fin

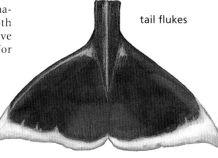

tail flukes

has a hooked tip, and it is usually light grey or white on the trailing half. The small flippers lie close to the head and are black on both sides. When viewed from above, the flukes are shaped like a wide ginkgo leaf and have white or grey trailing edges.

BLOW: This porpoise does not make a visible blow. As it swims and breaks the surface, however, a V-shaped cone of water, often called a "rooster tail," comes off its head. Many boaters look for this splash because it can be seen from a much greater distance than the porpoise itself.

dive sequence

OTHER DISPLAYS: The Dall's Porpoise does not leap out of the water, but it is exceptionally fast and even seems hyperactive as it darts and zigzags about. It is among the fastest cetaceans and has been clocked at speeds of up to 55 km/h. The Dall's Porpoise appears to love bow riding and will zoom toward a fast-moving boat like a black-and-white torpedo. If one of these porpoises comes to the bow of your boat, do not slow down for a better look because it will quickly lose interest in any vessel that is going slower than about 20 km/h.

GROUP SIZE: Dall's Porpoises are commonly found in groups of 10 to 20 individuals, though meetings of hundreds or even thousands may occur in some waters.

FOOD: These porpoises feed at the surface or in deep water, and their primary foods include squid, lanternfish, hake, mackerel, capelin and other schooling fish. Their maximum feeding depth is estimated to be 500 m.

REPRODUCTION: Two peaks in calving seem to occur, one in February or March and another in July or August. Peak mating must have a similar split, because gestation is about 11.5 months, and lactation is believed to last less than one year. Males reach sexual maturity when they are about 1.8 m long (four to five years old), and females when they are 1.7 m long (three to four years old).

SIMILAR SPECIES: A young **Orca** (p. 106) is larger and would never be seen unattended by its mother. The **Pacific White-sided Dolphin** (p. 102) is much greyer overall.

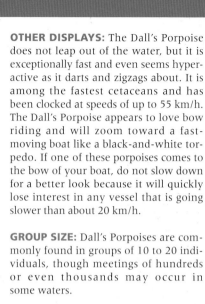

Pacific White-sided Dolphin

Atlantic White-sided Dolphin
Lagenorhynchus acutus

This large, colourful dolphin inhabits cool to temperate waters in the North Atlantic. Common around the Maritimes and Newfoundland, the Atlantic White-sided Dolphin can be seen in pods of a couple of dozen individuals. Larger pods, numbering about 60, are found near the shores of Newfoundland. Sometimes pods of several hundred dolphins are reported.

Another common dolphin in the same region is the White-beaked Dolphin (*L. albirostris*), and these two species are often misidentified. The White-beaked Dolphin lacks the yellow streak along its sides. Both species are playful and quick to bow ride.

This dolphin tends to prefer cool waters with low salinity. Although it is not strongly migratory, the Atlantic White-sided Dolphin moves between inshore and offshore waters in response to food availability. Normally, it is considered an open ocean species, but it comes to nearshore waters in spring in search of mackerel.

DESCRIPTION: The Atlantic White-sided Dolphin actually appears tricoloured: the belly and lower sides are white, the sides are grey and the back is dark grey or black. Its facial markings are quite distinctive; it has a black eye ring and a bicoloured beak, which is black above and grey below. A faint, grey stripe may connect the leading edge of the flipper with the eye ring or the rear margin of the lower jaw. There is a diagnostic lateral, yellow stripe to the rear of the dorsal fin, sometimes with a leading portion of white. The fins are all strongly pointed. The pectoral fin is about 30 cm in length, the dorsal fin may be up to 50 cm in height, and the tail flukes are up to 60 cm across. Females are a bit smaller than males.

BLOW: There is no discernible blow, but this dolphin often clears the water every 10 to 15 seconds to breathe.

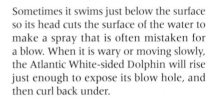

Sometimes it swims just below the surface so its head cuts the surface of the water to make a spray that is often mistaken for a blow. When it is wary or moving slowly, the Atlantic White-sided Dolphin will rise just enough to expose its blow hole, and then curl back under.

OTHER DISPLAYS: Playful and acrobatic, Atlantic White-sided Dolphins are known to breach high out of the water and bow ride with fast boats. They often

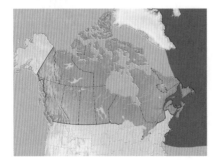

RANGE: This dolphin is found in Subartic waters of the North Atlantic Ocean. The western edge of its range extends from Greenland to New England, and the eastern edge stretches from the United Kingdom and Ireland to western Norway.

Total Length: 2.1–2.9 m

Total Weight: 180–230 kg

Birth Length: 0.9–1.1 m

Birth Weight: about 20 kg

swim just below the surface, creating a spray off their heads and dorsal fins.

GROUP SIZE: Pods of dolphins numbering several dozen are the most common.

DID YOU KNOW?

The total population of Atlantic White-sided Dolphins is about 300,000 individuals. Fewer than 1000 of these dolphins are hunted every year, but accidental drownings caused by fishing nets account for many more deaths.

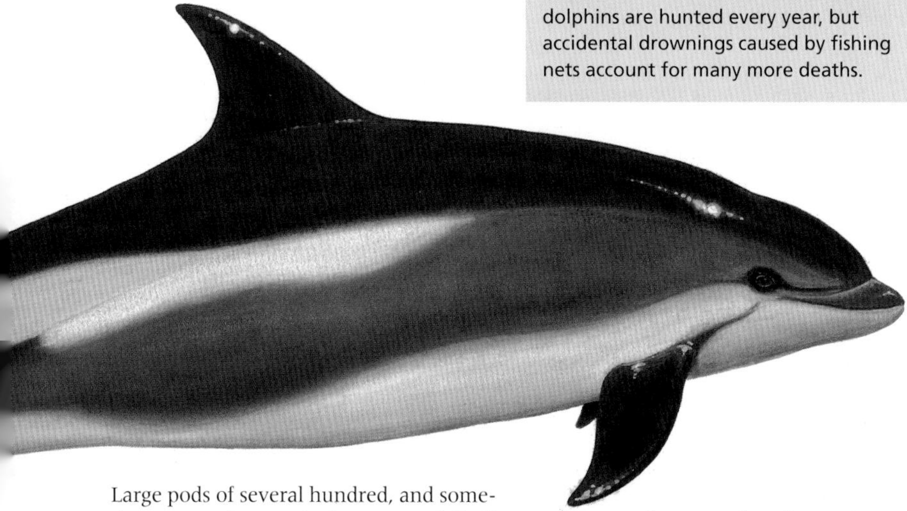

Large pods of several hundred, and sometimes even thousands, form, especially if they are following large whales that are feeding. Social dynamics also influence pod segregation. Young, sexually immature dolphins sometimes form their own pods, much like teenagers hanging out with each other.

FOOD: Lanternfish, herring, mackerel, squid, smelt, hake and shrimp appear to make up the bulk of the diet. Atlantic White-sided Dolphins are not considered deep divers. Although they favour waters between 70 and 300 m deep, they do not dive to the bottom and rarely stay underwater for more than four minutes. Sometimes they cooperate in their efforts to surround and attack schools of fish.

REPRODUCTION: Females give birth to one calf every two years. The peak calving period is in June and July but can extend from May to August. A female will mate again when her calf is about a year old, and gestation is about 11 months. Calves are weaned when they are 1.5 years old. Females reach sexual maturity at six to eight years of age, males at eight or nine. These dolphins can live 25 to 30 years, but most do not survive to 20 years of age.

SIMILAR SPECIES: The **White-beaked Dolphin** has a grey saddle behind the dorsal fin and a white tailstock and lacks the yellow lateral stripe. Sometimes these two species form mixed pods, so identification can be difficult. The **Harbour Porpoise** (p. 92) is smaller and grey overall.

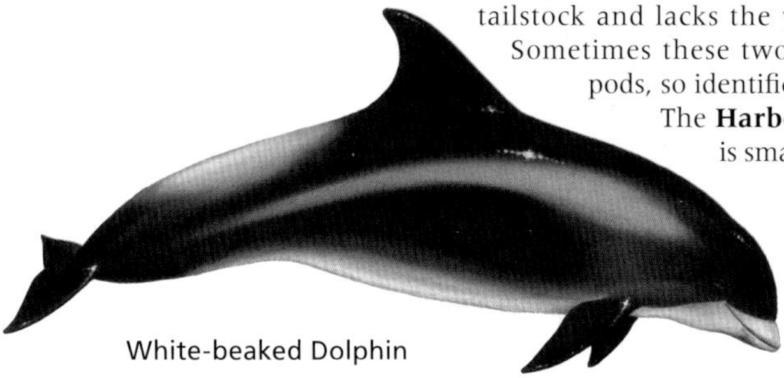

White-beaked Dolphin

Pacific White-sided Dolphin
Lagenorhynchus obliquidens

The acrobatic Pacific White-sided Dolphin is a favourite of whale watchers on the West Coast. This boisterous dolphin is so inquisitive and entertaining that it frequently "steals the show" from larger, less-engaging cetaceans.

Do these dolphins enjoy entertaining? It would seem so, because they often step up their antics when boats full of eager spectators are around. To the astonishment of the viewers on one occasion, an overly zealous individual leaped more than 3 m out of the water and accidentally landed on the deck of a large research boat. The researchers quickly returned the exhibitionist to the water, of course, but the event remains a testament to the impressive antics of these dolphins.

As a group, Pacific White-sided Dolphins are both acrobatic and sociable. Together they surf ocean waves, catch wakes, ride bow waves and "porpoise" in unison. Sometimes groups of 1000 to 2000 of these dolphins gather in offshore waters. They also socialize with other dolphin and marine mammal species, most notably the Northern Right Whale Dolphin (*Lissodelphis borealis*), Risso's Dolphin (*Grampus griseus*), seals and sea lions.

Despite their gentle-looking faces and intensely social behaviour, Pacific White-sided Dolphins can be aggressive, both with other marine mammals and with each other. They have been seen pestering larger whales, much like crows

and magpies bother dogs. Pacific White-sided Dolphins have been seen clustering around the heads of Orcas and Humpback Whales while they swim. Sometimes the larger whales appear to get fed up and dive deeply to get away. Although Orcas are fearsome predators that are known to feed on dolphins, the dolphins seem to know which Orcas are safe to harass. Transient Orcas, for example, would kill and eat the

RANGE: Pacific White-sided Dolphins are found only in the northern portion of the Pacific Ocean.

Total Length: up to 2.4 m; average 2.1 m

Total Weight: up to 180 kg; average 95 kg

Birth Length: about 1 m

Birth Weight: about 14 kg

DID YOU KNOW?

Pacific White-sided Dolphins and other members of the same genus are often referred to as "lags," a diminutive of their genus name, *Lagenorhynchus*.

dolphins if given a chance. Resident Orcas, however, are mainly fish eaters, and the Pacific White-sided Dolphins are generally safe around them. The exact reason why these dolphins pester larger cetaceans is not known, nor is it understood how the dolphins know which Orcas they can bother without being harmed. Some researchers speculate that the dolphins may be trying to get the larger cetaceans to change direction away from a certain area. This may be related to feeding areas where fish are plentiful. Yet another theory suggests that the pestering is really play, and that both the dolphins and the larger cetaceans enjoy the game. As more video documentation of these mixed species encounters becomes available, researchers may gain a better understanding of what is really occurring.

Recently, the number of Pacific White-sided Dolphins has been increasing around Vancouver Island and in the region between the island and the mainland. No one can adequately explain this because the species was previously thought to prefer the open ocean. Some evidence shows that capelin are now found in increasing numbers where previously there were none, and the dolphins appear to love eating these little fish.

Like most dolphins, the Pacific White-sided Dolphin has acute senses through which it perceives its marine environment in an extremely sophisticated manner. Its sense of touch is many times greater than our own, and it can feel subtle changes in the pressure of the water around it. If another creature approaches a dolphin from outside its field of vision, the dolphin can detect the animal's presence from the pressure wave that the animal creates.

Because their sense of touch and their echolocation are so highly developed, it is a common misconception that dolphins have poor eyesight. In fact, they have exceptionally good eyesight both in and

out of the water. When a dolphin leaps into the air, it can clearly see all of its surroundings. Humans, by comparison, are faced with blurry images when we take our goggles off in water—a medium that is about 800 times denser than air.

ALSO CALLED: Lag, Pacific Striped Dolphin, White-striped Dolphin, Hook-finned Dolphin.

DESCRIPTION: The Pacific White-sided Dolphin has a distinctive and beautiful colour pattern of white, grey and nearly black. Its back is mainly dark, and a large, greyish patch begins in front of the eyes and extends down each side to below the dorsal fin. Along the sides of the tail-stock, another similarly coloured patch may thin into a streak that runs ahead of the dorsal fin. A distinct, dark lateral line borders the pure white undersides. The eyes are dark, as is the tip of the barely discernible beak. This dolphin's most distinguishing feature is the rearward-pointing, bicoloured dorsal fin, which is

dark on the leading edge and pale grey on the trailing edge. The flippers may be similarly coloured, but they are often dark all over. The flukes are pointed, slightly notched in the middle and dark above and below.

BLOW: Pacific White-sided Dolphins do not make a distinct blow, but they often splash about and produce sprays that resemble a blow.

OTHER DISPLAYS: These acrobats perform dazzling breaches, somersaults and an array of other abovewater displays. They often swim just under the surface with their dorsal fins exposed to slice through the water.

GROUP SIZE: Pacific White-sided Dolphins are commonly seen in groups of 10 to 50, but larger groups may form temporarily.

FOOD: These dolphins eat a variety of creatures such as squid, anchovies, hake and other small fish. They feed in groups to better herd the fish, and each adult consumes about 9 kg per day.

REPRODUCTION: Calving and mating occur from late spring to autumn, and gestation is estimated to be 9 to 12 months. A mother nurses her calf for up to 18 months, and she gives birth again soon after weaning the previous calf. Both females and males reach sexual maturity when they are about 1.8 m long. Social maturity influences the age at first mating.

Striped Dolphin

SIMILAR SPECIES: The **Striped Dolphin** (*Stenella coeruleoalba*) has prominent eye stripes. **Saddleback dolphins** (*Delphinus* spp.) have tawny or yellowish sides. The **Harbour Porpoise** (p. 92) is smaller and much greyer overall.

Orca
Orcinus orca

The Orca, with its striking colours and intelligent eyes, has fascinated humankind for centuries. Revered by indigenous peoples of the West Coast, this black-and-white giant is now used as a symbol for everything from biodiversity protection to nonhuman intelligence.

Orcas are among the most widely distributed mammals on earth, and they live in every ocean of the world, from cold polar seas to warm equatorial waters. Warming polar temperatures and decreasing ice cover has permited Orcas to dominate the food chain even in waters where they were previously absent. Uncontested as the top marine predator, they feed on a wider variety of creatures than any other cetacean and are regarded as intelligent yet fearsome creatures—the wolves that rule the seas.

Studies on the North Pacific Coast indicate that there are three distinct forms of Orcas. The two common groups are the "transients" and the "residents," distinguishable by appearance and behaviour. Transient Orcas tend to be larger, with taller, more pointed dorsal fins and saddle patches that are larger and more uniformly grey. They also make erratic direction changes while travelling, whereas residents travel along more predictable routes.

The feeding and socializing behaviours of the two groups also differ: transients are more likely to feed on other sea mammals such as Harbour Seals and sea lions, live in smaller pods of one to seven individuals, dive for up to 15 minutes and vocalize less than residents. By contrast, resident Orcas feed mainly on fish and cephalopods, live in groups of 10 to 25 individuals, rarely dive longer than three to four minutes and are highly vocal.

Recently, researchers have identified a new group of Orcas. These "offshore" Orcas resemble the residents in appearance, but they usually live

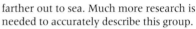

farther out to sea. Much more research is needed to accurately describe this group.

Unlike rorqual whales, Orcas have never been hunted heavily by humans. Some hunting has taken place in the past several decades, but it has not threatened the total population. Unfortunately, live capture for the aquarium trade has taken many Orcas and their close cousins, the Bottlenose Dolphins (*Tursiops* spp.), from the wild. These activities cause much controversy because whales are intelligent animals, and many people feel that to keep them confined in an aquarium is unjust. Much of

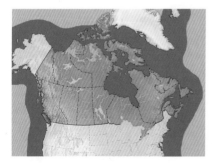

RANGE: Orcas are found in all the oceans and seas of the world.

Total Length: up to 9.8 m; average 8.2 m

Total Weight: up to 7200 kg; average 5700 kg

Birth Length: 1.8–2.1 m

Birth Weight: about 200 kg

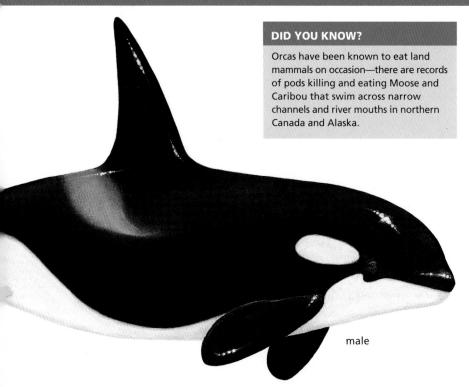

male

what we have learned about cetacean intelligence and biology comes from studies of captive animals, and this knowledge can help us better understand and protect these creatures in the wild.

ALSO CALLED: Killer Whale, Grampus.

DESCRIPTION: The Orca is unmistakable—its body is jet black with a white lower jaw and underparts, as well as white patches behind the eyes and on the sides. Its large flippers are paddle-shaped, and the dorsal fin is tall and triangular. An old male may have a fin as tall as 2 m, and the fins of some old individuals may be wavy when seen from the front. The female's dorsal fin is smaller and more curved than the male's. Behind the dorsal fin, there is often a grey or purplish "saddle." Each whale's dorsal fin and saddle patch has a unique shape, and the fin often bears scars. The eye is below and in front of the white facial spot, and the snout tapers to a rounded point. The flukes are dark on top and whitish below, with pointed tips, concave trailing edges and a distinct notch in the middle. A male Orca is commonly more than 1 m longer than a female.

BLOW: In cool air, the Orca makes a low, bushy-shaped blow.

OTHER DISPLAYS: Orcas are extremely acrobatic for their size. They are often seen breaching clear out of the water, and they also engage in lobtailing, logging,

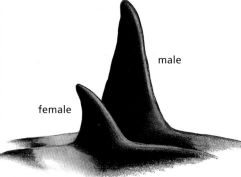

male

female

flipper-slapping and spyhopping. They may speed-swim, or "porpoise," with their entire body leaving the water at each breath. Orcas are also inquisitive and often approach boats, apparently to get a better look at the humans on board.

Subsurface beaches of rounded pebbles attract Orcas—they seem to enjoy rubbing their bodies on the smooth stones.

GROUP SIZE: Orcas travel in pods of 3 to 25 individuals. Certain social gatherings

dive sequence

may attract several pods at a time. The most interesting feature of resident Orcas is that they are one of very few described species in which the family unit is the matriline (all surviving offspring from the female lineage stay together for their entire lives, usually an adult female, her offspring and her daughter's offspring).

FOOD: Orcas feed on a wider variety of prey than any other cetacean, partly because of their global distribution. Several hundred species are potential prey to these top predators, including, but not limited to, fish, seals, other cetaceans, dugongs, sea turtles and birds.

REPRODUCTION: Mating takes place between individuals within a pod and rarely outside the social group. Males reach maturity when they are about 6 m long, and females when they are about 5 m long. Females first give birth at about the age of 15. Winter appears to be the peak calving season, and gestation is believed to be 12 to 16 months.

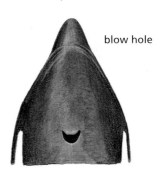

blow hole

SIMILAR SPECIES: The **Dall's Porpoise** (p. 96) is frequently mistaken for a baby Orca. The **False Killer Whale** (*Pseudorca crassidens*) lacks the distinctive white markings of the Orca.

Dall's Porpoise

Beluga

Delphinapterus leucas

In the cold northern waters of the world, the elegant white Beluga lives among the blue-white icebergs. This whale is unmistakable, but passing boaters may overlook it amid the floating chunks of ice. The Beluga is the only cetacean with good numbers in Hudson Bay.

Often nicknamed "Sea Canary" by sailors, this loquacious whale has a vast repertoire of whistles, warbles, squeaks, clucks and squeals that is unlike that of any other cetacean. The Beluga produces these sounds in the air passages in its head, and the sounds are probably modified and amplified in the forehead "melon." The Beluga seems to have the unique ability to bulge or shrink its melon at will, which may account for the large variety of sounds it can produce. The melon of the Beluga, like that of other toothed whales, functions primarily for sound amplification and detection for echolocation. Given the size of its melon, the Beluga probably has one of the most sophisticated sonar systems of any whale. It can navigate in water that is barely deep enough to cover its body, as well as in the open ocean.

Despite research efforts, we do not know the significance of the Beluga's calls. We know that certain sounds are associated with courtship and familial greetings, but whether these whales just chatter noisily or whether they actually communicate with language remains a mystery. In cases where a juvenile has become stranded and must wait six hours for a new high tide—risking sunburn, dehydration and attack by a Polar Bear—the sounds made by the pod members in the water are distinctive. In one documented case, when a stranded juvenile survived the tide and returned to the bay, the cacophony of squeaks and trills from its pod were distinctly different from their earlier sounds.

An anomalous population of Belugas survives in the St. Lawrence River and the Saguenay River. This population commonly inhabits the St. Lawrence only as far inland as Québec City, but individuals have been seen as far west as Montréal. It is believed that these Belugas remained there after the retreat of the last ice age. This population is in a slow decline, and the whales may

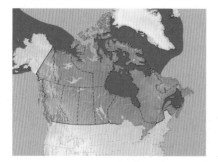

RANGE: The Beluga has a circumpolar distribution in Subarctic and High Arctic seas. Its movements follow the formation and retreat of seasonal ice. A small population lives in the St. Lawrence River and the Saguenay River.

Total Length: up to 5.5 m; average 4 m

Total Weight: *Male:* 450–1000 kg (maximum 1400 kg); *Female:* 250–700 kg

Birth Length: about 1.5 m

Birth Weight: about 60 kg

suffer ill health from the pollutant burdens that contaminate their bodies. When the Belugas of the St. Lawrence die, their bodies must be removed and treated like toxic waste because of the toxins stored in their fat.

ALSO CALLED: White Whale, Sea Canary.

DESCRIPTION: The white body of an adult Beluga is unmistakable. This whale moults annually,

DID YOU KNOW?

The skin of a Beluga is very thick—at least 10 times thicker than that of dolphins and 100 times thicker than that of terrestrial mammals.

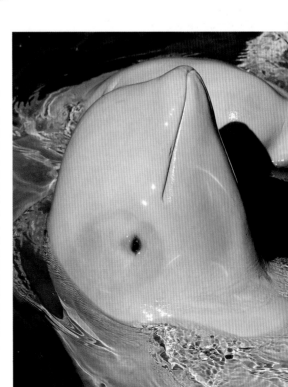

and just prior to moulting, the old skin appears slightly yellow. After the moult, the new skin is gleaming white. At birth, calves are pink or brown but become dark grey within a few weeks. A Beluga does not turn white until it is about five or six years old, and as a juvenile, it is dark brown, slate blue or pinkish grey. The Beluga has a small, bulbous head with a very short but prominent beak. The flippers may curl upward slightly and so may the tips of the flukes. The trailing edge of the flukes and the dorsal ridge may have a brown tinge. The flukes are notched in the middle and may appear convex in shape.

Despite its robust appearance, the Beluga is an extremely flexible creature, and it can twist and turn as it dives underwater. Males are generally larger than females.

BLOW: The blow of the Beluga is low and steamy and can only be seen on calm, cool days. You may have more luck if you listen for the loud, puffing sound instead.

OTHER DISPLAYS: Belugas do not breach, but they frequently spyhop and lobtail. When they rise above the surface, they turn their heads left and right as they survey their surroundings—a range of motion that is possible because the neck vertebrae are not fused as in most other whales. Because Belugas are gleaming white, their underwater activities are visible provided the whales are not too deep. When they are moulting, they roll and rub themselves on the bottom. In shallow river deltas where hundreds of Belugas can be seen together, their frolicking and playful behaviour is visible from the surface. In many cases,

Belugas underwater near a whale-watching boat will turn on their sides or even swim upside down so they can look up at the human observers on the deck above them.

GROUP SIZE: Belugas are commonly seen in small groups of 5 to 20 members, but at certain times of the year, hundreds or even a few thousand of these whales may gather together at river deltas, straits and sounds.

FOOD: The Beluga feeds on a variety of sea creatures such as squid, fish, molluscs and other invertebrates. Although it may feed near the surface and has been characterized as a coastal species, it frequently dives to depths of 400 to 800 m, where it is thought to feed on deepwater creatures such as squid.

REPRODUCTION: When a female is ready to give birth, she swims into deltaic calving waters that are a couple of degrees warmer than the open sea. A calf stays with its mother for at least two years and sometimes as long as four years if the mother does not get pregnant immediately after the calf is weaned. Normally, a female gives birth every three years. Peak calving periods are in late summer, mating is in early spring or summer, and gestation is about 14 months. Males become sexually mature when they are about eight years old and 3.6 to 4.2 m long and females when they are four to seven years old and 2.7 to 3.2 m long.

SIMILAR SPECIES: No other whale is all white. The **Narwhal** (p. 114) is light grey, and older individuals are whiter. Male Narwhals and some females have a tusk.

Narwhal

Narwhal
Monodon monoceros

The extraordinary Narwhal has one of the most distinguishing characteristics of any mammal—a horn like a unicorn. This "horn" is really a tooth, so it is more aptly called a tusk. Unlike the tusk of a Walrus, the Narwhal's tusk does actually resemble that of the fabled unicorn—it is long and twisted and comes to a tapered point. Rarely (perhaps one in 500 whales), a narwhal may have two tusks. Normally only the male has a tusk, though the female sometimes has one as well. Surprisingly, there is even a record of a female with two tusks—truly one in a million.

Many theories have come and gone as to the purpose of the Narwhal's tusk, including to attract a mate, for courtship battles, to stir up food from the bottom, to stab food, to break through ice and to enhance sonar. Recent findings indicate that the tusk is a very sensitive appendage, with millions of nerves radiating outward from its long, central nerve. With its tusk, a Narwhal is capable of determining water temperature and pressure, as well as particle concentration, including salinity. All these factors are vital for life under the ice and determining which waters are rich in prey. If the tusk breaks, it repairs itself with new dentine.

The worldwide Narwhal population likely numbers more than 150,000, many of which live in northern Canadian waters and the fjords of western Greenland.

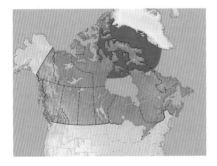

RANGE: Narwhals are found in Canadian Arctic and Greenlandic waters, and east to Scandinavia and Siberia.

Total Length (without tusk):
Male: up to 4.7 m; *Female:* up to 4.1 m

Tusk Length: up to 3 m

Total Weight: *Male:* up to 1600 kg; *Female:* up to 1000 kg

Birth Length: 1.5–1.7 m

Birth Weight: about 80 kg

DESCRIPTION: The Narwhal is mainly greyish white with dark grey or brown spots concentrated over the back and head. Some may be completely dark over their backs, though most become lighter with age. Like the Beluga, the Narwhal has no dorsal fin but has a dorsal ridge about 5 cm high. The lack of a dorsal fin is believed to be an adaptation to the sea ice; sometimes this whale uses its back to break through ice that forms over its breathing holes. The pectoral fins measure 30 to 40 cm, and the width of the tail flukes is about 3 m.

DID YOU KNOW?

Narwhals are among the deepest diving of all marine mammals. They regularly dive to depths of 800 m and can go as deep as 1500 m. These deep dives last about 25 minutes.

BLOW: The blow is small and puffy, and the gasping sound can be easily heard.

OTHER DISPLAYS: Narwhals may spend several minutes at breathing holes with their heads and tusks out of the water. They are very vocal, and their clicks, squeaks and whistles are used daily. Like dolphins, they release streams of bubbles underwater, which may be another kind of communication.

GROUP SIZE: Narwhals normally live in small pods of three to eight individuals, though groups of 20 or more may form. On rare occasions, there have been reports of hundreds and even thousands of individuals congregating in open waters. The reason for such large gatherings is not understood.

FOOD: Narwhals primarily feed on squid, fish, crustaceans and molluscs. The neck vertebrae are not fused, allowing these whales a great range of motion as they hunt. Because Narwhals only have a few teeth, they likely catch their prey by suction and swallow it whole.

REPRODUCTION: Mating tends to be from March to May, and after a gestation of about 15 months, a single calf is born, in July or August. Very rarely twins are born. The calf is a dark brownish grey colour, and it lightens and gains spots as it matures. Calves may be nursed for as long as 20 months, though usually less. Narwhals may live more than 50 years.

SIMILAR SPECIES: The **Beluga** (p. 110) is similar in shape but is all white and has no tusk.

Beluga

North Atlantic Right Whale
Eubalaena glacialis

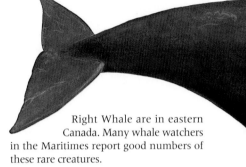

Considerable debate is still ongoing in the scientific community as to the number of species of right whales. Many scientists believe there is only one species worldwide, *Eubalaena glacialis,* whereas others feel strongly that this name represents only the northern population, and that the Southern Hemisphere has a different species, the Southern Right Whale (*E. australis*). To further complicate the classification, some researchers split the northern species into the North Pacific Right Whale (*E. japonica*) and the North Atlantic Right Whale (*E. glacialis*). At this time, the three populations are geographically isolated, and they are treated as three different species within this book.

The highly endangered right whales were the most heavily pursued species during the height of whaling. The name "right whale" is derived from mariners referring to it as the "right" whale to hunt—it is large, has thick blubber, long baleen and swims slowly. The northern species were the first whales to be hunted commercially, and hundreds of thousands were killed to feed the demand for whale oil and other whale products. By 1920, the northern right whales were nearly extinct. Even today, there are fewer than 100 North Pacific Right Whales right whales and 300 North Atlantic Right Whales. In the Southern Hemisphere, there are as many as 3000 Southern Right Whales. Some of the best chances of seeing a North Atlantic Right Whale are in eastern Canada. Many whale watchers in the Maritimes report good numbers of these rare creatures.

When North Atlantic Right Whales are seen in the wild, they can be quite approachable and inquisitive. They are sometimes playful, bumping and pushing floating objects in the water. As a cautionary note, these qualities of inquisitiveness and playfulness can result in a whale playing near or with your boat. Some kayakers have been surprised when their boats were bumped or lifted in friendly jest.

As everyone knows, the largest animal in the world is the Blue Whale (*Balaenoptera musculus*). The right whale, however, wins a few titles of its own. Although it is not a very noble record to hold, the right whale may well be the slowest-swimming whale. Also, its large tail is the broadest relative to body length. The title that wins this whale some true prestige, however, is having the largest testicles. With testes weighing around 500 kg each, no other

RANGE: North Atlantic Right Whales are found in polar and temperate waters of the Northern Hemisphere. They favour offshore waters rather than the open ocean.

Total Length: up to 18 m; average 14 m

Total Weight: up to 86,000 kg; average 54,000 kg

Birth Length: 4.6–6.1 m

Birth Weight: about 900 kg

creature is as well endowed. By comparison, the testes of the Blue Whale weigh a mere 68 kg.

ALSO CALLED: Black Right Whale, Biscayan Right Whale.

DESCRIPTION: The robust North Atlantic Right Whale is easy to distinguish because of the callosities on its head and continues over the back. The baleen plates can be as long as 2.4 m. The flippers are large, spatula-shaped and dark on both sides. At close range, the finger bones in the flippers can be seen as distinct ridges. The flukes, also dark both above and below, are smooth and pointed and have a noticeable notch in the middle. Females are generally larger males.

its lack of a dorsal fin. Distinct callosities grow on the chin and rostrum, in front of the blow hole and above the eyes, and the pattern of these callosities is unique to each individual. The body is blackish, dark blue or dark brown and may show slight mottling or even patchy white spots on the belly. The mouth line arches strongly upward, and then drops nearly vertically to meet the eye. There is a distinct indentation behind the blow hole and then a bulge that

BLOW: When viewed from the front or the rear, the right whale's distinctive blow is widely V-shaped. The sound is loud, and the spray can reach a height of 7 m.

OTHER DISPLAYS: The North Atlantic Right Whale is quite acrobatic considering its bulk. It often breaches, lobtails, flipper-slaps and spyhops. When it lobtails, it may

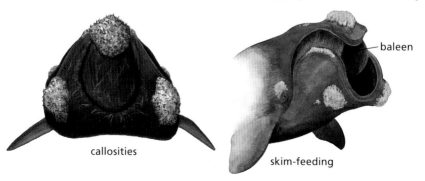

callosities

baleen

skim-feeding

invert in the water to such a great extent that it appears to be doing a headstand.

GROUP SIZE: North Atlantic Right Whales are typically found in groups of two or three individuals. Sometimes a group numbering up to 12 members is seen.

Seasonal feeding waters attract larger numbers of whales at certain times.

FOOD: A remarkably efficient filter-feeder, the North Atlantic Right Whale feeds on some of the smallest zooplankton in the sea. Its primary food is tiny krill and

dive sequence

other copepods. To feed, the whale swims slowly with its mouth open. The unique shape of the baleen makes the mouth look like a cave. As the whale progresses through a concentration of copepods, the water enters the "cave" and passes through the baleen. The creatures are trapped against the silky hairs of the baleen plates and eventually swallowed. This whale skim-feeds wherever food is present—at the surface or at depth.

REPRODUCTION: North Atlantic Right Whale populations have never managed to recover the way other whale species have. The reasons are not clear, but the combination of isolated populations and a slow reproductive rate may be responsible. The female gives birth only once every three to four years. Although this whale courts and mates year-round, peak conception appears to be in winter, as is calving, which indicates a gestation period of about one year. Both genders reach sexual maturity when they are 75 to 80 percent of the adult length.

blow

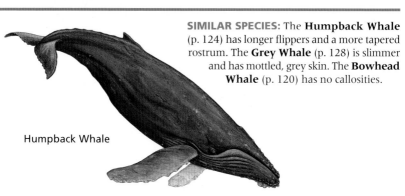

SIMILAR SPECIES: The **Humpback Whale** (p. 124) has longer flippers and a more tapered rostrum. The **Grey Whale** (p. 128) is slimmer and has mottled, grey skin. The **Bowhead Whale** (p. 120) has no callosities.

Humpback Whale

Bowhead Whale
Balaena mysticetus

The large Bowhead Whale lives in cold Arctic waters, forced to move southward only by the winter ice. As the Arctic ice contracts and expands, so, too, does the Bowhead's range. In winter, this whale stays close to the southern edge of the ice.

The Bowhead Whale has an extremely thick layer of blubber that insulates it from its freezing environs. At 30 cm thick, a Bowhead's blubber is thicker than that of any other whale. This whale can even be called fat, since its weight is second only to that of the Blue Whale (*Balaenoptera musculus*), but its length is only average among whales. In addition to its thick blubber, the Bowhead has adaptations in its complex circulatory system that help to conserve heat.

A long-held belief is that humans and elephants are the longest-lived of all

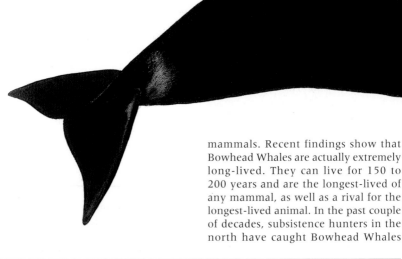

mammals. Recent findings show that Bowhead Whales are actually extremely long-lived. They can live for 150 to 200 years and are the longest-lived of any mammal, as well as a rival for the longest-lived animal. In the past couple of decades, subsistence hunters in the north have caught Bowhead Whales

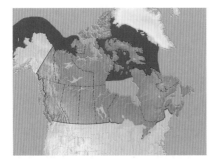

RANGE: The Bowhead Whale is found primarily in Arctic waters. It has a short, nomadic migration that matches the advance and retreat of the sea ice.

Total Length: up to 20 m; average 17 m

Total Weight: up to 109,000 kg; average 73,000 kg

Birth Length: 3.7–4.6 m

Birth Weight: about 900 kg

that showed evidence of having survived previous hunting attempts. In one case, a female had the tip of an explosive harpoon embedded in the blubber of her neck. This particular kind of harpoon was used for whaling over 100 years ago. Another Bowhead had an ivory spear point embedded in its blubber. These discoveries prompted researchers to study the eyes of Bowheads, because

DID YOU KNOW?

Bowhead Whales are slowly increasing in number and recovering their range, though this whale is still listed as a species of special concern. Although populations in the eastern Arctic were nearly decimated by whaling, there are now stable—though still small—populations in all Arctic seas.

certain tissues in the eye can help determine an animal's age.

Like several other whales, Bowheads have a large vocal repertoire. Some of their calls are used in long songs, which may be a part of courtship. Other sounds are used daily, indicating that vocalizations play an important role in their lives. Many reports of individuals coordinating activities even though separated by considerable distances suggest that Bowheads can communicate effectively in spite of the distance. Long-distance communication has been shown in other large whales as well, such as the Blue Whale.

Bowhead Whales are currently protected from commercial whaling, but subsistence hunters in specific regions are permitted to kill a small number annually. The Inuit of Canada, Alaska and Greenland take 10 to 25 whales per year, but this number is believed to be sustainable. Bowheads are not just important as food to Native peoples, but they are culturally significant as well. The Inuit have a strong belief in the spirit of the Bowhead Whale, and their stories and legends reflect this relationship.

ALSO CALLED: Arctic Whale, Greenland Right Whale.

DESCRIPTION: The Bowhead Whale is a large, rotund whale with a strongly

bowed mouth line. It is mainly black, though its chin may be white with a series of black spots. The body is generally smooth, free of both barnacles and callosities. The upper jaw is very narrow when viewed from above, and there is a prominent indentation behind the blow hole. There is no dorsal fin. The flippers are dark on both sides. There may be a pale grey ring around the tailstock. The flukes, which are up to 7 m wide, are also dark on both sides. The Bowhead's baleen is longer than that of any other whale—usually 3 to 4 m long, with a maximum recorded length of 5.2 m—and there is no baleen at the front of the mouth. Females are generally larger than males.

BLOW: This whale has two widely separated blow holes. The resulting blow is broadly V-shaped and can rise up to 7 m.

OTHER DISPLAYS: Before a dive, a Bowhead Whale may raise its tremendous flukes into the air. It also breaches

blow

dive sequence

occasionally, with half to three-quarters of its body leaving the water vertically and falling on one side with a tremendous splash.

GROUP SIZE: Bowheads are seen either singly or in small groups of up to six members. Good feeding sites may attract up to 15 whales. Rarely, during migration events, Bowheads may form loose groups of up to 60 individuals.

FOOD: Bowhead Whales are often seen skim-feeding through waters rich in krill and other copepods. Several whales may form a U-shaped group while skim-feeding in the same direction at the same speed.

REPRODUCTION: In summer, these whales are found in Arctic seas at higher latitudes, where they mate. Gestation is about 13 months, and females give birth to one calf in spring or summer, with at least two years between pregnancies. Calves nurse for about one year, and both sexes become sexually mature when they reach 12 m in length.

SIMILAR SPECIES: The **North Atlantic Right Whale** (p. 116) has callosities over its head, and at present, encountering right whales and Bowheads in the same waters is unlikely.

North Atlantic Right Whale

Humpback Whale
Megaptera novaeangliae

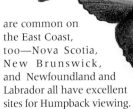

Humpback Whales are renowned for their extensive migrations and haunting songs. These whales are popular among whale watchers and seem to enjoy performing for their boat-bound admirers.

Both the east and west coasts of Canada offer some excellent opportunities to see Humpbacks. Some Humpbacks may spend the entire summer off the west coast of Vancouver Island, but the best times to see them are during their spring and autumn migrations. The Humpbacks of this region breed in the warm waters of either Hawaii or Mexico, and they typically pass by British Columbia between late March and early May and again from late August to October. Humpbacks are common on the East Coast, too—Nova Scotia, New Brunswick, and Newfoundland and Labrador all have excellent sites for Humpback viewing.

A Humpback's impressive song can last from a few minutes to half an hour, and an entire performance can go on for several days, with only short breaks between each song. The complex underwater vocalizations are composed of trills, whines, snores, wheezes and sighs, and they are among the loudest and most mysterious sounds produced by any animal.

Although the true meaning of their songs elude us, we do know that only the males sing and that they perform mainly during the breeding season, implying that the main purpose of the vocalizations is courtship. Male Humpbacks also act aggressively toward one another in their tropical breeding waters and battle to determine dominance. A dominant male becomes the escort of a female with a calf. Presumably, a female with a calf is, or will soon be, receptive to mating.

Other than the brief bouts of fighting that occur between males during the breeding season, Humpback Whales are gentle and docile. They feed primarily on schooling

bubble-netting

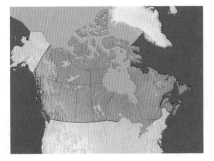

RANGE: Humpback Whales are found in all the world's oceans, migrating seasonally between polar and tropical waters.

Total Length: up to 19 m; average 14 m

Total Weight: up to 48,000 kg; average 27,000 kg

Birth Length: 4–5 m

Birth Weight: 900–1800 kg

fish or krill, using "lunge-feeding" or "bubble-netting" to concentrate prey. When lunge-feeding, a whale approaches a school of fish and surges forward, gulping a large volume of fish and water into its greatly stretched throat. The thick baleen permits water to be squeezed out of the whale's mouth, while the fish remain inside to be

DID YOU KNOW?

As much as 8000 km can lie between a Humpback Whale's high-latitude summer feeding waters and its tropical mating and calving waters.

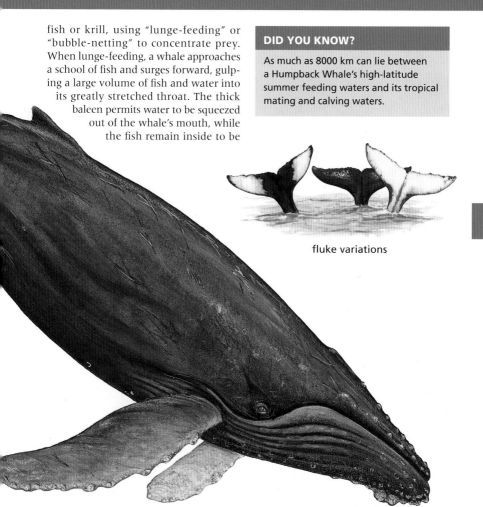

fluke variations

swallowed. When bubble-netting, one or more whales circle a school of fish or krill from below while releasing a constant stream of bubbles. The bubbles rise, confusing and momentarily trapping the fish, then the whales surge up inside the cylindrical "net" and gulp the fish into their mouths.

DESCRIPTION: This whale is slightly more robust in the body than other rorquals. Its body colour is either dark grey or dark slate blue, and the underside may be the same colour as the back or nearly white. A Humpback's head is slender, with numerous knobs and projections around the snout. The mouth line arches downward to the eye, and 12 to 36 grooves are visible on the pale throat. The flippers are distinctively long and knobby, with variable patterns of white markings. The tail flukes are strongly swept back and have irregular trailing edges. Like the flippers, the flukes have unique white markings that can be used to identify individuals. The dorsal fin can be small and stubby or high and curved, and several small knuckles are visible on the dorsal ridge between the fin and tail. The Humpback often carries barnacles and whale lice. Females are typically longer than males.

BLOW: The Humpback Whale makes a thick, orb-shaped blow that can reach up to 3 m high and is visible at a great distance. From directly in front or behind, the blow may appear slightly heart-shaped.

OTHER DISPLAYS: An acrobatic whale, the Humpback dazzles whale watchers with high breaches that finish in a tremendous splash. Other behaviours that it may repeat several times include lobtailing, flipper-slapping and spy-hopping. This whale is often inquisitive and sometimes approaches boats. When it breathes and dives, it rolls through the water and shows a strongly arched back. The tail flukes are lifted high only on deep dives.

dive sequence

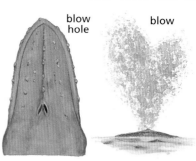

blow
hole

blow

GROUP SIZE: Humpbacks commonly live in groups of two to nine members. Some groups may have as many as 15 members, and occasionally, a single whale is seen on its own. Good feeding and breeding waters usually draw large numbers, and groups of up to 100 individuals have been known to form.

FOOD: Humpbacks feed only in summer, and after their winter in the tropics, they are slim and hungry. A whale may feed by either lunging or bubble-netting, with much individual variation enhancing each technique. Major foods include krill, herring, sand lance and capelin.

REPRODUCTION: Courtship between Humpbacks is elaborate and involves lengthy bouts of singing by the males. Mating usually occurs in warm waters, and a single calf is born following a gestation of about 11.5 months. The calf stays close to its mother and nurses for about one year. Both sexes reach sexual maturity when they are about 12 m long (four to nine years of age).

SIMILAR SPECIES: The **Grey Whale** (p. 128) is slimmer and has mottled grey skin with large numbers of barnacles and whale lice.

Grey Whale

Grey Whale
Eschrichtius robustus

Grey Whales, among the most frequently observed of the large whales, are famous for their extensive migrations, which are believed to be the longest of any mammal. In their voyages, these whales travel back and forth between the cold Arctic seas where they spend summer and the warm Mexican waters where they overwinter. Each year, almost the entire world population of Grey Whales performs this cycle, travelling about 20,000 km along the western coast of North America.

During their summers in Arctic or near-Arctic seas, Greys feed on abundant, bottom-dwelling crustaceans known as amphipods. These whales eat enormous quantities of food throughout their five to six months in the North. During migration and especially during their stay in southern waters, they eat very little and may even fast. Having lost as much as 30 percent of their body weight, they are slim and hungry when they return to the food-rich Arctic waters.

The journey to winter waters takes approximately three months—the whales leave the Arctic by late September and arrive in the warm waters off California

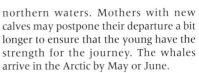

and Mexico by late December. This southward migration coincides with the species' reproductive activity, and once in warm waters, a female either mates or gives birth. If she mates, her journey south the next year will be to give birth, because gestation takes about 13.5 months. Conversely, if she gives birth, she will court and mate the following year.

By late February or March, Grey Whales begin their return to northern waters. Mothers with new calves may postpone their departure a bit longer to ensure that the young have the strength for the journey. The whales arrive in the Arctic by May or June.

The best time to view passing Grey Whales in British Columbia is between late March and early May and again from late August to October. Some may spend the entire summer off the west coast of Vancouver Island.

RANGE: Grey Whales are now found only in coastal areas of the North Pacific, mainly in North American waters. A small population spends summers in the Sea of Okhotsk off Siberia and migrates to the southern tip of Korea for the winter.

Total Length: up to 15 m; average 13 m

Total Weight: up to 40,000 kg; average 32,000 kg

Birth Length: about 4.5 m

Birth Weight: about 450 kg

Grey Whales, which once inhabited both the Atlantic and the Pacific oceans, have been close to extinction at least twice in history and now live only in Pacific waters. Greys are particularly vulnerable to whalers because they live mainly in shallow waters. As a result of many years of protection, these whales now number some 21,000 in the eastern Pacific. Their only natural predator is the Orca, which might take young or weak individuals.

DID YOU KNOW?

Grey Whales are favourites among whale watchers because they can be very friendly and may even approach boats. In extraordinary encounters, Greys seem to enjoy the occasional back rub from willing admirers.

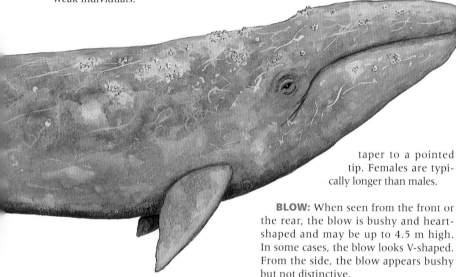

taper to a pointed tip. Females are typically longer than males.

BLOW: When seen from the front or the rear, the blow is bushy and heart-shaped and may be up to 4.5 m high. In some cases, the blow looks V-shaped. From the side, the blow appears bushy but not distinctive.

ALSO CALLED: Devilfish, Scrag Whale, Mussel-digger.

DESCRIPTION: The Grey Whale is easily distinguished from other whales by its mottled grey appearance and narrow, triangular head. The head is slightly arched between the eye and the tip of the snout, and the jaw line usually has a similar arch. There are many yellow, orange or whitish patches of crusted barnacles or lice over most of the body, but especially on the head. There is no dorsal fin, but this whale does have one large bump where the dorsal fin should be and a series of smaller bumps, or "knuckles," that continue along the dorsal ridge to the tail. In a dive, these bumps are visible, as are the pointed flukes and distinctly notched tail. The small flippers are wide at the base but

OTHER DISPLAYS: Grey Whales exhibit breaching, spyhopping and fluking. They will breach anywhere in their range but do so most often in breeding lagoons in the south. They rise nearly vertically out of

blow

the water, come down with an enormous splash and then repeat the breach two or three times in a row. Spyhopping is also common, and they may keep their heads out of the water for 30 seconds or more. In shallow water, Grey Whales may "cheat" and rest their flukes on the bottom so they can keep their heads above water with minimal effort. Before a deep dive, they raise their flukes clear of the water's surface.

GROUP SIZE: Generally these whales are seen in groups of only one to three individuals. They may migrate in groups of up to 15, and food-rich areas in the

dive sequence

north can attract dozens or hundreds of Grey Whales at a time.

FOOD: Unlike other baleen whales, the Grey Whale is primarily a bottom-feeder. Its food consists of benthic amphipods and other invertebrates. A feeding whale dives down to the bottom and rolls onto one side, sticking out its lower lip and sucking in great volumes of food, water and muck. Once its mouth is full, it uses its powerful tongue to push the silty water out through the baleen, trapping the crustaceans inside and swallowing them whole. Most Grey Whales are "right-lipped," the way humans are mainly right-handed, and they prefer to feed using the right side of the mouth. A close-up look at a Grey's face will reveal its "handedness," because the side it uses to feed will have numerous white scars and no barnacles. Inside the whale's mouth, the same uneven wear is evident—on right-lipped whales, the baleen plates on the right side are shorter and more worn than the plates on the left.

REPRODUCTION: Male Grey Whales are sexually mature when they are just over 11 m long, and females when they are nearly 12 m long (from 5 to 11 years of age for both sexes). Mating occurs in December or January, and a single calf is born 13.5 months later, in the following January or February. The young start their journey northward with their mothers when they are only two months old and continue nursing until they are six to nine months old.

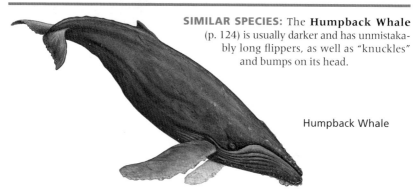

SIMILAR SPECIES: The **Humpback Whale** (p. 124) is usually darker and has unmistakably long flippers, as well as "knuckles" and bumps on its head.

Humpback Whale

CARNIVORES

This group of mammals is aptly named because, though some members of the order Carnivora are actually omnivorous and eat a great deal of plant material, most of them prey on other vertebrates. These "meat eaters" vary greatly in size, from the small Least Weasel to the large and muscular Polar Bear.

Cat Family (Felidae)

Excellent and usually solitary hunters, all wildcats except the cheetah have long, curved, sharp, retractile claws. Like dogs, cats walk on their toes—they have five toes on each forefoot and four toes on each hindfoot—and their feet have naked pads and furry soles. As anyone who has a house cat knows, the top of a cat's tongue is rough, with spiny, hard, backward-pointing papillae, which are useful to the cat for grooming its fur.

Skunk Family (Mephitidae)

Biologists previously placed skunks in the weasel family, but recent DNA research has led taxonomists to group the North American skunks (together with the stink badgers of Asia) in a separate family. Unlike most weasels, skunks are usually boldly marked, and when threatened, they can spray a foul-smelling musk from their specialized anal glands.

Weasel Family (Mustelidae)

All weasels are lithe predators with short legs and elongated bodies. They have anal scent glands that produce an unpleasant-smelling musk, but, unlike skunks, they use it to mark territories rather than in defence. Most species have been trapped for their thick, long-lasting fur.

Raccoon Family (Procyonidae)

Raccoons are small to midsized omnivores that, like bears (and humans), walk on their heels. They are very good climbers and are best known for their long, banded, bushy tails and distinctive black masks.

Hair Seal Family (Phocidae)

The hair seals are also known as "true seals," and they are believed to share a common ancestor with the mustelids. These seals have hindflippers that permanently face backward, so they cannot rotate their hips and hindlegs to support their body weight. Hair seals and eared seals are collectively referred to as pinnipeds.

Walrus Family (Odobenidae)

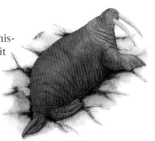

The Walrus—the only member of this family—is unmistakable. Like a hair seal, all of its digits have claws and it has no external ears, but like a member of the eared seal family (Otariidae), the Walrus is able to rotate its hindlimbs under its body to help it move on land. Perhaps the most distinguishing feature of the Walrus is its tusks. Both sexes have tusks, though the male's tend to be much longer.

Eared Seal Family (Otariidae)

Eared seals include the fur seals and sea lions, and all are believed to share a common ancestor with bears. These seals can rotate their hindlegs forward to help support their weight when they are on land.

Bear Family (Ursidae)

The three North American members of this family are the world's largest terrestrial carnivores. All bears are plantigrade—they walk on their heels—and have powerfully built forelegs and a short tail. Although most bears sleep through the harshest part of winter, they do not truly hibernate—their sleep is not deep and their body temperature drops only a couple of degrees.

Dog Family (Canidae)

This family of dogs, wolves, coyotes and foxes is one of the most widespread terrestrial, nonflying, mammalian families. The typically long snout houses a complex series of bones associated with the sense of smell, which plays a major role in finding prey and in communication. Members of this family walk on their toes, and their claws are blunt and non-retractile.

Mountain Lion
Puma concolor

A pug mark in the snow or a heavily clawed tree trunk are powerful reminders that Mountain Lions are still found in much of Canada. This large cat once occurred throughout much of North America, but conflicts with settlers and their stock animals resulted in the widespread removal of this great feline. Today, loss of habitat and human settlement are the most serious issues affecting Mountain Lion populations. Nevertheless, this secretive wildcat is adept at surviving, even where we least expect it. In spite of population declines, it is the most widespread carnivore in the Americas, ranging from northern Canada south to Patagonia. Its common names reflect this distribution: "Puma" is derived from the name used by the Incas of Peru, and "Cougar" comes from Brazil. Mountain Lions are recovering parts of their former range in eastern Canada, especially in Ontario and Québec.

The Mountain Lion is generally a solitary hunter, except when a mother is accompanied by her young. When the young are old enough, they follow their mother and sometimes even help her kill—a process

RANGE: The Mountain Lion formerly ranged from northern British Columbia east to the Atlantic and south to Patagonia. It has been extirpated from much of southern Canada and most of the United States except the western mountains. A tiny population remains in the Everglades of Florida, and there are occasional reports from Maine and New Brunswick.

Total Length: 1.5–2.7 m

Shoulder Height: 65–80 cm

Tail Length: 50–90 cm

Weight: 30–90 kg

that teaches them how to hunt for themselves. Although the Mountain Lion is capable of great bursts of speed and giant bounds, it often opts for a less energy-intensive hunting strategy. Silent and nearly motionless, a cat will wait in ambush in a tree or on a ledge until prey approaches. By leaping onto the shoulders of its prey and biting deep into the back of the neck while attempting to knock its victim off balance, the Mountain Lion can take down an adult deer or a small Moose. This big cat needs the equivalent of about one large animal per week to survive, and Mountain Lion densities in the wild tend to correlate with ungulate densities. It is an adaptable creature that may hunt by day or night, though most activity is crepuscular.

One of Canada's most charismatic animals, the Mountain Lion is a creature many wildlife lovers hope to see…from a safe distance. This elusive cat is a master of living in the shadows, but if you spend enough time in the wilderness, you may one day see a streak of burnished brown flash through your peripheral vision. If this streak was actually a Mountain Lion, you can count yourself among the extremely lucky. Few people—even biologists—ever get more than a fleeting glimpse of this graceful feline.

If you startle one—which is quite improbable because it usually knows of your presence long before you realize that it is nearby—it will quickly disappear from sight. Only a young cat might approach for a closer look. A juvenile Mountain Lion, like most young carnivores, is extremely curious and may not yet realize that humans are best avoided.

ALSO CALLED: Cougar, Puma, Panther.

DESCRIPTION: This handsome feline is the only large, long-tailed native cat in Canada. It is mainly buffy grey to tawny or cinnamon in colour, with a pale buff or nearly white underside. Its body is long and lithe, and its tail is more than half the length of the head and body. The head, ears and muzzle are all rounded. The tip of the tail, sides of the muzzle and backs of the ears are black. Some individuals have prominent facial patterns of black, brown, cinnamon and white. Males are larger than females.

HABITAT: Mountain Lions are found most frequently in remote, wooded, rocky places, usually near an abundant supply of prey species. They normally inhabit areas away from people, though in many parts of British Columbia, their range

overlaps considerably with areas of human activity.

FOOD: Mountain Lions prey on a wide variety of animals, but they specialize in hunting deer and other ungulates. Other prey species include beavers, porcupines, rabbits and birds. In bitter winters, animals weakened by starvation are easy prey for Mountain Lions.

DEN: A cave or crevice between rocks usually serves as a den, but a Mountain

DID YOU KNOW?

During an extremely cold winter, a Mountain Lion can starve if the carcasses of its prey freeze solid before it can get more than one meal. This cat's jaws are designed for slicing, and it has trouble chewing frozen meat.

Lion may also den under an overhanging bank, beneath the roots of a downed tree or even inside a hollow tree.

REPRODUCTION: A female Mountain Lion may give birth to a litter of one to six (usually two to four) kittens at any time of the year after a gestation of about three months. The black-spotted, tan kittens are blind and helpless at birth, but their blue eyes open at two weeks of age. Their mottled coats help camouflage them when their mother leaves to find food. As the kittens mature, they lose their spots and their eyes turn brown or hazel. They are weaned at about two to three months, by which time they weigh about 3 kg. Young Mountain Lions may stay with their mother for up to two years.

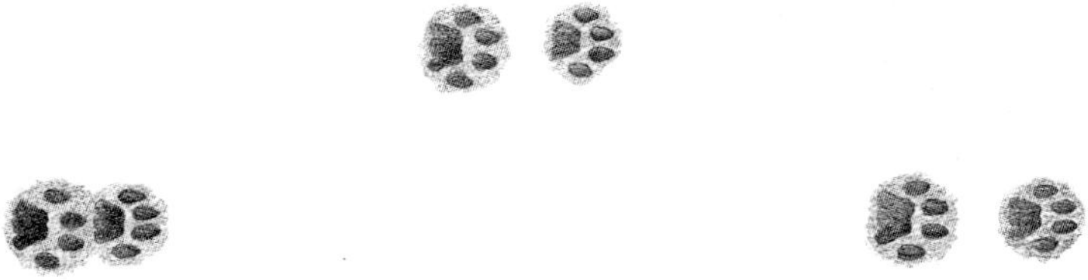

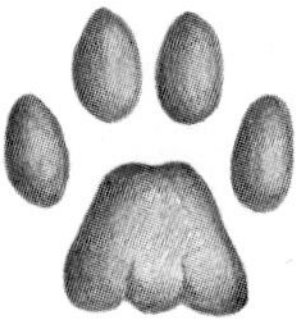

foreprint

hindprint

fast walking trail

SIMILAR SPECIES: The two other native cats in the region, the **Canada Lynx** (p. 138) and the **Bobcat** (p. 142), are smaller and have mottled coats and bobbed tails.

Canada Lynx

Canada Lynx
Lynx canadensis

Meat is on the nightly menu for the Canada Lynx, and the meal of choice is Snowshoe Hare. The classic predator-prey relationship of these two species is now well known to all students of zoology, but extensive field studies were required to determine how and why these species interact to such a great extent.

Periodic fluctuations in Canada Lynx numbers in local areas have been observed for decades—when hares are abundant, Canada Lynx kittens are more likely to survive and reproduce, but when hares are scarce, many kittens starve and the lynx population declines, sometimes rapidly and usually one to two years after the decline in hares.

The reason why the Canada Lynx is so focused on the Snowshoe Hare as its primary prey may never be understood

RANGE: Primarily an inhabitant of boreal and mixed forests, the Canada Lynx occurs across much of Canada and Alaska. Its range extends south into the mountains of the western U.S. and into the northern parts of Wisconsin, Michigan, New York and New England.

Total Length: 80–100 cm

Shoulder Height: 45–60 cm

Tail Length: 9–12 cm

Weight: 7–18 kg

completely, but the forest community in which this cat lives certainly affects its lifestyle. Many other carnivores, as well as birds of prey, compete with the lynx for the same forest prey, few are as skilled at catching Snowshoe Hares as the Canada Lynx.

This resolute carnivore copes well with the difficult conditions of its wilderness home. Its well-furred feet impart nearly silent movement and serve as snowshoes when snow is deep. Like other cats, the Canada Lynx is not built for fast, long-distance running—it generally ambushes or silently stalks its prey. The ultimate capture of an animal relies on sheer surprise and a sudden, overwhelming rush. With its long legs, a lynx can travel rapidly while trailing evasive prey in the tight confines of a forest. It can also climb trees quickly to escape an enemy or to find a suitable ambush site.

The Canada Lynx is primarily a solitary hunter found in remote forests. During population peaks, however, young lynx may disperse into less-hospitable environs. In recent memory, lynx have been reported within the limits of many major cities. These incidents are unusual, however, because lynx typically avoid contact with humans. In Canada, you are more likely to see this cat in forested areas where there is an abundance of hares, but even in this prime territory, it is elusive. Each rare observation of a wild lynx undoubtedly surprises those who are more accustomed to the appearance of a domestic feline—the stilt-legged Canada Lynx is more than twice the size of a house cat and is gangly in appearance.

DESCRIPTION: This medium-sized, short-tailed, long-legged cat has large feet and protruding ears tipped with 5-cm-long black hairs. The long, lax, silvery grey to buffy fur bears faint, darker stripes on the

sides and chest and dark spots on the belly and the insides of the forelegs. There are black stripes on the forehead and long facial ruff, and the entire tip of the stubby tail is black. The long, buffy fur of the hindlegs makes the lynx look like it is wearing baggy trousers. Its large feet spread widely when it is walking, especially in deep snow. The footprint of this cat is big and round, but thick fur often obscures the detail of the print.

HABITAT: The Canada Lynx is closely linked to the northern boreal forest. Desired habitat components include numerous fallen trees and thickets of

young spruce and balsam fir that serve as effective cover and ambush sites. The Canada Lynx depends on its prey, and its prey depends on the twigs, grasses, leaves, bark and vegetation of the dense forest.

FOOD: Snowshoe Hares typically make up the bulk of the diet, but a lynx can sustain itself on squirrels, grouse, rodents

foreprint

hindprint

walking trail

and even small domestic animals when hares are scarce. When a lynx does not eat all of its kill, it caches the meat under snow or leafy debris.

DEN: The den is typically an unimproved space beneath a fallen log, in a willow thicket or in a dense tangle of fallen trees. Canada Lynx do not share dens, and adult contact is restricted to mating. A mother lynx shares a den with her young until they are mature enough to leave.

REPRODUCTION: Lynx breed from March to early May, and the female gives birth to one to five (usually two or three) kittens in May or June after a gestation of a little over two months. The kittens are generally grey, with indistinct longitudinal stripes and dark grey barring on the limbs. Their eyes open in about 12 days, and they are weaned at two months. They stay with their mother through the first winter and acquire their adult coat at 8 to 10 months of age. A female usually bears her first litter near her first birthday.

Bobcat

SIMILAR SPECIES: The only other native cat that resembles the Canada Lynx is the smaller, shorter-legged **Bobcat** (p. 142). The Bobcat has shorter ear tufts and much smaller feet, and the tip of its tail is black above and white below. The **Mountain Lion** (p. 134) is much larger and has a long tail.

Bobcat
Lynx rufus

For those of us who are naturalists as well as feline enthusiasts, seeing a Bobcat in the wild is a rewarding experience. This cat is much more common in southern Canada than the Canada Lynx because the Bobcat has small feet and cannot cope with the deep snow that covers most of northern Canada in winter. Nighttime drives in the southern parts of the provinces offer some of the best opportunities for seeing one, though, at best, the experience is a mere glimpse of this cat bobbing along in the headlights.

The Bobcat looks like a large version of a domestic cat, but it is a wildcat in every sense of the word. It impresses observers with its lightfootedness, agility and stealth, usually leaving the momentary experience forever etched in the viewer's mind.

Over the past two centuries, Bobcat populations have fluctuated greatly because of this cat's adaptability to human-wrought change and its vulnerability to our persecution. Less restricted in its prey choices than the Canada Lynx, the Bobcat may vary its diet of hares with any number of small animals, including an occasional turkey or chicken. Its farmyard raids did not go over well with early settlers, and for more than 200 years, the Bobcat was

RANGE: The Bobcat has the widest distribution of any native cat in North America. It occurs in the southern Rockies and the interior of British Columbia, across southern Canada and south through much of the U.S. to Mexico.

Total Length: 75–125 cm

Shoulder Height: 45–55 cm

Tail Length: 13–17 cm

Weight: 7–13 kg

considered vermin. Even today, this striking native feline remains on the "varmint" list in parts of the southern U.S.

Despite its small size, the Bobcat is rumoured to be a ferocious hunter that can take down animals much larger than itself, though some sources dispute these stories. This exceptional feat, however, is likely possible—a Bobcat waits motionless on a rock or ledge for a deer to approach and leaps onto the neck of the unsuspecting animal. It then manoeuvres to the lower side of the neck to deliver a suffocating bite to the deer's throat. The Bobcat may resort to such rough tactics when food is scarce, but it usually dines on simpler prey, such as rabbits, birds and rodents. Most of its prey, big or small, is caught at night by stalking or ambush. During the day, the Bobcat remains immobile in any handy shelter.

Finding Bobcat tracks in soft ground may be the easiest way to determine the presence of this small cat. Unlike Coyote or Red Fox prints, Bobcat prints rarely show any claw marks, and there is one cleft on the front part of the main footpad and two on the rear. A Bobcat's print is slightly larger than a house cat's, and this animal tends to be found much farther from human habitation. Like all cats, the Bobcat buries its scat, and its scratches can help confirm its presence.

DID YOU KNOW?

Most cats have long tails, which they lash out to the side to help them corner more rapidly while in pursuit of prey. The Bobcat and the Canada Lynx, however, which typically hunt in brushy areas, have short, or "bobbed," tails that will not get caught in branches.

DESCRIPTION: The coat is generally tawny or yellowish brown, though the colour varies slightly with the seasons. The winter coat is usually dull grey with faint patterns. In summer, the coat often has a reddish tinge to it—the source of the scientific name *rufus*. A Bobcat's sides are spotted with dark brown, and there are dark, horizontal stripes on the breast and outsides of the legs. There are two black bars across each cheek and a brown forehead stripe. The ear tufts are less than 2.5 cm long. The chin and throat are whitish, as is the underside of the bobbed tail. The upper surface of the tail is barred, and the tip of the tail is black on top.

HABITAT: The Bobcat occupies open, coniferous and deciduous forests and brushy areas. It especially favours willow stands, which offer excellent cover for hunting.

FOOD: The Bobcat's preferred food seems to be hares, but it will catch and eat squirrels, rats, mice, voles, beavers, skunks, wild turkeys and other ground-nesting birds. When necessary, it will scavenge the kills of other animals, and it may even take down its own large prey, such as deer. In areas close to human habitation, it may take small domestic animals.

DEN: The Bobcat does not keep a permanent den. During the day, it uses any available shelter. The female prefers rocky crevices for the natal den, but she may also use a hollow log or the cavity under a fallen tree. The mother does little den improvement other then scrape away large twigs and stones.

REPRODUCTION: Bobcats typically breed in February or March but may breed at other times of the year. The female gives birth to one to six (sometimes three) fuzzy grey kittens in April or May after a gestation of 60 to 70 days. The kittens' eyes open after nine days. They are weaned at two months but remain with their mother for three to five months. Female Bobcats become sexually mature at one year of age, and males at two.

foreprint

hindprint

walking trail

trotting to loping trail

Canada Lynx

SIMILAR SPECIES: The **Canada Lynx** (p. 138) is the only other native bob-tailed cat in North America. These two cats are nearly the same size, but the lynx's hindlegs are longer, making it appear taller. The Canada Lynx also has much longer ear tufts, and the tip of its tail is entirely black. The **Mountain Lion** (p. 134) is larger and has a long tail.

Western Spotted Skunk
Spilogale gracilis

When agitated and fearing for its safety, this skunk resorts to the practice that has made this family infamous. If foot stamping and tail raising do not convey sufficient warning, the next stage certainly will.

Unlike the more familiar Striped Skunk, which always sprays with its body in a "U" position with all four feet planted on the ground, the Western Spotted Skunk sometimes sprays in an unusual, upturned position. Like a contortionist in a sideshow circus, this little skunk faces the threat and performs a handstand, letting its tail fall toward its head. It can maintain this balancing act for more than five seconds, which is usually sufficient time to take aim and expel a well-placed stream of fetid scent into the face of the threat. Many animals may attempt to kill and eat the Western Spotted Skunk before it sprays, but few are successful.

One of the first signs of spring in the wild, and also in many urban and suburban areas, is the smell of skunk in the air. In the southwestern corner of British Columbia, road-killed Western Spotted Skunks are likely the cause because they are more numerous than Striped Skunks in that region. Road fatalities are a major cause of death among skunks, despite their weasel-like agility and dexterity. Western Spotted Skunks are especially nimble—with surprising ease, they can climb up to holes in hollow trees or to bird nests, where they find shelter or food.

DESCRIPTION: This small skunk is mainly black with a white forehead spot and a series of four or more white stripes broken into dashes on the back. The pattern of white spots is different on each

RANGE: The Western Spotted Skunk ranges from southwestern British Columbia to the southern tip of Texas and south through Mexico.

Total Length: 32–58 cm

Tail Length: 10–21 cm

Weight: 450–900 g

individual. The tail is covered with long, sparse hairs, and the tip is white with a black underside. The ears are small, rounded and black, and the face strongly resembles a weasel's. This skunk may walk, trot, gallop or make a series of weasel-like bounds. At night, its eyeshine is amber.

HABITAT: This skunk is found in woodlands, rocky areas, open prairie or scrublands. It does not occupy marshlands or wet areas, but farmlands make an excellent home. It is mainly nocturnal, and even in prime habitat, this skunk is seldom seen.

FOOD: This omnivorous mammal feeds on insects, berries, eggs, nestling birds, small rodents, lizards and frogs. Animal matter usually accounts for the larger part of its diet. An opportunistic forager, the Western Spotted Skunk will eat nearly anything that it finds or can catch. Insects, especially grasshoppers and crickets, are the most important food in summer, and small mammals comprise the winter diet.

DEN: The Western Spotted Skunk is nomadic in comparison to the Striped Skunk. It rarely makes a permanent den,

> ### DID YOU KNOW?
>
> Spotted skunks become sexually mature at a very young age. A male may be able to mate when he is just five months old, and a female usually mates in September or October of her first year.

preferring to hole up temporarily in almost any safe spot—in a rock crevice, under a fallen log, building or woodpile, in the abandoned burrow of another mammal or even in a tree cavity. The natal den is used for a longer period than other den spaces, and it differs primarily in the grass and leaves with which the female lines the inside for comfort. In harsh winters, several skunks may den together to conserve energy and wait out inclement weather.

REPRODUCTION: Females exhibit delayed implantation, and two to six (usually four) young are born in May or June after a gestation of about 60 days. The eyes and ears are closed at birth. The young skunks are covered with a fine fur that reveals their future coat pattern. The eyes open after one month, and by two months, the young are weaned. The family frequently stays together through autumn and may overwinter in the same den, not dispersing until the following spring.

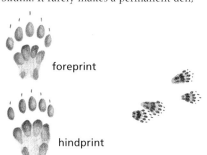

foreprint

hindprint

walking trail

Striped Skunk

SIMILAR SPECIES: The **Striped Skunk** (p. 148) is the only other animal in the same range with a black-and-white coat. As its name suggests, however, its white markings are in broad stripes.

Striped Skunk
Mephitis mephitis

In nature, many of the most strikingly marked creatures are poisonous or unsafe in some way. This is true for skunks—their bold, black-and-white patterns convey a clear warning to stay away. Anyone not heeding it will likely receive a faceful of foul, repugnant fluid.

Butylmercaptan is responsible for the skunk's stink. Seven different sulfide-containing "active ingredients" have been identified in the musk, which not only smells bad, but also irritates the skin and eyes. Prior to spraying, a distressed Striped Skunk will twist its body into a "U" shape with all four feet on the ground, so that both its head and tail face the threat. If a skunk successfully targets an intruder's eyes, there is intense burning, copious tearing and sometimes a short period of blindness. As well, the musk stimulates nausea in humans.

Surprisingly, this species is tolerant of observation from a discreet distance, and watching a skunk can be very rewarding—its gentle movements contrast with the hyperactive behaviour of its weasel cousins. The Striped Skunk's activity begins at sundown, when it emerges from its daytime hiding place. It usually forages among shrubs, but it often enters open areas, where it can be seen with relative ease. An opportunistic predator, the Striped Skunk feeds on whatever animal matter is available, even digging for grubs, worms and other invertebrates. During winter, its activity is much reduced, and this skunk spends the coldest periods in communal dens.

The most frequent predator of the Striped Skunk is the great horned owl. Lacking a highly developed sense of smell, this owl does not seem to mind the skunk's odour—nor do the few other birds that commonly scavenge road-killed skunks.

RANGE: The Striped Skunk is found across most of North America, from Nova Scotia to Florida in the East and from the southwestern Northwest Territories to northern Baja California in the West. It is absent from parts of the deserts of southern Nevada, Utah and eastern California.

Total Length: 55–80 cm

Tail Length: 20–35 cm

Weight: 1.9–4.2 kg

DESCRIPTION: This cat-sized, black-and-white skunk is familiar to most people. Its basic colour is glossy black. A narrow, white stripe extends up the snout to above the eyes, and two white stripes begin at the nape of the neck, run along the back on either side of the midline and meet again at the base of the tail. The white bands often continue on the tail, ending in a white tip, but there can be much variation in the white markings. The foreclaws are long. A pair of musk glands on either side of the anus discharges the foul-smelling, yellowish liquid for which the skunk is famous.

HABITAT: In the wild, the Striped Skunk seems to prefer streamside woodlands, groves of hardwood trees, semi-open areas, brushy grasslands and valleys. It also regularly occurs around farmsteads and even in the hearts of cities, where it can be an urban nuisance that eats garbage and raids gardens.

FOOD: All skunks are omnivorous. Insects, including bees, grasshoppers, beetles and various larvae, make up the largest portion, about 40 percent, of the spring and summer diet. To get at bees, a skunk will scratch at a hive entrance until the insects emerge,

foreprint

hindprint

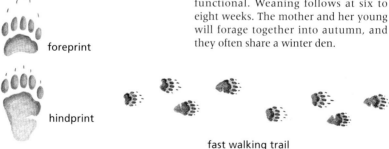

fast walking trail

> **DID YOU KNOW?**
>
> Fully armed, the Striped Skunk's scent glands contain about 30 ml of noxious, smelly stink. The spray has a maximum range of about 6 m, and a skunk is accurate for half that distance.

and then chew up great gobs of mashed bees, thus incurring the beekeeper's wrath. The rest of the diet is composed of bird eggs and nestlings, amphibians, reptiles, grains, green vegetation and, particularly in autumn, small mammals, fruits and berries. Along roads, carrion is often an important component of a skunk's diet.

DEN: In most instances, the Striped Skunk builds a bulky nest of dried leaves and grasses in an underground burrow or beneath a building. Winter and maternal dens are underground.

REPRODUCTION: A female Striped Skunk gives birth to 2 to 10 (usually five to seven) blind, helpless young in April or May, after a gestation of 62 to 64 days. The typical black-and-white coat pattern is present on the skin at birth. The eyes and ears open at two to four weeks, and at five to six weeks, the musk glands are functional. Weaning follows at six to eight weeks. The mother and her young will forage together into autumn, and they often share a winter den.

Striped Skunk

SIMILAR SPECIES: The **Western Spotted Skunk** (p. 146) has white spots instead of stripes. The **American Badger** (p. 170) has a white stripe running up its snout, but it is larger and squatter and has a grizzled, yellowish grey body.

American Marten
Martes americana

The quick, active, agile American Marten is equally at home on the forest floor or among branches and tree trunks. Its fluid motions and attractive appearance contrast with its swift, deadly hunting tactics.

Unfortunately, this animal's antics are not easily observed because it tends to inhabit wilder areas. The American Marten has been known to occupy human structures for short periods of time, should a food source be near, but more typical sightings are restricted to flashes across roadways or trails.

A close relative of the Eurasian Sable (*M. zibellina*), the American Marten is widely known for its soft, lustrous fur, and it is still targeted on traplines in remote wilderness areas. As with many species of forest mammals, populations seem to fluctuate markedly every few years—a cyclical pattern revealed by trappers' records. Some scientists attribute these cycles to changes in prey abundance and a corresponding population decline, whereas others suggest that marten populations simply migrate from one area to another.

The American Marten is often used as an indicator of environmental conditions because it depends on food found in mature coniferous forests. The loss of such forests has led to declining populations and even extirpation from parts of Canada where the species was once numerous. Hopefully, modern methods of forest management will maintain adequate habitat for the American Marten and prevent its decline.

ALSO CALLED: Pine Marten, American Sable.

RANGE: The range of the American Marten coincides almost exactly with the distribution of boreal and mixed coniferous forests across North America. It may repopulate where mature forests have developed in areas that were formerly cut or burned.

Total Length: 50–68 cm

Tail Length: 18–23 cm

Weight: 0.5–1.2 kg

DESCRIPTION: This slender-bodied, fox-faced mustelid has a beautiful tan to chocolate brown coat and a long, bushy tail. The feet are well furred and equipped with strong, nonretractile claws. The conspicuous ears are 3.5 to 4.5 cm long, and the almond-shaped eyes are dark. The breast spot, when present, is pale cream to bright amber and varies in size from a small dot to a large patch that occupies the entire region from the chin to the belly. There is a scent gland on the centre of the abdomen. Males are 20 to 40 percent larger than females.

HABITAT: The American Marten prefers mature, particularly coniferous, forests that contain numerous dead trunks, branches and leaves to provide cover for its rodent prey. It does not occupy recently burned or cut areas.

FOOD: Although rodents such as squirrels, mice and voles make up most of the diet, the American Marten is an opportunistic feeder that will eat hares, bird eggs and chicks, insects, fish, snakes, frogs, carrion and occasionally berries and other vegetation. This active predator hunts both day and night.

DID YOU KNOW?

Although the American Marten, like most members of the weasel family, is primarily carnivorous, it has been known to consume an entire apple pie left cooling outside a window and to steal doughnuts off a picnic table.

DEN: The preferred den site is a hollow tree or log that is lined with dry grass and leaves. American Martens are small enough to refurbish and occupy abandoned woodpecker holes. The female bears her litter in a nest either in a hollow tree, on the ground under protective cover or in an underground burrow.

REPRODUCTION: Breeding occurs in July or August, but with delayed implantation of the embryo, the litter of one to six (usually three or four) young is not born until March or April. The young are blind and almost naked at birth and weigh just 28 g. The eyes open at six to seven weeks, at which time the young are weaned from a diet of milk to one of mostly meat. The mother must quickly teach her young to hunt, because when they are only about three months old, she will reenter estrus, and, with mating activity, the family group disbands. Most females first mate at 15 months of age and have their first litters at two years of age.

print

walking trail

Fisher

SIMILAR SPECIES: The **Fisher** (p. 152) is twice as long, with a long, black tail and fur that is frosted or grizzled grey to black. It seldom has an amber breast patch. The **American Mink** (p. 164) has a white chin and irregular white spots on the chest, but it has shorter ears and legs (it does not climb well) and a smaller, much less bushy, cylindrical tail.

Fisher
Martes pennanti

For the lucky naturalist, meeting a Fisher in the wild is a rare pleasure. The rest of us must content ourselves with the knowledge that this reclusive animal remains a top predator in coniferous wildlands. Historically, the Fisher was more numerous, and it was once found throughout the northern boreal forest, the northeastern hardwood forests and the forests of the Rocky Mountains and the Pacific ranges. Its numbers are still very good, but chance sightings are limited to secluded areas.

RANGE: Fishers occur across the central part of Canada except the Prairies and into the northeastern U.S. In the West, they are found from the Northwest Territories through most of British Columbia and into the Cascades and the Sierra Nevada, as well as through the Rockies to Wyoming.

Total Length: 80–120 cm

Tail Length: 30–40 cm

Weight: 2.0–5.5 kg

The Fisher prefers intact wilderness, often disappearing shortly after development begins in an area. Forest clearing, habitat destruction, fires and overtrapping resulted in its decline or extirpation over much of its range, but there have been a few reintroductions and a gradual recovery in some areas over the past two decades.

The Fisher is among the most formidable predators in Canada. It is particularly nimble in trees, and the anatomy of its ankles is such that it can rotate its feet and descend trees headfirst. Making full use of its athleticism during foraging, the Fisher incorporates any type of ecological community into its extensive home range, which can reach 120 km across. According to Ernest Thompson Seton, a legendary naturalist of the 19th century, as fast as a squirrel can run through the treetops, a marten can catch and kill it, and as fast as the marten can run, a Fisher can catch and kill it.

The Fisher is a good swimmer, but despite its name, it rarely eats fish. Perhaps this misnomer arose because of confusion with the similar-looking—though much smaller—American Mink, which regularly feeds on fish. These two members of the weasel family are most easily distinguished by their preferred habitats: the American Mink inhabits riparian areas, whereas the Fisher prefers deep forests. In winter, both species can be found in marshy areas, so size and general appearance become the best keys for identification.

Few of the animals on which the Fisher preys can be considered easy picking. The most notable example is the Fisher's famed ability to hunt the North American Porcupine. What the porcupine lacks in mobility, it more than makes up for in defensive armament, and a successful attack requires all the Fisher's speed, strength and agility. This hunting skill is far less common than folklore would suggest, however, and the Fisher does not exclusively track porcupines. Most of a Fisher's diet consists of rodents, rabbits, grouse and other small animals.

DID YOU KNOW?

The scientific name *pennanti* honours Englishman Thomas Pennant (1726–98). In the late 1700s, he predicted the decline of the American Bison and postulated that Native Americans entered North America via a Bering Strait land bridge.

DESCRIPTION: The Fisher has a foxlike face, with rounded ears that are more noticeable than those of other large weasels. In profile, its snout appears distinctly pointed. The tail is dark and more than half as long as the body. The colouration over its back is variable, ranging from frosted grey or gold to black. The underside, tail and legs are dark brown. There may be a white chest spot. Males have a longer, coarser coat than females and are typically 20 percent larger.

HABITAT: Preferred habitats include dense coniferous and mixed forests. In winter, Fishers may inhabit marshy areas. They are not found in young forests or in areas where logging or fire has thinned or removed the trees. Fishers have extensive home ranges and may only visit a particular part of their range once every two to three weeks. They are most active at night and thus are seldom seen.

FOOD: Like other members of the weasel family, the Fisher is an opportunistic hunter, killing squirrels, hares, mice, Muskrats, grouse and other birds. Unlike almost any other carnivore, however, the Fisher may hunt porcupines, which it kills by repeatedly attacking the head. It also eats berries and nuts, and carrion can be an important part of its diet.

DEN: Hollow trees and logs, rock crevices, brush piles and cavities beneath boulders all serve as den sites. Most dens are only temporary lodging because the Fisher is always on the move throughout its territory. The natal den is more permanent and is usually located in a safe place, such as a hollow tree. A Fisher may excavate a winter den in the snow.

foreprint

REPRODUCTION: A litter of one to six (usually two or three) young is born in March or April. The mother will breed again about a week after the litter is born, but implantation of the embryo is delayed until January of the following year. During mating, the male and female may remain coupled for up to four hours. The helpless young nurse and remain in the den for at least seven weeks, after which time their eyes open. When they are three months old, they begin to hunt with their mother, and by autumn, they are independent. The female usually mates when she is two years old.

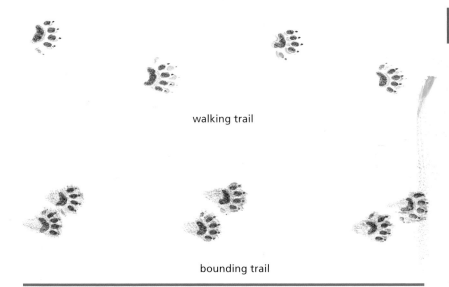

walking trail

bounding trail

American Marten

SIMILAR SPECIES: The **American Marten** (p. 150) is smaller and lighter in colour, and it usually has a buff or amber chest spot. The **American Mink** (p. 164) is smaller, with shorter ears and legs and a cylindrical, much less bushy tail. The Fisher typically has a more grizzled appearance than either the American Marten or the American Mink.

Short-tailed Weasel
Mustela erminea

summer
coat

The Short-tailed Weasel is common throughout almost all of Canada and may even be the most abundant land carnivore. Despite its numbers, the Short-tailed Weasel is not commonly seen because, like all weasels, it is most active at night and inhabits areas with heavy cover.

A foraging Short-tailed Weasel explores every hole, burrow, hollow log and brush pile for potential prey. In winter, it travels both above and below the snow in its search for food. Once a likely meal is located, the weasel seizes its prey with a rush and then wraps its body around the animal, driving its needle-sharp canines into the back of the skull or neck. If the weasel catches an animal larger than itself, it seizes the prey by the neck and strangles it.

The Short-tailed Weasel's dramatic change between its winter and summer coats led Europeans to give it two different names: an animal wearing the dark summer coat is called a "stoat," and in the white winter pelage, it is known as an "ermine." In Canada, three weasel species are white in winter and brown in summer, so the stoat and ermine labels are best avoided to prevent confusion. The trigger for the weasel's colour change is actually decreasing daylight rather than the onset of cold temperatures and snow.

DESCRIPTION: The short summer coat has brown upperparts and creamy white underparts, often suffused with lemon yellow. The feet are snowy white, even in summer, and the last third of the tail is black. The short, oval ears extend noticeably above the elongated head. The eyes are black and beady. The long neck and narrow thorax give the weasel a snakelike

winter
coat

RANGE: In North America, this weasel occurs throughout most of Alaska and Canada, south to northern California and northern New Mexico in the West and northern Iowa and Pennsylvania in the East.

Total Length: 20–35 cm

Tail Length: 4–9 cm

Weight: 45–105 g

appearance. Starting in October and November, the weasel becomes completely white, except for the black tail tip. The lower belly and inner hindlegs may retain some yellow highlights. In late March or April, the weasel moults back to its summer coat.

HABITAT: This weasel is most abundant in coniferous or mixed forests and streamside woodlands. In summer, it may often be found on alpine tundra, where it hunts on rockslides and talus slopes.

FOOD: The diet appears to consist almost entirely of animal prey, including mice, voles, shrews, chipmunks, rabbits, bird eggs and nestlings, insects and even amphibians. The weasel often eats every part of a mouse except the filled stomach, which may be excised with precision and left on a rock.

DEN: The Short-tailed Weasel commonly takes over the burrow and nest of a mouse, ground squirrel, chipmunk, pocket gopher

DID YOU KNOW?

Short-tailed Weasels typically mate in late summer, but after little more than a week, the embryos stop developing. In early spring, up to eight months later, the embryos implant in the female's uterus, and the young are born about one month later.

or lemming (often after eating the original occupant), and modifies it for weasel settlement. The nest is lined with dried grasses, shredded leaves and the fur and feathers of prey. Sometimes a weasel accumulates the furs of so many animals that the nest grows to a diameter of 20 cm. The nest may also be located in a hollow log, under a building or in an abandoned cabin that has a sizable mouse population.

REPRODUCTION: In April or May, the female gives birth to 4 to 12 (usually six to nine) blind, helpless young that weigh just 1.8 g each. Their eyes open at five weeks, and soon thereafter, they accompany the adults on hunts. At about this time, a male has typically joined the family. In addition to training the young to hunt, he impregnates the mother and all her young females, which are sexually mature at two to three months. Young males do not mature until they are about one year old—a reproductive strategy that prevents interbreeding among littermates.

print

bounding trail

Least Weasel

SIMILAR SPECIES: The **Least Weasel** (p. 162) is generally smaller, and, though there may be a few black hairs at the end of its short tail, the tail tip is not entirely black. The **Long-tailed Weasel** (p. 158) is generally larger and has orangey underparts, generally lighter upperparts and yellowish to brownish feet in summer.

Long-tailed Weasel
Mustela frenata

summer
coat

O n a sunny winter day, there may be no better wildlife experience than to follow the tracks of a Long-tailed Weasel. This curious animal zigzags as though it can't make up its mind which way to go, and every little thing it encounters seems to offer a momentary distraction. Its bountiful energy is easily read in its tracks as it leaps, bounds, walks and circles through its range.

The Long-tailed Weasel hunts wherever it can find prey—on and beneath the snow, along wetland edges, in burrows and even occasionally in trees. It overpowers small animals such as mice and snakes, killing them instantly. Larger prey species, up to the size of a rabbit, are grabbed by the throat and neck and then wrestled to the ground. As the weasel wraps its snakelike body around its prey in an attempt to throw it off balance, it tries to kill the animal with bites to the back of the neck and head.

Unlike the Short-tailed Weasel and the more northerly Least Weasel, the Long-tailed Weasel occurs only in North America. With the conversion of native grasslands to farmland, this species has been declining throughout much of its range. Still, in some native pastures that teem with rodents, this weasel can be found bounding about during the day, continuously hunting throughout its wakeful hours.

DESCRIPTION: The summer coat has rich cinnamon brown upperparts and usually orangey underparts. The feet are brown in summer. The tail is half as long as the body, and the terminal quarter is black. The winter coat is entirely white, except for the black tail tip and sometimes an orangey wash on the belly. As in all weasels, the body is long and slender—the forelegs appear to be positioned well back on the body—and the head is barely wider than the neck.

HABITAT: The Long-tailed Weasel is an animal of open country and open woodlands. It may be found in agricultural areas and on grassy slopes. Sometimes it forages in valleys and open forests.

RANGE: From a northern limit in central British Columbia and Alberta, this weasel ranges south through most of the U.S. (except the southwestern deserts) and Mexico into northern South America.

Total Length: 28–42 cm
Tail Length: 12–29 cm
Weight: 85–400 g

FOOD: Although the Long-tailed Weasel can successfully subdue larger prey than either of its smaller relatives, voles and mice still make up most of its diet. It also preys on ground squirrels, Red Squirrels, rabbits and shrews, and it takes the eggs and young of ground-nesting birds when it encounters them.

DEN: This weasel usually makes its nest in the burrow of a small mammal it has eaten. The nest cavity is lined with soft materials such as fur from prey or dried grasses. The female makes a maternal den in the same manner, either in a purloined burrow, in a hollow log or under an old tree stump.

REPRODUCTION: Long-tailed Weasels typically mate in midsummer, but through delayed implantation of the embryos, the young are not born until April or May. The four to nine (usually six to eight) blind, helpless young are born with sparse, white hair, which becomes a fuzzy coat after one week and

> **DID YOU KNOW?**
>
> Weasel signs are not uncommon if you know what to look for. The tracks typically follow a paired pattern, and the twisted, black, hair-filled droppings, which are about the size of your pinkie finger, are often left atop a rock pile.

a sleek coat in two weeks. At 3.5 weeks of age, the young begin to supplement their milk diet with meat; they are weaned when their eyes open, just after seven weeks. By six weeks, there is a pronounced difference in size, with young males weighing about 99 g and females about 78 g. At about this time, a mature male weasel typically joins the group to breed with the mother and the young females as they become sexually mature. The group travels together, and the male and female teach the young to hunt. The group disperses when the young are 2.5 to 3 months old.

winter coat

print

bounding trail

Short-tailed Weasel

SIMILAR SPECIES: The **Short-tailed Weasel** (p. 156) is typically smaller, has a relatively shorter tail and has a lemon yellow (not orangey) belly and white feet in summer. The **Least Weasel** (p. 162) is much smaller and may have a few black hairs on the tip of its short tail, but it never has an entirely black tip.

Black-footed Ferret
Mustela nigripes

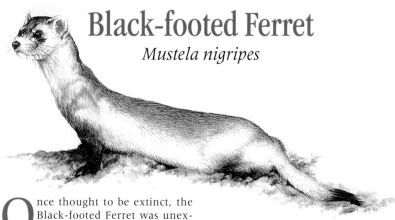

O nce thought to be extinct, the Black-footed Ferret was unexpectedly rediscovered in Wyoming when a rancher's dog laid a limp corpse in his yard in 1981. Since that time, small populations of North America's only ferret have been reestablished near prairie dog colonies in several places in the grasslands of the U.S. The current wild population of these ferrets numbers only about 800 adults. The Black-footed Ferret once occurred throughout the Great Plains, but it preys almost exclusively on prairie dogs, and the widespread eradication of these rodents by ranchers led to the collapse of the ferret population.

The reintroduction of Black-footed Ferrets into the Canadian Prairies has been in the making for many years. Ferrets in the Toronto Zoo and other participating conservation agencies are bred with the intention of reintroducing them to the Prairies. Grasslands National Park in southern Saskatchewan has been chosen as the prime site for reintroduction because it has intact native grasslands and the last remaining population of Black-tailed Prairie Dogs. Before being released into the wild, however, the captive-bred ferrets need to be taught how to survive. To achieve this, the ferrets are sent to "bootcamp" in Colorado. The U.S. Fish and Wildlife Service operates the Black-footed Ferret Conservation Breeding Centre in Fort Collins, where captive-bred ferrets are taught how to interact with other, wild ferrets and to hunt prairie dogs. The first of several planned annual releases into Grasslands National Park occurred in October 2009. Happily, researchers were able to locate many of the ferrets the following spring. Having survived their first winter, there is a good chance that the population will thrive and grow, marking 2010 as the first year in Canada with resident Black-footed Ferrets in more than 70 years. With additional annual releases, the Black-footed Ferret may soon become an icon for conservation success stories.

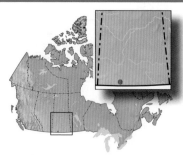

RANGE: The Black-footed Ferret formerly inhabited most of the Great Plains, from southern Alberta and Saskatchewan to northern Texas. Current populations are discontiguous. The species has been successfully reintroduced into southern Saskatchewan.

Total Length: 49–58 cm

Tail Length: 11–14 cm

Weight: 480–900 g

DESCRIPTION: The Black-footed Ferret has an elongated head, broadly oval, furred ears, short legs and a fairly cylindrical tail. The body is heavier than that of other weasels but still has the typical long, thin shape. The short, glossy hairs of the upperparts vary in colour from creamy white to cinnamon brown to yellowish grey. There are distinct white patches on either side of the snout and between the ears and eyes. The forehead is dark, and a dark brown mask extends across the eyes. The throat and belly are white. The legs, feet and tail tip are dark brown or black.

HABITAT: This species inhabits open, arid grasslands and shrubby areas, mostly in the vicinity of prairie dog "towns."

FOOD: Black-tailed Prairie Dogs are the primary food. When they are scarce, the Black-footed Ferret will eat other rodents, including ground squirrels,

DID YOU KNOW?

The Black-footed Ferret is nocturnal, and at night, it will slip into a prairie dog burrow to prey on the sleeping inhabitant.

mice and pocket gophers, as well as small birds, eggs and even reptiles.

DEN: A ferret typically makes its den in a prairie dog burrow, after having eaten the inhabitant. It may enlarge the burrow and add new chambers. Unlike prairie dogs, ferrets do not tamp down the excavated dirt, but instead drag it away, leaving two parallel grooves of dirt leading from the burrow entrance. A prairie dog burrow with the telltale loose dirt and grooves around it could be the home of this endangered mammal.

REPRODUCTION: Black-footed Ferrets mate in March or April. The female bears one to five (usually three or four) young after a gestation of about 42 days. The young are born helpless, and, like other weasels, they grow rapidly.

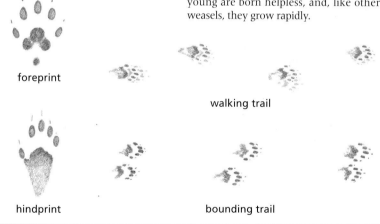

foreprint

walking trail

hindprint

bounding trail

Long-tailed Weasel

SIMILAR SPECIES: The **Long-tailed Weasel** (p. 158) lacks the black mask and legs, is somewhat smaller and tends to have orangey underparts.

Least Weasel
Mustela nivalis

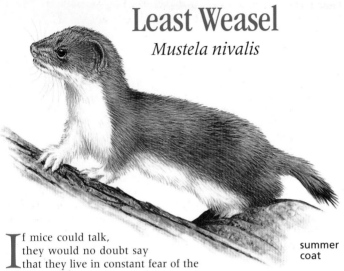

summer coat

If mice could talk, they would no doubt say that they live in constant fear of the Least Weasel. As it hunts, this weasel enters and fully explores every burrow it encounters. This pint-sized carnivore is small enough to squeeze into the burrows of mice and voles, and any small, moving creature seems to warrant attack.

The Least Weasel is the smallest weasel—in fact, the smallest member of the carnivore order—in the world, but it has a monstrous appetite. It needs to consume about half of its weight each day (about two voles), but it can only consume this amount at a rate of a little less than 2 g of vole per hour. The result is that a Least Weasel, with its high metabolism but low feeding rate, needs to eat up to 10 times per day. If it finds a group of voles or other small rodents, it is quick enough to kill them all within seconds. It begins to eat its prey immediately, but stores leftovers to be eaten later. A Least Weasel may have many caches of food to help sustain its frequent need to feed.

This weasel can be active at any time, day or night, though few people ever see it in action. Most human encounters with a Least Weasel result from lifting plywood, sheet metal or hay bales. These sightings are understandably brief because the weasel wastes little time finding the nearest escape route, and it will enter any hole larger than 2.5 cm across—much to the dismay of the current resident.

Because the Least Weasel changes colour in response to shorter day lengths, an unseasonably early snow in autumn or an early melt in spring can make a weasel stand out against its environs. In spite of

RANGE: In North America, the Least Weasel's range extends from western Alaska through most of Canada and south to Nebraska and Tennessee. It is absent from the Maritimes, New York State and New England.

Total Length: 15–22 cm

Tail Length: 2–4 cm

Weight: 30–55 g

this visual disadvantage, this weasel is still an efficient hunter.

DESCRIPTION: In summer, this small weasel is walnut brown above and white below. The short tail may have a few black hairs at the end but never has an entirely black tip. The ears are short, scarcely extending above the fur. In winter, the entire coat is white, including the furred soles of the feet. Only a few black hairs may remain at the tip of the tail.

HABITAT: The Least Weasel usually does not inhabit dense coniferous forests, preferring aspen parkland, open, grassy areas, forest edges or tundra. It sometimes occupies abandoned buildings and rock piles. Prey abundance seems to influence the distribution of this weasel more than habitat does.

FOOD: Voles, mice and insects are the usual prey, but amphibians, birds and eggs are taken when they are encountered.

DEN: The burrow and nest of a vole that has fallen prey to a Least Weasel is a typical den site. The nest is usually lined with rodent fur and fine grasses, which may become matted like felt and reach a thickness of 2.5 cm. In winter, frozen, stored mice may be dragged into the nest to thaw prior to consumption.

DID YOU KNOW?

During the autumn moult, white fur first appears on the animal's belly and spreads toward the back. The reverse occurs in spring—the brown coat begins to form on the weasel's back and moves toward its belly.

REPRODUCTION: Unlike many weasels, the Least Weasel does not exhibit delayed implantation of embryos, and a female may give birth in any month of the year. After a gestation of 35 days, 1 to 10 (usually four or five) wrinkled, pink, hairless young are born. The young begin to eat meat at three weeks. After their eyes open at 26 to 30 days, their mother begins to take them hunting. They disperse at about seven weeks, living solitary existences except for brief mating encounters.

winter coat

print

bounding trail

Short-tailed Weasel

SIMILAR SPECIES: The **Short-tailed Weasel** (p. 156) is generally larger, has a longer tail with an entirely black tip and usually has lemon yellow highlights on the belly. The **Long-tailed Weasel** (p. 158) is larger, has a much longer, black-tipped tail and has orangey underparts, generally lighter upperparts and brown feet in summer.

American Mink
Neovison vison

To many people, watching the fluid undulations of a bounding American Mink is more valuable than its much-prized fur. The mink is a lithe weasel that was once described by naturalist Andy Russell as moving "like a brown silk ribbon." Indeed, like most weasel species, the American Mink seems to move with the unpredictable flexibility of a toy Slinky in a child's hands.

Mink are tenacious hunters, following scent trails left by potential prey over all kinds of obstacles and terrain. Almost as aquatic as otters, these opportunistic feeders routinely dive to depths of several metres in pursuit of fish. Their fishing activity tends to coincide with breeding aggregations of fish in spring and autumn, and with times when low oxygen levels force fish to congregate in oxygenated areas in winter. It is along watercourses, therefore, where mink are most frequently observed, and their home ranges often stretch out in linear fashion, following rivers for up to 5 km.

The American Mink is active throughout the year, and it is often easiest to follow by trailing its winter tracks in snow. The paired prints left by its bounding gait betray this inquisitive animal's adventures as it comes within sniffing distance of burrows, hollow logs and brush piles. This active predator always seems to be on the hunt and rarely passes up a feeding opportunity. A mink may kill more than it can eat, and surplus kills are stored for later consumption. Food caches are often tucked away in the mink's overnight den, which are typically dug into a riverbank, beneath a rock pile or in the home of a permanently evicted Muskrat.

DESCRIPTION: The sleek coat is generally dark brown to black, usually with white spots on the chin, the chest and sometimes the belly. The legs are short.

RANGE: This wide-ranging weasel occurs across most of Canada and the U.S., except for the High Arctic tundra and the dry southwestern regions of the U.S.

Total Length: 47–70 cm

Tail Length: 15–20 cm

Weight: 0.6–1.4 kg

The tail is cylindrical and only somewhat bushy. The anal scent glands produce a rank, skunklike odour. Males are nearly twice as large as females.

HABITAT: The American Mink is rarely found far from water. It is comfortable in a variety of habitats from coniferous or hardwood forests to open grassland, provided that a lake, pond or river is available.

FOOD: American Mink are fierce predators of Muskrats, but in their desire for nearly any type of animal flesh, they also take frogs, fish, waterfowl, eggs, mice, voles, rabbits, snakes and even crayfish and other aquatic invertebrates.

DEN: The den is usually in a burrow close to water. A mink may dig its own

DID YOU KNOW?

The name "mink" comes from a Swedish word that means "stinky animal." Although not as aromatic as a skunk, the American Mink is one of the smelliest weasels. Its anal musk glands can release a stinky liquid when the animal is threatened, but a mink cannot aim the spray.

burrow, but more frequently, it takes over the bank den of a Muskrat or American Beaver and lines the nest with grasses, feathers and other soft materials.

REPRODUCTION: Mink breed any time between late January and early April, but because the period of delayed implantation varies in length from one week to 1.5 months, the female almost always gives birth in late April or early May. The actual gestation is affected by temperature but lasts about two months, and a litter consists of 2 to 10 (usually four or five) helpless, blind, pink, wrinkled young. Their eyes open at 24 to 31 days, and weaning begins at five weeks. The mother teaches the young to hunt for two to three weeks, after which they hunt for themselves but stay with their mother until fall.

foreprint

hindprint

bounding trail

American Marten

SIMILAR SPECIES: The **American Marten** (p. 150) has a bushier tail, longer legs and an orange or buffy throat patch, and it is not as sleek looking. The **Fisher** (p. 152) is much larger, with longer ears and a much bushier tail. The **Northern River Otter** (p. 174) is larger and has a longer, tapered tail and webbed feet.

Wolverine
Gulo gulo

The largest member of the weasel family, the Wolverine is one of the most poorly understood mammals in North America. It is an elusive animal and historically was the subject of many myths and tall tales. More recently, the Wolverine has become a symbol of remote, pristine wilderness. Although most of us will never see a Wolverine, the knowledge that it maintains a hold in remote forests may reassure us that expanses of wilderness still exist.

Tales of the this animal's gluttony—its reputation rivals that of the African hyena—have lingered in wilderness lore for centuries. Pioneers warned their children against the dangers of the forests, and often they meant Wolverines.

The Wolverine is an efficient and agile predator—it can crush through bone in a single bite, it has long, semi-retractile foreclaws that allow it to climb trees, and it is ferocious enough to challenge a lone bear or wolf. What we rarely hear about is this animal's intelligence, its uniqueness among its weasel relatives and its sheer stealth and beauty.

A few behavioural studies of the Wolverine have demystified this animal; instead of a vicious glutton, it becomes a clever and adaptable creature. Nevertheless, some of the Wolverine's reputation is well deserved. True to its nickname "Skunk Bear," it produces a stink that

RANGE: In North America, the Wolverine is a species of the boreal forest and tundra of Alaska and northern Canada. It follows the montane coniferous forests from Alaska as far south as California and Colorado.

Total Length: 70–110 cm

Tail Length: 17–25 cm

Weight: *Male:* 11–18 kg; *Female:* 6–12 kg

rivals a skunk's odour. The abundant scent is produced in anal glands and is used primarily to mark territory.

The Wolverine's habitat preferences seem to vary as its diet shifts with the seasons. In summer, it eats mostly rodents and other small mammals, as well as birds and berries. In winter, it scavenges carrion, mainly hoofed mammals left by wolves, as well as roadkills or winter die-offs. Like a vulture, the Wolverine can detect carcasses from far away.

The Wolverine some of the most powerful jaws in the mammal world and uses them to tear meat off frozen carcasses or to crunch through bone to get at the rich, nourishing marrow inside. Few other large animals are able to extract as much nourishment from a single carcass.

DESCRIPTION: Although the Wolverine's weasel-like head is small relative to its body size, the long legs and long fur look like they belong on a small bear. Unlike a bear,

DID YOU KNOW?

The Wolverine's lower jaw is more tightly bound to its skull than that of most other mammals. The articulating hinge that connects the upper and lower jaws cannot dislocate because it is wrapped by bone in an adult Wolverine, and it is this characteristic that gives this animal such a powerful bite.

however, the Wolverine has an arched back and a long, bushy tail. The shiny coat is mostly dark cinnamon brown to nearly black. There may be yellowish white spots on the throat and chest. A buffy or pale brownish stripe runs down each side from the shoulder to the flank, where it becomes wider. The stripes meet just before the base of the tail, leaving a dark saddle.

HABITAT: The Wolverine predominantly occupies large areas of remote, forested

wilderness and sometimes also tundra. Its enormous territory encompasses a great variety of habitats, and this agile, determined predator can likely conquer almost any wilderness terrain.

FOOD: Wolverines prey on mice, other small mammals, birds, beavers and fish. Caribou and even Moose have been attacked, often successfully. In winter, Wolverines scavenge carcasses of malnourished animals or the remains left by other predators. Depending on location and season, they may also eat berries, fungi and nuts.

foreprint

hindprint

On the West Coast, they will scavenge whale and seal carcasses.

DEN: The Wolverine's den may be among the roots of a fallen tree, in a hollow tree butt, in a rocky crevice or even in a long-lasting snowbank. The natal den is often underground, and the mother lines it with leaves. A Wolverine may maintain several dens throughout its territory that range in quality from makeshift cover under tree branches to a permanent underground dugout.

REPRODUCTION: Wolverines breed between late April and early September, but the embryos do not implant in the uterine wall of the female until January. Between late February and mid-April, the female gives birth to a litter of one to five (generally two or three) cubs. The stubby-tailed cubs are born with their eyes and ears closed and with a fuzzy, cream-coloured coat that sets off the darker paws and mask. They nurse for 8 to 10 weeks, then they leave the den and their mother teaches them to hunt. The mother and her young typically stay together through the first winter. The young disperse when they become sexually mature in spring.

loping trail

American Badger

SIMILAR SPECIES: The **American Badger** (p. 170) is squatter, with a distinctive vertical stripe on its forehead, and it lacks the lighter side stripes. The **Black Bear** (p. 216) is larger, has a shorter tail and lacks the light buffy stripe along each side.

American Badger
Taxidea taxus

In tall-grass prairies and open pastures, a lucky observer may encounter an American Badger. With a flair for remodelling, the badger is nature's roto-tiller and backhoe. The large holes that it leaves are of critical importance as den sites, shelters and hibernacula for dozens of species, from Coyotes to black-widow spiders. Even toads can be found inside the cool, moist burrows.

The American Badger enjoys a reputation for fierceness and boldness that was acquired in part from a distantly related European weasel bearing the same name. While it is true that a cornered badger will put up an impressive show of attitude, like most animals, it prefers to avoid a fight. When it is severely threatened or in competition, the American Badger's claws, strong limbs and powerful jaws make this animal a dangerous opponent. In spite of its impressive arsenal, the badger routinely kills only rodents and other small mammals. However, on rare occasions, American Badgers have taken Coyote pups. Conversely, a group of Coyotes can defeat a badger.

Pigeon-toed and short-legged, the American Badger is not much of a sprinter, but its heavy front claws enable it to move large quantities of earth in short order. Although a badger's predatory nature is of benefit to landowners, its natural digging skills have led many to be killed—cattle and horses have been known (very rarely) to break their legs when stepping carelessly into badger excavations.

RANGE: The American Badger is found in Canada in south-central British Columbia, the Prairies and southern Ontario, continuing through much of the United States to Baja California and the central Mexican highlands.

Total Length: 80–85 cm

Tail Length: 13–16 cm

Weight: 5–11 kg

American Badgers tend to spend a great part of winter sleeping in their burrows, but they do not enter a full state of hibernation like many mammals do. Instead, badgers emerge from their slumber to hunt whenever winter temperatures are moderate.

In spite of low population densities, almost all sexually mature female badgers are impregnated during the nearly three months that they are sexually receptive. As with most members of the weasel family, once the egg is fertilized, further embryonic development is put off until the embryos implant, usually in January, which results in a spring birth.

DESCRIPTION: Long, grizzled, yellowish grey hair covers this short-legged, muscular member of the weasel family. The hair is longer on the sides than on the back or belly, which adds to the flattened appearance of the body. A white stripe originates on the nose and runs back onto the shoulders and sometimes slightly beyond. The top of the head is otherwise dark. A dark, vertical crescent or triangle is apparent on each side of the face between the short, rounded, furred ears

> **DID YOU KNOW?**
>
> American Badgers make an incredible variety of sounds. Adults hiss, bark, scream and snarl, whereas young badgers grunt, squeal, bark, meow, chirr and snuffle. Also, the badger's front claws clatter when it runs on hard surfaces.

and the eyes. A whitish or pale buff stripe originates at the edge of the mouth on each side and passes beside the eye and up to just above the ear. The short, bottlebrush tail is more yellowish than the body, and the lower legs and feet are very dark brown, becoming blackish on the extremities. The three central claws on each forefoot are greatly elongated for digging. Skulls from older badgers may have the "wrap-around" jaw articulation seen in Wolverine skulls.

HABITAT: Essentially an animal of open places, the American Badger shuns thick forests. It is usually found in association with ground-dwelling rodents, typically in grasslands, agricultural areas and open parkland.

FOOD: Burrowing mammals fill most of the American Badger's dietary needs, but it also eats eggs, young ground-nesting birds, mice and sometimes carrion, insects and snails.

DEN: A badger may dig its own den or take over another animal's burrow. The den may approach 9 m in length and have a diameter of about 30 cm. A bulky grass nest is built in an expanded chamber near or at the end of the burrow. A large pile of excavated earth is generally found to one side of the burrow entrance.

REPRODUCTION: Delayed implantation of the embryo is characteristic, and one to five (usually four) naked, helpless young are born in March or April. Their eyes open after a month, and at two months, their mother teaches them to hunt. They leave the burrow in early evening, trailing their mother. The young investigate every grasshopper or beetle they encounter, but the mother directs the expedition to small mammal burrows. She often cripples a rodent and then leaves it for her young to kill. The young disperse in autumn, when they are three-quarters grown. Some of the young females may mate when they are only four to five months old, but most American Badgers are not sexually mature until 14 months of age.

foreprint

hindprint

walking trail

SIMILAR SPECIES: The American Badger's facial pattern and overall appearance are unique. The **Raccoon** (p. 182) has a long, ringed tail and horizontal, black-and-white facial markings, rather than vertical ones. The **Wolverine** (p. 166) has a different range. The **Striped Skunk** (p. 148) also has black-and-white markings, but it is smaller and has a mainly black body.

Raccoon

Northern River Otter
Lontra canadensis

Although it may seem too good to be true, all those playful characterizations of Northern River Otters are founded on truth. Otters often amuse themselves by rolling, sliding, diving or "body surfing," and they may also push and balance floating sticks with their noses or drop and retrieve pebbles for minutes at a time. They seem particularly interested in playing on slippery surfaces and will leap onto snow or mud with their forelegs folded close to their bodies for a streamlined toboggan ride. Unlike most members of the weasel family, river otters are social animals, and they will frolic together in the water and take turns sliding down banks.

With their streamlined bodies, rudder-like tails, webbed toes and valved ears and nostrils, Northern River Otters are well adapted for aquatic habitats. When they emerge from the water to clamber over rocks, there is a serpentine appearance to their movements. The large amount of playtime they seem to have results from their efficiency at catching prey when it is plentiful. Although otters generally cruise along slowly in the water by paddling with all four feet, they can dart after prey with the ease of a seal whenever hunger strikes. When otters swim quickly, they propel themselves mainly with vertical undulations of their body, hindlegs and tail. They can hold their breath for as long as four minutes.

RANGE: The Northern River Otter occurs from near treeline across Alaska and Canada south through forested regions to northern California and northern Utah in the West and Florida and the Gulf Coast in the East. It is largely absent from the Midwest and the Prairies and Great Plains.

Total Length: 0.9–1.4 m

Tail Length: 30–50 cm

Weight: 5–11 kg

Because of all its activity, the Northern River Otter leaves many signs of its presence when it occupies an area. Its slides are the most obvious and best-known evidence—but be careful not to mistake the slippery beaver trails that are common around beaver ponds for otter slides. Despite its other aquatic tendencies, an otter always defecates on land. Its scat is simple to identify—it is almost always full of fish bones and scales.

A Northern River Otter may make extensive journeys across land, even through deep snow. Although this animal looks clumsy on land, with its humped, loping gait, it can easily outrun a human. On slippery surfaces such as wet grass, snow and ice, the otter glides along, usually on its belly with its legs tucked either back or forward to help steer and push. On flat ground, snow slides are sometimes pitted with blurred footprints where the otter has given itself a push for momentum.

In the past, the Northern River Otter's thick, beautiful, durable fur led to excessive trapping that greatly diminished its numbers. Trapping has since been reduced, and the otter seems to be slowly recolonizing parts of North America from which it has been absent for decades. Even in areas where it is known to occur, however, it is infrequently seen. Water quality can influence otter densities because much of its prey is aquatic.

DESCRIPTION: This large, weasel-like carnivore has dark brown upperparts that look black when wet. It is paler below, and the throat is often silver grey. The head is broad and flattened, with small eyes and ears and prominent, whitish whiskers. The feet are webbed. The long tail is thick at the base,

DID YOU KNOW?

When a troupe of agile Northern River Otters travels single file through the water, the creatures' undulating, lithe bodies combine to form a very serpentlike image—perhaps with enough similarity to give rise to rumours of lake-dwelling sea monsters.

gradually tapering to the tip. Males are larger than females. The otter does not hibernate, and in winter, it will chase fish under the surface of the ice.

HABITAT: Year-round, Northern River Otters live primarily in and along wooded rivers, ponds and lakes, but they sometimes roam far from water. They may be active day or night but tend to be more nocturnal close to human activity. In winter, otters are found on lakes with beaver lodges or on bog ponds with steep banks containing old beaver burrows, through which they can enter the water. Also, riffles and waterfalls with pools provide important access to water in winter.

FOOD: Fish, crayfish, turtles and frogs form the bulk of the diet, but otters occasionally depredate bird nests and also eat small mammals such as mice, young Muskrats and young beavers, and sometimes even insects and earthworms.

DEN: The permanent den is often in a bank, with both underwater and abovewater entrances. When roaming on land, an otter rests under roots or overhangs, in hollow logs, in the abandoned burrows of other mammals or in

fast running trail

abandoned American Beaver or Muskrat lodges. Natal dens are usually abandoned Muskrat, American Beaver or Woodchuck dens.

REPRODUCTION: The female bears a litter of one to six blind, fully furred young in March or April. The young weigh about 135 g at birth. They first leave the den at three to four months of age and leave their parents at six to seven months. Otters become sexually mature at two years of age, but may not become accomplished breeders until they are five to seven years old. The mother breeds again soon after her litter is born, but delayed implantation of the embryos puts off the birth until the following spring.

foreprint

hindprint

SIMILAR SPECIES: The **American Beaver** (p. 305) is stouter and has a wide, flat, hairless, scaly tail. The **American Mink** (p. 164) is much smaller, its feet are not webbed, and its shorter tail is cylindrical, not tapered.

American Beaver

Sea Otter
Enhydra lutris

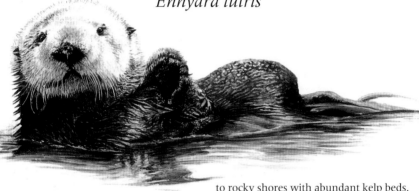

For many people, playful and intelligent Sea Otters are among the most desired animals to see when visiting the West Coast. These fully aquatic carnivores have such a buoyant body and curious demeanour that watching them is not only comical, but also mesmerizing. When two Sea Otters are playing, they turn somersaults at the surface and wrestle together as if trying to dunk each other. When they are resting, they lounge on their backs at the surface and rub their faces with curled-up paws, much like cats do when grooming. Sea Otters are even neighbourly, regularly hobnobbing with sea lions and seals.

It is easy to tell the difference between Sea Otters and Northern River Otters on the West Coast—with the right information. Sea Otters do not venture more than 2 km from shore or into water more than 30 m deep. Typically, they stay close

to rocky shores with abundant kelp beds. They also prefer open coastlines and are therefore rarely seen along the Inside Passage. River otters, on the other hand, are frequently seen in the sheltered waters of the Inside Passage. Also, river otters are well known for their travels, and an otter seen several kilometres from shore or one that is swimming long distances from island to island is likely a Northern River Otter. Any otter seen moving or eating on land is, again, a river otter. Sea Otters are very clumsy on land, and their locomotion is limited to an ungainly, heavy lope, an awkward, slow walk or an even slower, body-dragging slide. Being so limited on land, Sea Otters rarely come out of the water, though they may haul out onto rocks to rest during rough or stormy weather.

At night or for daytime rest, a Sea Otter wraps itself in kelp at the water's surface. The kelp is attached underwater

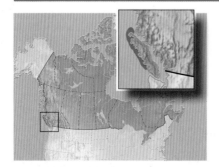

RANGE: Sea Otters are found in scattered locations along the West Coast from southern California to the Aleutian Islands of Alaska.

Total Length: 0.8–1.5 m

Tail Length: 25–41 cm

Weight: 23–45 kg

and prevents the otter from drifting while it sleeps.

Unlike most other marine mammals, Sea Otters drink seawater. Whales and pinnipeds are able to meet most of their water needs from the metabolism of food, and although they may safely drink seawater, they rarely do so unless they are in a period of fasting. Sea Otters, however, drink about a cupful of seawater daily, and their relatively large kidneys can concentrate the salts in their urine by up to 30 percent. This means they can excrete salt in a higher concentration than the surrounding seawater, resulting in a net gain of free water. By comparison, when humans drink seawater, dehydration occurs. Our kidneys cannot cope with the salt concentration of seawater, and therefore water leaves every cell of the body to dilute the salt—basically, in order to remove the excess salt, more water gets urinated than what was ingested in the first place.

The Sea Otter does not have a layer of insulating blubber like other marine mammals, relying on its high metabolism and thick coat to keep warm. The coat has two layers, and the insulating undercoat traps air next to the skin, keeping the skin warm and dry. The lush coat is both a blessing and a curse because Sea Otters were

DID YOU KNOW?

Sea Otters caught in oil spills have an extremely poor chance of survival. The oil slicks the animals' coats and destroys the insulating and waterproofing qualities of the fur.

once hunted almost to extinction for their pelts. The otters that now inhabit the West Coast are probably all descendants of reintroduction programs that took place in 1969. Populations are slowly increasing, but the species is still listed as threatened.

DESCRIPTION: The Sea Otter is the largest member of the weasel family and the smallest completely marine mammal. This stout-bodied creature has a short tail and rounded head. Its slightly flattened tail is no more than one-third of the length of its body. Its fur may be a variety of colours, including light brown, reddish brown, yellowish grey or nearly black, and is very thick, especially on the throat and chest. The head is often lighter than the body; in old males, the head may be nearly white. This otter has two "pockets" formed by a fold of skin between its chest and each underarm. Its tiny ears may appear pinched but are

otherwise inconspicuous. All four feet are webbed, and the hindlegs resemble flippers. Males are larger than females.

HABITAT: This otter lives almost its entire life in shallow coastal waters, favouring areas with kelp beds or reefs with nearby or underlying rocks.

FOOD: The Sea Otter feeds primarily on sea urchins, crustaceans, shellfish and fish. When foraging, it dives underwater for up to five minutes and returns to the surface with its prey and a stone, using the pockets of skin that run from its chest to its underarms to carry its load. One otter was seen unloading six urchins and three oysters from its "cargo holds." To unload

walking trail

its pockets and eat, the otter rests on its back. It places the stone on its chest and uses it as an anvil on which to bash the shell of an urchin, shellfish or crustacean repeatedly until the shell breaks, exposing the flesh inside. Unlike the Northern River Otter, the Sea Otter does not come to shore to eat its meal.

REPRODUCTION: Mating occurs in the water, usually in late summer, and the female gives birth to a single pup 6.5 to 9 months later, a gestation that probably includes a period of delayed implantation. On rare occasions, a female has two pups. A pup is fully furred and has open eyes at birth. The female gives birth in the water, and to nurse her young, she floats on her back and allows the pup to sit on her chest. The pup also plays and naps on its mother's chest. It is weaned at one year of age but may stay with its mother for several more months, even if she gives birth again. If the mother senses danger, such as an approaching shark or Orca, she will hold her pup under her forelegs and dive into a kelp bed until the danger passes.

foreprint
(in sand)

hindprint

SIMILAR SPECIES: The **Northern River Otter** (p. 174) has a longer tail, its hindlegs are not flipper-like, and it is almost invariably the type of otter seen on land. **Seals** and **sea lions** (pp. 186–203, 208–10) have different body shapes, and their forelegs are flippers instead of paws.

Northern River Otter

Raccoon
Procyon lotor

The Raccoon is famous for its black "bandit" mask and ringed tail. The mask suits the Raccoon, because it is well known as a looter of people's gardens, cabins, campsites and, of course, garbage cans. This animal is likely to investigate every tasty bit of food and shiny object it finds. Despite its roguish behaviour, however, the Raccoon has never been associated with ferociousness or savagery—it is mainly a curious, docile animal unless cornered or threatened. Testing a Raccoon's ferocity is an unnecessary and simple-minded act, and it has been known to wound and even kill attacking dogs.

Raccoons are among the most frequently encountered wild carnivores in the southern parts of most provinces, especially British Columbia and Ontario. When a Raccoon is seen, which is usually at night, it quickly bounds away, effectively evading flashlight beams by slipping into a burrow or climbing to a tree retreat. Should its tree sanctuary be found, the raccoon will remain still, waiting for the invasive experience to end.

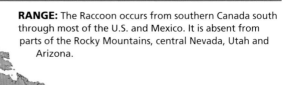

RANGE: The Raccoon occurs from southern Canada south through most of the U.S. and Mexico. It is absent from parts of the Rocky Mountains, central Nevada, Utah and Arizona.

Total Length: 65–100 cm

Tail Length: 19–40 cm

Weight: 5–14 kg

Raccoons tend to frequent muddy, aquatic environments, a characteristic that allows people to find their diagnostic tracks along the edges of wetlands and waterbodies. Like bears and humans, Raccoons walk on their heels, so they leave unusually large tracks for their body size. They will methodically circumnavigate wetlands in the hope of finding duck nests or unwary amphibians upon which to dine.

The way that the Raccoon typically feels its way through the world has long been recognized. In fact, our word "raccoon" is derived from the Algonquian name for this animal, *aroughcoune*, which means, "he scratches with his hands." One of the best-known characteristics of the Raccoon is its habit of dunking its food in water before eating it. It had long been thought that the animal was washing its food—the scientific name *lotor* is Latin for "washer"—but biologists now believe that a Raccoon's sense of touch is enhanced by water, and that it is actually feeling for inedible bits to discard.

Long, cold winters are an ecological barrier to the dispersal of these animals because they do not hibernate and so require year-round food availability. They may sleep for extended periods in parts of their range, but they still come out on warmer nights. Over the past century, however, Raccoons have been moving into colder climes, perhaps because of increasing human habitation in these areas. When Raccoons first appeared in Winnipeg, Manitoba, in the 1950s, many people were quite surprised and took them to the local zoo, thinking they were escapees rather than a new species of urban "wildlife."

DESCRIPTION: The coat is blackish to brownish grey overall, with lighter, greyish brown underparts. The bushy tail, with its four to six alternating, blackish rings on a yellowish white background, makes the Raccoon one of the most recognizable North American carnivores.

> **DID YOU KNOW?**
>
> Raccoons have thousands of nerve endings in their "hands" and "fingers." It is an asset they constantly put to use, probing under rocks and in crevices for food.

There is a black mask across the eyes, bordered by white "eyebrows" and a mostly white snout, and a strip of white fur separates the upper lip from the nose. The ears are relatively small. The Raccoon is capable of producing a wide variety of vocalizations—it can purr, growl, snarl, scream, hiss, trill, whinny and whimper.

HABITAT: Raccoons are most often found near streams, lakes and ponds. They favour woodlands near agricultural areas and are not found in vast open grasslands or tundra.

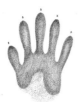

foreprint

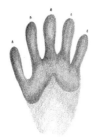

hindprint

running group

FOOD: The Raccoon fills the role of medium-sized omnivore in the food web. Besides eating fruits, nuts, berries and insects, it avidly devours clams, frogs, fish, eggs, young birds and rodents. Just as a bear does, the Raccoon consumes large amounts of food in autumn to build a large fat reserve that will help sustain it over winter.

DEN: The den is often located in a hollow tree, but Raccoons are increasingly using sites beneath abandoned buildings or under discarded construction materials. In rougher terrain, dens are sometimes found in rock crevices, where grasses or leaves carried in by the female may cover the floor.

REPRODUCTION: After about a two-month gestation, the female bears two to seven (typically three or four) young in late spring. The young weigh just 57 g at birth. Their eyes and ears open at about three weeks, and when they are six to seven weeks old, they begin to feed outside the den. At first, the mother carries her young by the nape of the neck, as a cat carries kittens. About a month later, she starts taking them on extended nightly feeding forays. Some young disperse in autumn, but others remain until their mother forces them out when she needs room for her next litter.

walking trail

SIMILAR SPECIES: Only the **American Badger** (p. 170) could possibly be confused with a Raccoon, but the badger is much squatter, its facial markings are vertically oriented (unlike the horizontal mask of a Raccoon), and its shorter, thinner tail does not have the Raccoon's distinctive rings.

American Badger

Harbour Seal
Phoca vitulina

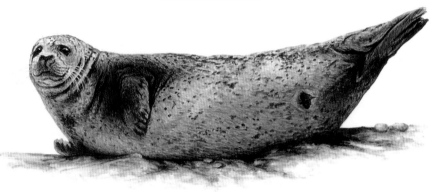

Inquisitive Harbour Seals are well-known residents of Canada's coastal waters. These seals are frequently referred to as being sociable or gregarious, but this perception is not entirely true. Although many seals may bask on rocks together, they pay very little attention to their neighbours and seldom interact. Only during the pupping season is there interaction, and it is primarily between females. Mothers with newborn pups may congregate in a "nursery" in shallow water where the pups can sleep. These nursery groups form solely as a protective measure against possible predation. While most of the females and pups sleep, some are likely to be awake and watchful for danger. The same is true for hauled-out seals. Where several seals are together, chances are good that there is always at least one individual awake and vigilant.

Harbour Seals cannot sleep at the surface like sea lions and sea otters. During the day, they sleep underwater in shallow coastal water by resting vertically just above the bottom. Young pups commonly sleep in this manner. They can go without breathing for nearly 30 minutes, and though they sometimes wake up to breathe, they frequently rise to the surface and take a breath without awakening, then sink back to the bottom. When the tide is out, these seals sleep high and dry in their preferred haul-out sites. They frequently rest with their heads and rear flippers lifted above the rock.

Harbour Seals tend to be wary of humans, and if you approach one, it is likely to dive immediately into the water.

RANGE: Harbour Seals are found along the western coast of North America from western Alaska to California and in the East from southeastern Ellesmere Island and Hudson Bay to the Carolinas. They also inhabit parts of Greenland, Europe and Asia.

Total Length: 1.2–1.8 m

Tail Length: 9–11 cm

Weight: 50–140 kg

However, many kayakers and boaters have enjoyed watching inquisitive individuals that approach their boats for a better look. This kind of experience is controlled by the seal—if it wants to see you, it will come closer, but if the seal is afraid of you, it will leave. Do not approach a seal that is trying to avoid you because it can cause unnecessary stress to the animal.

ALSO CALLED: Common Seal.

DESCRIPTION: A Harbour Seal is typically dark grey or brownish grey with light, blotchy spots or rings. The reverse colour pattern is also common—light grey or nearly white with dark spots. The underside is generally lighter than the back. The outer coat is composed of stiff guard hairs about 1 cm long, and this characteristic is what gives seals in this family the name "hair seals." The guard hairs cover an insulating undercoat of sparse, curly hair that is about 0.5 cm long. Pups bear a spotted, silvery or grey-brown coat at birth. The head is large and round, and there are no visible ears. Each of the short front flippers bears long, narrow claws. The male is the larger gender in this species.

DID YOU KNOW?

Sometimes these seals follow fish several hundred kilometres up major rivers; there are even residents in some inland lakes. Harbour Seals can be found in Harrison Lake, BC, more than 180 km upriver from the coast.

HABITAT: This nearshore species is frequently found in bays and estuaries. Common haul-out sites include intertidal sandbars, rocks and rocky shores, and favoured spots are used by Harbour Seals generation after generation.

FOOD: Harbour Seals feed primarily on fish, including rockfish, cod, herring, flounder and salmon. To a lesser extent, they also feed on molluscs such as clams, squid and octopus, and crustaceans such as crayfish, shrimp and crabs. Newly weaned pups seem to consume more shrimp and molluscs than do adults. Adult Harbour Seals have been seen taking fish from nets, and some have even entered fish traps to feed, making a clean getaway afterward.

seal is ı
w it—eve
lifts its hea
langer. Even
vatchful and
ar that has
the snow to

d in the Arctic
in the
ador

beach trail

REPRODUCTION: The breeding season for Harbour Seals varies geographically—the farther north a population, the later the pupping and breeding. Gestation lasts 10 months, and a single pup is born between April and August. Pupping can be as early as February in the southern extremes of this species' range and as late as July in the northern extremes. The pups are weaned at four to six weeks of age—after they have tripled their birth weight on their mother's milk, which is more than 50 percent fat. Within a few days of weaning their pups, females mate again in August and September. Harbour Seals become sexually mature at three to seven years. Captive seals have lived more than 35 years, though the typical lifespan in the wild is no more than 25 years.

flipper print

SIMILAR SPECIES: The **Ringed Seal** (p. 190) is smaller and has irregular, pale rings on its sides. The **Grey Seal** (p. 192) generally has fewer spots, and its nostrils are set farther apart. The **Harp Seal** (p. 194) has irregular, wishbone- or harp-shaped markings. The **Hooded Seal** (p. 198) is more strongly spotted, and the male has an inflatable forehead sac. The **Bearded Seal** (p. 196) is larger, uniformly coloured and bears a distinct "beard" of long whiskers.

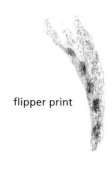

Ringed Seal

Ringed Seal
Pusa hispida

Studies of the Ringed Seal have provided information not only about the biology of this small seal, but also about the biogeological history of the continent. Ringed Seal fossils have been found near Ottawa, and we know that seas covered that region more than 10,000 years ago. By identifying the variety of sea creatures the Ringed Seal feeds on today, we can draw inferences about the animals living in the ancient seas that once covered the province of Ontario.

When a Ringed Seal dives for crustaceans, it can reach a maximum depth of about 91 m. It can stay under for up to 18 minutes, though a dive usually lasts for five minutes or less. Unlike its close cousin the Bearded Seal, the Ringed Seal cannot tolerate the pressure at great sea depths, nor can it tolerate being without air for much longer than 15 minutes.

As a result, the Ringed Seal remains in shallower areas, whereas the Bearded Seal occupies deeper waters.

Although it spends most of its time in the water, the Ringed Seal digs small chambers under the snow but above the ice, where it rests when it needs to reoxygenate its blood after a long dive. Sometimes many seals are found together in chambers that are connected by short passageways.

The Walrus is the biggest threat to the Ringed Seal in the water, while the Polar Bear, the Arctic Fox and humans are the dangerous creatures on land and ice. There is no place where this ____ truly safe, and it seems to kno___ ___ry minute or so a Ringed Seal ___ ___ to scan its surroundings for ___ in its snowy den, this seal is ___ ready to flee—a Polar Be___ sniffed it out can dig through ___

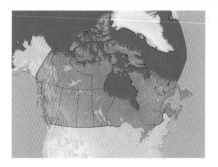

RANGE: Ringed Seals are foun_ seas from Point Barrow, Alaska, West and to Hudson Bay and Lab_ in the East.

Total Length: 1.0–1.7 m

Tail Length: 7–10 cm

Weight: 50–100 kg

reach it, and the Arctic Fox, which preys on seal pups, is small enough to walk right into the den.

DESCRIPTION: One of the smallest aquatic carnivores, this little seal reaches a maximum length of 1.7 m, though it is usually only 1.2 m long. Its fur is variegated brown or even bluish black above and pale yellowish or grey below, with irregular, doughnut-shaped rings of light-coloured fur on the sides and back. It has long whiskers and "eyebrows," a short muzzle and dark eyes. Males are typically larger than females.

HABITAT: The Ringed Seal spends most of its life underwater in the northern polar seas. It maintains breathing holes under the ice by digging the ice away with its sharp claws. This seal usually stays under or near landfast ice, and it is seldom seen on shifting or broken ice floes.

FOOD: In shallow water, these seals feed mainly on shrimp, small fish and crustaceans. In deeper water, their diet shifts to larger zooplankton such as amphipods.

REPRODUCTION: A female typically mates annually, in March or April, and gives birth

DID YOU KNOW?

The presence of PCBs and DDT in the environment is believed to have reduced the pregnancy rates of female Ringed Seals to as low as 28 percent in certain areas—a percentage low enough to threaten the survival of some local Ringed Seal populations.

to a single pup the following March, after three months of delayed implantation and nine months of gestation. The Ringed Seal is unique among seals because of the maternal den it constructs under the snow. A pregnant female may hollow out a shallow chamber in the snow above a breathing hole, or she may use a natural snow cave. Maternal chambers are typically about 2 × 3 m wide and only 60 cm deep. The pup is born helpless, but with all its teeth, and it is covered in lanugo (soft, downy fur). The pup nurses for about two months, and when its mother leaves the den to feed, the pup remains by itself and waits quietly. Some pups may stay near their mothers for the remainder of the year, whereas other pups disperse soon after weaning.

SIMILAR SPECIES: The **Harbour Seal** (p. 186) is larger, with spotty or patchy markings rather than the distinctive doughnut-shaped markings on the Ringed Seal. The **Grey Seal** (p. 192) is larger and has fewer spots. The **Harp Seal** (p. 194) has irregular, wishbone- or harp-shaped markings. The **Hooded Seal** (p. 198) is more strongly spotted, and the male has an inflatable forehead sac. The **Bearded Seal** (p. 196) is larger, uniformly coloured and bears a distinct "beard" of long whiskers.

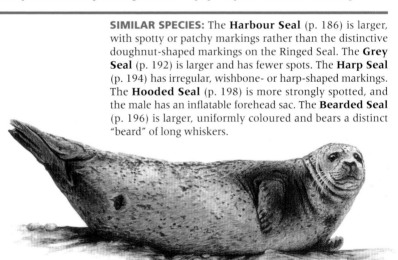

Harbour Seal

Grey Seal
Halichoerus grypus

Every year, off the coast of Nova Scotia, tens of thousands of Grey Seal pups are born on sandy Sable Island. The breeding colony on this island is the largest in the world for Grey Seals. Because they are thought to be interfering with recovering fish stocks, these seals are at risk of widescale population culls. The issue is extremely controversial because Sable Island is also a designated wildlife preserve, with a small population of Feral Horses. One proposed method of reducing the population is to give a contraceptive to at least 15,000 females every year for five years. Surprisingly, this would cost about the same as culling the population of about 200,000 seals over four years. These methods seem drastic, and many naturalists maintain that they are unnecessary. Fishermen and businessmen of the industry are quick to blame Grey Seals for the collapse of cod fishing and the poor recovery of cod in the last few years. In fact, Grey Seals are opportunistic feeders and, though they eat cod when it is available, they also eat many other fish, including species that prey on cod. Ultimately, Grey Seals and cod have always coexisted in the local marine ecosystem; the fishing industry likely has only itself to blame for destabilizing the balance by regularly overharvesting the fish.

ALSO CALLED: Atlantic Seal, Horsehead Seal.

DESCRIPTION: This medium-sized seal has a very straight profile and nostrils that are quite far apart. It is mainly silvery grey or brown, with darker

RANGE: Grey Seals are found in the coastal waters of Canada's East Coast, as well as Iceland, the United Kingdom, Ireland, Scandinavia and northern Europe.

Total Length: *Male:* 2.0–2.7 m; *Female:* 1.6–2.2 m

Tail Length: 9–13 cm

Weight: *Male:* 240–320 kg; *Female:* 150–260 kg

DID YOU KNOW?

Grey Seals are extremely vocal, and in addition to bleats and gurgling sounds, they sing eerie songs. Their combined voices sound like a group of school children pretending to be ghosts.

spots and blotches. An older male may have heavy scarring around its neck. Males are as much as 30 percent larger than females.

HABITAT: Grey Seals inhabit the northern Atlantic Ocean, specifically coastal waters where they can hunt for benthic fish at depths of 70 to 100 m. There are haul-out sites and rookeries on both Atlantic shores, in North America and northern Europe.

FOOD: Grey Seals feed on a wide variety of fish, especially those found on or near the sea floor ones. Eels, Atlantic cod, flatfish, herring and skates are commonly eaten, as well as octopus and lobster. These seals fast during the pupping and breeding season.

REPRODUCTION: Females come to shore in September, and pups are born from late September through November. Within a few weeks of giving birth, females may mate again. Gestation is a little over 11 months, including a period of delayed implantation. Like Harp Seal young, pups are born with white, fuzzy lanugo. Pups weigh 14 kg at birth but grow quickly because their mother's milk is about 60 percent fat. At about one month of age, the pups shed their lanugo, which is replaced by waterproof adult fur. Soon after, they head out to sea and are able to catch their own prey. Sometimes yearling pups are seen on islands in spring, perhaps lost or resting. Grey Seals become sexually mature between three and five years old, and they can live for a long time—females up to 40 years and males up to 30 years.

SIMILAR SPECIES: The **Harp Seal** (p. 194) is lighter overall, and mature adults have an irregluar, wishbone- or harp-shaped pattern on their backs. The **Harbour Seal** (p. 186) also inhabits Sable Island. It lacks the straight profile and has nostrils set closer together.

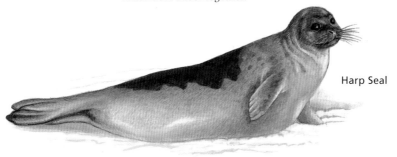

Harp Seal

Harp Seal
Phoca groenlandica

Among colonies of Harp Seals, "survival of the fattest" seems to aptly describe the mechanism guiding their population dynamics. Pups weigh about 11 kg when they are born, but they are capable of putting on an astonishing 25 kg in their first 10 to 12 days. They require this fat to survive—the pups are weaned when they are only 9 to 12 days old, but they cannot swim until they are at least four weeks old. During these initial weeks, the pups rely on their newly stored fat to survive.

Harp Seals are strongly migratory and swim about 5000 km per year. Once they leave their rookeries, they head north to spend the summer feeding in High Arctic waters. When the ice starts to advance in autumn, these seals begin their southward migration ahead of the ice. In winter, they haul out to their rookeries again.

Harp Seals were extensively hunted in the past and are still hunted today. Canada's annual seal hunt focuses on this species, and these seals are killed for meat, fur, oil and their genitals, which are sold on the black market as an aphrodisiac. International boycotts of seal fur and pressure by environmental groups to stop the seal hunts have met with limited success. Other than subsistence hunting by the Inuit, seal hunting continues partly for sport and partly because seals are often blamed for contributing to the collapse of the East Coast fishing industry even though cod makes up only three to five percent of the diet of these animals. Canada continues to permit the killing of hundreds of thousands of Harp Seals per year. The hunt in 2010 had a quota of 330,000 seals, making Canada one of the world leaders of marine mammal hunting.

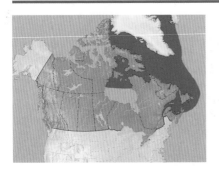

RANGE: Harp Seals are circumpolar and migrate between Arctic and Subarctic waters. In Canada, their winter range is around Newfoundland, the Gulf of St. Lawrence and the Maritimes; in summer, they can reach as far north as Ellesmere Island and west into northern Hudson Bay.

Total Length: 1.6–1.9 m

Tail Length: 8–11 cm

Weight: 85–180 kg

ALSO CALLED: Greenland Seal, Saddle-backed Seal, Jumping Seal.

DESCRIPTION: Silvery white or greyish overall, this seal gets its name from the dark wishbone- or harp-shaped patch of fur over its back. Some individuals (called "Spotted Harps") never fully develop the harp pattern and have only spots or blotches. A Harp Seal pup is covered in fuzzy white fur for its first 12 days, then moults after weaning into a silver coat with sparse dark spots.

HABITAT: During the winter months, Harp Seals haul out on ice in northeastern Canada, where they give birth and later mate. The rest of the year is spent at sea; the peak of summer finds them in High Arctic seas as far north as Ellesmere Island. When feeding, they prefer waters not more than 300 m deep.

FOOD: Harp Seals regularly dive to depths of 150 to 200 m to find prey. They consume many kinds of small fish, including capelin, cod, herring, sculpin, halibut, redfish and plaice. In fact, they are known to eat more than 65 different kinds of fish and at least 70 kinds of invertebrates, including crab, lobster and shrimp.

REPRODUCTION: Pups are born in rookeries in late February or March. They are 80 to 90 cm long and have a fuzzy, yellow coat. After two or three days, the yellow coat is shed

DID YOU KNOW?

The mortality rate of Harp Seal pups is as high as 30 percent, but if they survive their first year, they have a good chance of living for 30 to 35 years.

and the pups are covered in their famous white lanugo. The lanugo only lasts for 12 days, the period of the pups' most intense growth. After the lanugo is shed, the pups have a silvery coat with spots for up to four years. The young become sexually mature sometime after four years of age, at which time the adult colour pattern is achieved. Females mate soon after giving birth, and males stay around the pupping colony for weeks to improve their chances of finding receptive females. Females have a delayed implantation of about four months, and their gestation is 7.5 months. This results in pups being born at the same time every year.

SIMILAR SPECIES: The **Grey Seal** (p. 192) and **Harbour Seal** (p. 186) are usually darker and lack the harp pattern.

Grey Seal

Bearded Seal
Erignathus barbatus

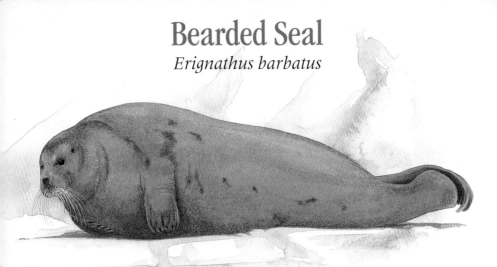

The nondescript Bearded Seal is one of the most vocal of all seals and has an extensive repertoire of unusual and haunting calls. In spring, the male Bearded Seal sings melodious yet eerie songs beneath the ice. Each song can last for up to one minute and carries for a long distance underwater. To court a female, a bull hauls out on the ice and serenades the cow with his echoing, warbling calls.

When humans hold their breath, the increasing carbon dioxide levels in the blood trigger them to breathe again in a short time. This trigger is greatly desensitized in seals, and they are able to postpone breathing for about 30 minutes. When a seal is underwater, its nostrils are closed and its heartbeat slows from the normal 100 beats per minute to about 10 beats per minute. During a dive, the seal's muscles function mainly anaerobically, creating an oxygen debt in the body. The seal stores oxygen in its blood and in certain molecules of the muscles, and this stored oxygen is used, or "spent," during the dive. The longer the dive lasts, the greater the body's oxygen debt becomes. When the dive is over, the seal must rest and rebuild, or "pay back," its body's oxygen stores. In other words, when a Bearded Seal is basking, it is not just enjoying the sun, it is allowing its body to slowly reoxygenate after a long dive.

DESCRIPTION: As its name suggests, this seal has a beard of long, flat bristles on either side of the muzzle. Other seals have whiskers, but none compare to the long, bristly beard of the Bearded Seal. The head of this large seal is noticeably small for its body size. The fur varies in

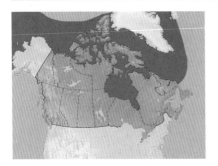

RANGE: Bearded Seals inhabit northern coastal waters and shallow Arctic seas from Point Barrow, Alaska, to Labrador.

Total Length: 2.0–2.6 m

Tail Length: 9–13 cm

Weight: 250–360 kg

DID YOU KNOW?

The long bristles of this seal's "beard" are extremely sensitive, and they help the Bearded Seal locate its favourite foods on the muddy sea bottom.

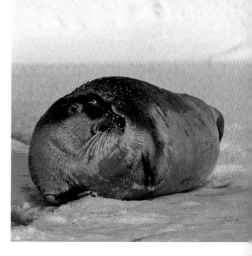

colour from uniformly light brown to tawny golden to greyish. The front flippers are distinctively squarish, and the third digit is longer than the others. Unlike many other pinnipeds, female Bearded Seals are slightly larger than males.

HABITAT: The Bearded Seal spends most of its time in the waters of cold Arctic seas. It is primarily found in water no deeper than 150 m, though it can dive to depths of 200 m. The Bearded Seal is only rarely seen on land—it prefers open water and rests on drift ice and ice floes. This seal is not migratory, but it does follow the ice as it moves seasonally.

FOOD: Using its powerful claws and keenly sensitive whiskers, the Bearded Seal locates bottom-dwelling fish and shellfish to eat. Favourite foods include shrimp, bivalves and gastropods, as well as crabs, worms and small fish such as Arctic cod, sand lance and sculpins.

REPRODUCTION: During the mating season in May, these normally solitary seals form groups of up to 50 individuals. The males are marvellously vocal, and they emit warbling calls on the ice surface to serenade the cows. Females are not vocal. Sometimes the bulls fight each other to establish hierarchy, but the dominant bulls do not form harems. Females give birth on the ice to one pup in April or May, after a gestation of about 11 months. The pup is well developed at birth and is nursed for no more than 12 to 18 days. The pup puts on blubber during this short nursing period and goes from weighing about 35 kg at birth to about 90 kg by day 18, at which time the pup is weaned and the mother leaves.

SIMILAR SPECIES: The **Harbour Seal** (p. 186) and **Grey Seal** (p. 192) are smaller and have spotty or patchy markings. The **Ringed Seal** (p. 190) is also much smaller and has distinctive doughnut-shaped markings on its sides. The **Harp Seal** (p. 194) has irregular, wishbone- or harp-shaped markings. The **Hooded Seal** (p. 198) is strongly spotted, and the male has an inflatable forehead sac.

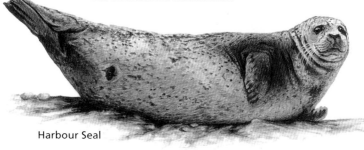

Harbour Seal

Hooded Seal
Cystophora cristata

The unusual Hooded Seal gets its name from the inflatable skin sac above the male's nostrils. During courtship, the male exhales air into this sac, which becomes larger with age. The male can also inflate the membrane between his nostrils, which then looks like a red balloon. When shaken, the red nasal sac makes a sound similar to flicking a balloon, and this sound is probably involved in courtship as well. The inflated sacs may be used to attract a female or threaten other males, or both. When the sac above the nose is not inflated it looks like a loose "hood."

In late winter, Hooded Seals gather in great number on the ice, where they give birth and later mate. There are four major pupping areas, three of which are in Canada: in the Gulf of St. Lawrence near the Magdalen Islands, just off the coast of Newfoundland and Labrador, and in the Davis Strait. The fourth site is just off Greenland's east coast. Males begin fighting to establish territories as soon as the females begin to give birth in spring. After a pup is born, the male waits for the pup to be weaned. The appearance of a family—mother, father, pup—is misleading, because the male is not the pup's father and will likely never even see that particular female again. Once the seals leave the ice, they are solitary as they migrate northward into good feeding waters. They remain at sea until they haul out to moult, usually from June to August. Mostly they moult on ice, but in some parts of their range, they may use rocky shorelines. Juveniles can be seen in the St. Lawrence River, sometimes as far west as Montréal.

Like the Harp Seal, the Hooded Seal is regularly hunted, but far fewer animals are taken. The total population of Hooded

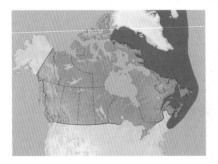

RANGE: The Hooded Seal is found in the central and western North Atlantic.

Total Length: *Male:* 2.3–2.9 m; *Female:* 2.0–2.3 m

Tail Length: 10–13 cm

Weight: *Male:* up to 435 kg; *Female:* up to 350 kg

Seals is only about 650,000. Canada continues to permit sealers a quota of about 10,000 Hooded Seals annually but has prohibited the killing of the pups.

ALSO CALLED: Crested Seal, Bladder-nosed Seal.

DESCRIPTION: An adult is silvery grey with dark spots that are more concentrated on the head and face, contributing to the seal's "hooded" appearance. The male is about twice as big as the female. The male develops his inflatable nasal sac at sexual maturity, sometime after he is four years old. Females do not have nasal sacs, but they still have the hooded appearance owing to the darker colour of their heads. Hooded Seal pups are fuzzy and bluish grey, usually with a darker back and lighter underparts.

HABITAT: These seals make extensive migrations through Arctic and Subarctic waters. They haul out on ice twice a year, once for the pupping season and again soon after to moult. Their range and abundance is strongly influenced by yearly changes in ice cover.

FOOD: Hooded Seals feed on a wide variety of fish, molluscs and crustaceans. Their diet seems to very regionally but mainly includes halibut, redfish, cod,

capelin, herring, mussels, starfish, squid, shrimp and octopus. They feed at depths of 100 to 600 m.

REPRODUCTION: Females give birth to a single pup from mid-March to early April. When they are ready to give birth, females find a secluded spot well away from the water on the most stable ice possible. Newborn pups are covered in a bluish grey lanugo and are nicknamed "bluebacks." Pups are procolial and can even swim soon after birth, if necessary. Females have some of the richest milk of all mammals, at least 60 percent fat. Some pups are nursed for a little as four days, the shortest lactation period of any mammal, but they are usually not weaned until they are four to eight days old. This short period of nursing is sufficient for the pup to double its weight, from 24 kg at birth to about 47 to 50 kg when weaned. A few days after the pup is weaned, the male that has been "guarding" the female will mate with her in the water. The female has delayed implantation of the embryo for about 3.5 months, so her eight month gestation means that pups are born at the same time every year.

SIMILAR SPECIES: A **Bearded Seal** (p. 196) is similar in size but has a long, bristly "beard." A male **Grey Seal** (p. 192) can be as large as a Hooded Seal but lacks the "hood."

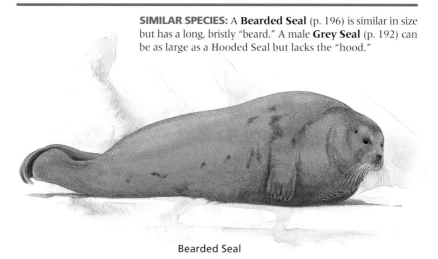

Bearded Seal

Northern Elephant Seal
Mirounga angustirostris

male

female

The enormous Northern Elephant Seal is one of the world's largest seals. Only its Southern Hemisphere counterpart, the Southern Elephant Seal (*M. leonina*), is slightly larger.

In the waters off British Columbia, this seal is mainly pelagic, and only rarely are individuals seen in inshore waters. If one is seen on shore (mainly on Vancouver Island), it is likely diseased or injured.

Elephant seals are famous for their incredible diving capabilities. When they dive for food, they can remain submerged for 80 minutes and reach depths of more than 1500 m. Among marine mammals, only the Sperm Whale (*Physeter macrocephalus*) and some of the beaked whales can dive deeper or for longer.

Northern Elephant Seals are also known for their long migrations—two return trips per year. Sometime between December and March, the adults arrive on sandy beaches in California or Mexico, where females give birth and mate. Afterward, the adults and the young of the year depart for good feeding waters. Adult males and some juveniles may venture as far north as the Gulf of Alaska

RANGE: These enormous seals are found from coastal Baja California north to the Gulf of Alaska. They probably disperse within a few hundred kilometres of the coast during the nonbreeding season.

Total Length: *Male:* 3.7–4.9 m; *Female:* 2.1–3.7 m

Tail Length: about 25 cm

Weight: *Male:* up to 2300 kg; *Female:* up to 900 kg

and the Aleutian Islands, where they feast on the abundant sea life. Most females and young of the year do not travel quite as far, preferring to feed in waters between 40° N and 45° N latitude.

Between April and August, after feeding for two to five months, Northern Elephant Seals return to the sandy shores of Mexico and California to moult. When these seals moult, they shed their short, dense, yellowish grey pelage along with large patches of old skin. While they lie on a beach, their skin starts to become itchy, and they begin scratching and biting to loosen the shedding fur. Once the fur and dead skin has been loosened, it dries and separates from the new, darker fur underneath. During this moulting phase, the patchy skin looks a bit like old paint peeling off wood. After the seals have scratched away the last of the old pelage, they look sleek and slate grey, often with wavy, circular markings in their coats.

During both the mating and moulting seasons, elephant seals fast, losing up to 36 percent of their body weight. After the moulting season, they once again venture north to food-rich waters to replenish their bodies before mating.

Adult males that travel twice a year to and from the waters around Alaska tally up more annual travel miles than almost any other mammal, about the same as the renowned Grey Whale. Each year, male Northern Elephant Seals may cover 21,000 km and spend more than 250 days at sea.

The commercial harvest of seals for oil at the height of the whaling years reduced the total number of Northern Elephant Seals to between 100 and 1000 individuals, with some local populations being completely extirpated. These seals are now fully protected under the U.S. Marine Mammal Protection Act, and their numbers have increased dramatically. More sightings off the British Columbia coast are now being reported.

DESCRIPTION: The sheer size of this seal usually gives away its identity. If you have any doubt, however, look at its nose—both sexes have a nose that extends past the mouth. An adult male has a pendulous, inflatable, 30-cm-long snout that resembles a trunk. The female's nose extends only a few centimetres beyond her mouth and cannot

be inflated. Both sexes have large, dark eyes, and the skin on the inside of their mouths is bright pink. This seal's skin is mainly grey or light brown with similarly coloured, sparse hair. The hindflippers appear to be lobed on either side and have reduced claws. The tough skin of the male's neck and chest is covered with creases, scars and wrinkles, a feature absent in the female. Pups are born black but moult to silver at one month of age.

HABITAT: The Northern Elephant Seal lives in the temperate waters off the West Coast. It migrates between northern feeding waters and southern breeding and moulting beaches twice a year. During the breeding and moulting seasons, this seal hauls out onto sandy beaches. It does not haul out onto rocks, but it may cross over rocks if necessary to reach a sandy beach. This seal rarely hauls out during the feeding season; instead, it rests on the water's surface and may remain offshore for weeks at a time.

FOOD: Northern Elephant Seals feed on a variety of sea creatures, including squid, octopus, small sharks, rays, pelagic red crabs and large fish. Adult males feed on larger prey than do females and pups.

DID YOU KNOW?

On land, Northern Elephant Seals are very noisy—the males produce a series of loud, rattling snorts, and females make sounds resembling monstrous belches.

REPRODUCTION: These seals are polygamous but are not strongly territorial. During the breeding season, from December to March, males arrive onshore first and battle fiercely for status in the social hierarchy; a high status means that a male can have a large harem. Duelling males will press their chests together and rise up vertically as they attempt to push each other down. During their sparring, they have their mouths wide open and will stab at each other's neck. They do not have large teeth, so the wounds inflicted are not too deep, but considerable bloodshed and injury does occur. The females come to shore a couple of weeks after the males, and within a few days, each cow gives birth to a pup conceived in the previous breeding season, after a gestation of 11 months. Pups are nursed for no more than one month, during which time the mothers fast. Just a few days before their pups are to be weaned, the females mate, and after weaning their pups, they leave. Females are sexually mature at two to five years of age, but most males cannot win a harem until they are 9 or 10 years old.

SIMILAR SPECIES: The **Harbour Seal** (p. 186), **Northern Sea Lion** (p. 210) and **California Sea Lion** (p. 214) are much smaller, and all lack the long, distinctive snout. The **Walrus** (p. 204) has a different range.

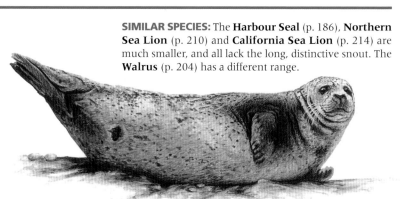

Harbour Seal

Walrus

Odobenus rosmarus

Truly one of a kind, the Walrus is the sole member of its family. It most likely evolved from an ancestral eared seal, but seven million years of adaptation has made the Walrus unlike any other living seal. Its unique tusks, which grow throughout an individual's life, are the its most distinguishing feature. The tusks are modified teeth and are coated in enamel when they first erupt. Over time,

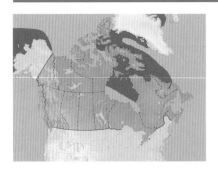

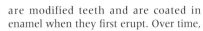

RANGE: The Atlantic race of the Walrus is found in the Arctic seas from Greenland to northern Canada, south into Hudson Bay and rarely along Labrador. The Pacific race is found from the Bering Sea to the western Beaufort Sea.

Total Length: *Male:* up to 3.7 m; *Female:* up to 2.9 m

Tail Length: no visible tail

Weight: *Male:* up to 1000 kg; *Female:* up to 800 kg

Tusk Length: *Male:* up to 76 cm; *Female:* up to 36 cm

the enamel wears off, leaving a tusk of pure ivory.

The exploitation of Walruses for their ivory was once extensive, but this has nearly been stopped, thanks to the involvement of the Canadian and U.S. governments. Only the Inuit, who have traditional links to this animal, continue to hunt the Walrus. They use all parts of the animal: the meat is eaten, the skin is used as leather for clothing or covers for boats, and the bones and tusks are used in sculptures and carvings that bring additional revenue into their communities.

There are two distinct Walrus races. The Pacific Walrus (*O. r. divergens*), which

DID YOU KNOW?

Although Walruses may sleep on land, they have the ability to sleep vertically in the water with their heads held up by inflated air sacs in the neck.

is the larger of the two, inhabits waters from the Chukchi Sea off northeastern Siberia east to Alaska. The Atlantic Walrus (*O. r. rosmarus*) lives in Hudson Bay and the Arctic seas east to Greenland. Both races live in groups of anywhere from a few dozen to up to 2000 individuals. The Atlantic Walrus is not migratory, unlike its Pacific counterpart, and it hauls out

either on land or on landfast ice that does not travel.

The Walrus is an excellent swimmer and can reach speeds of up to 24 km/h. When it dives for food, it usually descends to depths up to 80 m for about 20 minutes. Deeper and longer dives have been recorded, but Walruses have little need to do so as their preferred prey species inhabit benthic zones less than 80 m deep.

DESCRIPTION: Both the bull and the cow Walrus have tusks, but a cow's tusks are considerably shorter and more curled than a bull's. The Walrus has little hair over its body, and its skin colour varies from pale pink, yellow, reddish or brownish to nearly white. The hindflippers have nails on all five digits and can be rotated underneath so the walrus can "walk" on all fours on land. The foreflippers lack nails completely. The face is covered with about 400 whiskers, each up to 30 cm in length. The Walrus has no visible tail.

HABITAT: Historically, Walruses may have inhabited more of the rocky coasts and islets of mainland Canada, including Hudson Bay, than they do now, but the impact of human hunting has pushed them farther into the Arctic seas. Currently, they spend most of their time on Arctic pack ice over water that is about 18 m deep. Walruses spend more of their time basking than do other seals, usually on ice or on rocky outcroppings.

FOOD: Using its mouth like a powerful vacuum, the Walrus sucks molluscs out of their shells. Clams are the most common food, and a large, hungry Walrus can polish off up to 6000 of these bivalves in one day. Walruses are opportunistic, feeding on fish, crabs, shrimp, worms and a variety of other marine creatures, even other marine mammals.

REPRODUCTION: Unlike other seals that usually mate each year, a female Walrus mates once every three years. Gestation is 11 to 12 months, and the single calf, born in May, weighs 45 to 68 kg. The cow nurses her calf in the water in a vertical position. The calf remains with the mother and nurses from her for nearly two years. The mother separates from the calf before she has another baby.

SIMILAR SPECIES: No other seal grows tusks. The large **Northern Elephant Seal** (p. 200) has a different range. A large **Bearded Seal** (p. 196) can be as long as a Walrus, but a Walrus is much more rotund.

Northern Elephant Seal

Northern Fur Seal
Callorhinus ursinus

male

female

Completely at home in the ocean, the Northern Fur Seal almost never comes to shore, except during its breeding season. The rest of the year, this seal is pelagic off the coasts of British Columbia and neighbouring regions.

Northern Fur Seals are not gregarious. At sea, they are found either alone or in groups of no more than three individuals. Even during the breeding season, when large numbers come together on rocky islands, their interactions are limited to mating and courtship behaviours. The bulls savagely defend their harem territories, and though the females are less aggressive, they still keep to themselves. When the females have finished nursing their young, they depart the rocky shores and leave the pups to fend for themselves. Many pups die of disease within the first month or two after birth, but many of those that make it to weaning will survive.

The Northern Fur Seal travels farther at sea than almost any other seal or sea lion; only the Northern Elephant Seal covers more distance each year. Including its wanderings while feeding at sea and its migrations to and from breeding

RANGE: This wide-ranging species is found mainly offshore from California up the Pacific Coast and across the North Pacific to Japan. Rarely individuals have been seen at sea lion haul-out sites in British Columbia, such as Race Rocks at the southern tip of Vancouver Island.

Total Length: *Male:* 1.8–2.3 m; *Female:* 1.1–1.5 m

Tail Length: 5–12 cm

Weight: *Male:* 150–280 kg; *Female:* 38–54 kg

islands, a Northern Fur Seal tallies up about 10,000 km of travel per year.

DESCRIPTION: The Northern Fur Seal has a small head with long whiskers, small ears, large eyes and a short, pointed nose. Its tail is very short. The hind flippers are extremely large in relation to the seal's body size. When the seal is wet, it is sleek and black. When dry, the male is mostly dark greyish black, and the female shows a brownish or reddish throat and often some silvery grey on the underparts. The adult male has a thickened neck and is more than twice the weight of the female. Newborn pups are black, and male pups are larger than female pups.

HABITAT: These fur seals are pelagic for 7 to 10 months of the year. They come ashore only to breed, mainly on the rocky beaches of the Pribilof Islands in Alaska and the Commander Islands in Russia.

FOOD: Northern Fur Seals feed mainly on squid, along with herring, capelin and pollock up to 25 cm long. Almost all feeding takes place at night, when fish are closer to the surface. These seals may dive in search of food (the maximum recorded dive depth is 230 m), but most forage at depths of 70 m or less.

REPRODUCTION: Males come to shore in late May and June and battle to establish territories. Females come to shore in mid-June or July, and within two days, each female gives birth to a single pup conceived the previous summer. Mating occurs 8 to 10 days later. Pups nurse for four or five months.

SIMILAR SPECIES: In the same range, the **Harbour Seal** (p. 186) is a "true" seal that cannot rotate its hindflippers under its body, and it usually has spots. The **Northern Sea Lion** (p. 210) is larger. The **California Sea Lion** (p. 214) is usually larger and has shorter foreflippers with unequal toe lengths.

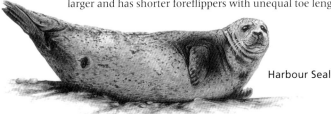

Harbour Seal

Northern Sea Lion
Eumetopias jubatus

The large Northern Sea Lion is a familiar sight to many people who frequent coastal areas. This gregarious creature is usually seen in groups of hundreds, and the rookeries at Cape St. James at the southern tip of Haida Gwaii (Queen Charlotte Islands) and North Danger Rocks, which lie south of Banks Island, contain up to 5000 sea lions.

The Northern Sea Lion's social system is more complex than that of the Harbour Seal, the other common pinniped on the West Coast. For example, when a group of sea lions is feeding, all individuals dive at the same time and surface together. This behaviour means that no single sea lion dives first and scares the fish away, ruining feeding opportunities for the others.

During the breeding and pupping season, hundreds of sea lions congregate at rookery sites that are used generation after generation. Two major rookeries exist in British Columbia, one on the northwest point of Vancouver Island, and the other near the southern end of Haida Gwaii. Mature bulls make a roaring

RANGE: Northern Sea Lions are found near shore from southern California up to Alaska and the Aleutian Islands, and across to Siberia and Japan.

Total Length: *Male:* 2.6–3.4 m; *Female:* 1.8–2.1 m

Tail Length: 12–15 cm

Weight: *Male:* up to 1000 kg; *Female:* 270–360 kg

sound, which, when combined with the grumbles and growls of the others, results in a cacophony that can be heard more than one kilometre away. Outside the pupping season, the "bachelors," the young of the year, some barren cows and the odd mature bull form loose colonies, feeding together and otherwise interacting.

Adult male Northern Sea Lions are the largest of the eared seals. There is great sexual dimorphism, with adult males being three to four times as heavy as females. A large difference in weight is characteristic of seals with territorial males that hold a harem. Females form loose aggregations with their pups within a male's territory, and they are far faster and more agile than the males.

Northern Sea Lions are well known for their curiosity and playfulness. They are very active, sometimes leaping clear out of the water and occasionally playing with rocks, seaweed and even feathers. They have even been seen jumping across surfaced whales. Their smaller cousins, the California Sea Lions, share

DID YOU KNOW?

Although Northern Sea Lions are saltwater mammals, they have been known to swim up major rivers in search of lamprey and salmon.

this playfulness and are commonly seen performing tricks in marine park shows.

For many years, sea lions were killed because it was believed that they fed on commercially valuable fish. Research has since indicated that they feed opportunistically on any readily available fish—commonly octopus, squid and "scrap" fish such as herring and greenling. Although the intentional killing of sea lions has decreased, their populations have declined by as much as 80 percent from historic numbers. The causes of this decline are unknown.

A unique characteristic of sea lions is that they frequently swallow rocks as large as 12 cm across. Although no one knows for sure, the most likely explanation for

this behaviour is that, as in birds, the stones help pulverize food inside their stomachs. Sea lions' teeth are ill suited for chewing, and they regularly swallow large chunks of meat and whole fish.

ALSO CALLED: Steller Sea Lion.

DESCRIPTION: The adult is light buff to reddish brown when dry, and brown to nearly black when wet. Its fur covers a thick blubber layer. The yellowish vibrissae may be as long as 50 cm. An adult male develops a huge neck that supports a mane of long, coarse hair. The female is sleek, without the massive neck. During the breeding season, adult males evicted from the colony often bear huge cuts and tears on the neck and chest, reminders of the vicious battles waged over a territory. The hindflippers are drawn forward under the body and—like all eared seals—the Northern Sea Lion can jump and clamber up steep rocky slopes at an amazing rate. Its foreflippers are strong and are used for propulsion underwater.

HABITAT: Northern Sea Lions live mainly in coastal waters near rocky shores, and they are seldom found more than 50 m from the water. During the breeding season, they occupy rocky, boulder-strewn beaches or rock ledges. Sea lions may rest in the water in a vertical position with their heads above the surface. They prefer to stay in the water during inclement weather, but when the sun shines, they usually haul out and bask on the rocks.

FOOD: These sea lions feed primarily on blackfish, greenling, rockfish, herring, Atka mackerel, Pacific cod, salmon and halibut. Other foods include squid, octopus, shrimp, clams and bottom fish. Males do not eat for one to two months while they are defending a territory.

REPRODUCTION: Males come to shore in early May and battle to establish their territories. Females come to shore in mid-May or June, and within three days, give birth to a single pup conceived during the previous summer. Females form loose aggregations with their pups in a male's territory, and mating occurs within two weeks after the pups are born. Pups nurse for about one year, but some have been known to nurse for as long as three years. Females are sexually mature at three to seven years of age and may live to be 30 years old. Males probably do not breed before age 10.

SIMILAR SPECIES: On the West Coast, the **Northern Fur Seal** (p. 208) and the **California Sea Lion** (p. 214) are both smaller and less common. The **Northern Elephant Seal** (p. 200) is larger and has a distinctively long snout.

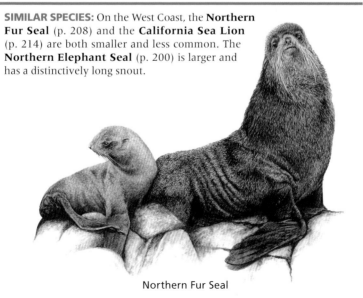

Northern Fur Seal

California Sea Lion
Zalophus californianus

California Sea Lions are among the most famous pinnipeds in the world. These amiable sea mammals regularly perform for the admiring visitors of circuses and marine aquariums. Balancing a ball on the nose, performing "handstands" and leaping through a hoop are just a few of the tricks that attract thousands of people and thousands of dollars each year. Many people, however, consider the live hunting required to populate marine aquarium stages to be unethical. They also think that teaching such an animal to do tricks merely for our entertainment diminishes the integrity of the creature's wild nature.

These sea lions come by their playfulness naturally—it is not something humans can take credit for teaching them. In the wild, females and young pups frequently play and cavort in the water, and they have even been known to play with other species. Flinging a piece of kelp around in the water and hitting the wild waves for some good bodysurfing are just part of the daily routine. Watching sea lions

RANGE: This sea lion inhabits coastal waters of the North Pacific from Vancouver Island to Mexico. There are three small, isolated populations in the Sea of Japan.

Total Length: *Male:* 2.0–2.5 m; *Female:* 1.5–2.0 m

Tail Length: about 12 cm

Weight: *Male:* 200–390 kg; *Female:* 45–110 kg

underwater is a treat because they turn sinuous loops and spirals in an aquatic ballet that belies their terrestrial ancestry.

California Sea Lions are generally rare in British Columbia. They do not breed in the province, and by May, they have gone south to establish their breeding territories. From October to February, however, at certain haul-out sites on the southeastern side of Vancouver Island, wintering males may actually be more common than wintering Northern Sea Lions.

Despite being much adored by children and tourists, California Sea Lions often suffer harsh treatment in the wild. In the 19th and early 20th centuries, they were killed in great numbers for oil and hides. Later in the 20th century, they were also slaughtered for the pet food industry.

DESCRIPTION: The California Sea Lion has a slender, elongated body, a blunt snout and a short but distinct tail. Adult males are brown, and females are tan with a slightly darker chest and abdomen. Coarse guard hairs cover only a small amount of the underfur. The male develops a noticeably raised forehead that helps distinguish it from the Northern Sea Lion. The front flippers are long and bear distinct claws. This species is noisy—males produce a honking bark, cows wail and pups bleat.

HABITAT: California Sea Lions are normally seen in coastal waters around islands

> **DID YOU KNOW?**
>
> The name "pinniped," which refers to all seals and sea lions, literally means "wing-footed"—an apt description of their fan-shaped flippers.

with rocky or sandy beaches. They tend to avoid the rocky islets preferred by Northern Sea Lions. Preferred haul-out sites include sandy or boulder-strewn beaches below rocky cliffs. In some places, they occupy sea caverns.

FOOD: These sea lions eat a wide variety of food, including at least 50 species of fish and many types of squid, octopus and other molluscs. Sea lions may feed on some commercially valuable fish species, but their diet is so varied that they probably do not contribute significantly to declining fish stocks.

REPRODUCTION: Males establish territories on rocky or sandy beaches along the coasts of California and Mexico in May, June or July. Females arrive in the breeding territories in May or June. If a female conceived the previous year, she will give birth to a single pup, and within a month, she will mate again. Most pups are weaned by eight months of age, but a few may nurse for a year or more. Pups begin eating fish before they are weaned.

SIMILAR SPECIES: In the same range, the **Northern Sea Lion** (p. 210) is larger and paler. The **Northern Fur Seal** (p. 208) is smaller, and its underparts are often more reddish. The **Harbour Seal** (p. 186) has spots and cannot rotate its hindlimbs under itself on land.

Northern Sea Lion

Black Bear
Ursus americanus

The Black Bear, an inhabitant of forests and open, marshy woodlands throughout most of Canada, is often feared by city dwellers who go to wilderness parks to appreciate the scenery. People who are more experienced with the wild forest and with animal behaviour may not fear the Black Bear but still treat it with healthy respect.

The Kermode Bear, a subspecies of the Black Bear, lives on several islands of coastal British Columbia. Up to 10 percent of this subspecies has a recessive colour characteristic that results in pale, cream-coloured fur. Also called Spirit Bears, these light-coloured bears feature prominently in the legends and beliefs of the indigenous peoples of the area. One legend says that Raven, the Creator, makes one in every 10 bears this snowy colour as a reminder of the hardships endured during the last ice age. The Kermode Bear is also a symbol of peace and harmony.

Contrary to popular belief and their classification as carnivores, Black Bears do not readily hunt larger animals. They are primarily opportunistic foragers and feed

RANGE: Across North America, the Black Bear occurs nearly everywhere there are forests, swamps or shrub thickets. It avoids grasslands and areas of dense human habitation.

Total Length: 1.4–1.8 m

Shoulder Height: 90–110 cm

Tail Length: 8–18 cm

Weight: 40–270 kg

on what is easy and abundant—usually berries, horsetails, other vegetation and insects, though they will not turn up their noses at fish, fawns or another carnivore's kill. Black Bear sows with young cubs are likely to attack young or sickly Moose, deer and Wapiti, but on the whole, large males are the more predatory of the two sexes.

In the past few decades, the ubiquitous dandelion has become increasingly abundant along roads and swaths cut into the forests, especially in mountain regions. In pursuit of its new favourite food, the Black Bear is now more frequently seen along roadsides. With dandelion leaves sticking out of its mouth and the puffy seeds stuck all over its face and muzzle, the bear looks like a little kid covered in its favourite ice cream. Unfortunately, together with an increase in bear sightings along roadsides, vehicle collisions that may claim bears' lives are also increasing.

Within its territory, a bear will have favourite feeding places and follow well-travelled paths to these sites. Keep in mind that the trails you hike in parks and wilderness areas may be used not only by humans, but also by bears en route to lush meadows or rich berry patches.

Normally, the Black Bear is a reclusive animal that will flee to avoid contact with humans if it hears you coming. If you surprise a bear, however, back away slowly, always facing the animal—do not run or climb a tree. In particular, heed its warning of a foot stamp, a throaty "huff" or the champing sound of its teeth. These signals indicate that the bear is agitated and probably does not like your presence, and it is giving you a clear warning to retreat from its territory. Research suggests that a bear may charge or attack when these warning signals are not understood by a person who instead remains frozen in place. The bear likely interprets such behaviour as a challenge.

Mismanagement of food and garbage in areas where both bears and people occur is the unfortunate cause of many bear deaths. Bears that become habituated to people and their garbage often become nuisances and are destroyed.

A grim threat to Black Bears throughout the world is the illegal trade in body parts. Bear paws and gallbladders bring high prices on the black market, and poaching occurs in both Canada and the U.S., but fortunately to a lesser extent than elsewhere. As the populations of many bears around the world shrink, however, North American bears face an increasing threat. With bear numbers dwindling and black-market values increasing, it is feared that the generally well-protected animals in Canadian parks may become prime targets of the trade.

DESCRIPTION: The long, shaggy coat ranges in colour from black to honey brown. The body is relatively short and stout, and the legs are short and powerful. The large, wide feet have curved, black claws. The head is large and has a straight profile. The eyes are small, and the ears are short, rounded and erect. The tail is very short. Adult males are about 20 percent larger than females.

HABITAT: Black Bears are primarily forest animals, and their sharp, curved foreclaws enable them to easily climb trees, even as adults. In spring, they often forage in natural or roadside clearings.

FOOD: Up to 95 percent of the Black Bear's diet is plant material and includes leaves, buds, flowers, berries, fruits and roots. This omnivore also eats animal matter such as

small mammals, small vertebrates and many kinds of invertebrates, including bees (and their honey) and other insects; even young hoofed mammals may be killed and eaten. Carrion and human garbage are eagerly sought out.

DEN: The den, which is only used during winter, may be in a cave, a hollow beneath a log, under the roots of a fallen tree or even in a haystack. A bear may even choose a depression in the ground and cover itself with leaves. The bear will not eat, drink, urinate or defecate during its time in the den. Hibernation is not deep; instead, it is as if the bear is very groggy or heavily drugged. This state is called

DID YOU KNOW?

During its winter slumber, a Black Bear loses 20 to 40 percent of its body weight. To prepare for winter, the bear must eat thousands of calories per day during late summer and autumn.

"torpor" and is technically not a true hibernation. Rarely, a bear will rouse and leave its den on mild winter days.

REPRODUCTION: Black Bears mate in June or July, but the embryos do not implant and begin to develop until the sow enters her den in November. The number of eggs that implant seems to be correlated with the female's weight and condition—fat mothers have more cubs. One to five (usually two or three) young are born in January, and they nurse while the sow sleeps. The cubs' eyes open, and they become active when they are five to six weeks old. The cubs leave the den with their mother when they weigh 2 to 3 kg, usually in April. The sow and her cubs generally spend the next winter together in the den, dispersing the following spring. Females typically bear young in alternate years.

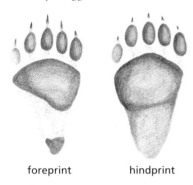

foreprint   hindprint

slow walking trail

Similar Species: The **Wolverine** (p. 166) looks somewhat like a small Black Bear but has a long tail, an arched back and pale side stripes. The larger **Grizzly Bear** (p. 220) has a blocky head and a shoulder hump.

Wolverine

Grizzly Bear
Ursus arctos

The mighty Grizzly Bear, more than any other animal, makes camping and travelling in the wild areas of western Canada an adventure. Since before the time of Sir Alexander Mackenzie's explorations, these creatures have had an almost mythical presence.

Fuelled by a mix of fear and curiosity, millions of visitors to mountain parks scan the roadsides and open meadows in the hope of catching a glimpse of this wilderness icon. Most people leave the parks without a personal Grizzly Bear experience, but when a bear is sighted, the human melee that ensues is unlike that which surrounds any other mountain animal. Crowds and "bear jams" are created, further contributing to the aura that surrounds this misunderstood species.

Grizzly Bears are indisputably strong—their massive shoulders and skull anchor muscles that are capable of

RANGE: In North America, this Holarctic species is largely confined to Alaska and northwestern Canada, with montane populations extending south into Idaho and Wyoming. It formerly ranged much farther, but European immigrants could not tolerate the threat it posed to livestock, so open-area populations were extirpated, leaving only remnant populations in the remote mountain wilderness.

Total Length: 1.8–2.6 m

Shoulder Height: 90–120 cm

Tail Length: 8–18 cm

Weight: 110–530 kg

rolling 90-kg rocks, dragging Wapiti carcasses and crushing some of the most massive ungulate bones. Ironically, Grizzly Bears do not commonly feed in this manner. Instead, their routine and docile foragings are concentrated on roots, berries and grasses.

An adaptable diner, the Grizzly Bear changes its diet to match the availability of foods. For instance, it eats huge quantities of berries when they are available in late summer. It swallows many of the berries whole, and its scat often ends up looking like blueberry pie filling. During this time of feasting, a Grizzly Bear's weight may increase by almost 1 kg per day, preparing it for the long winter ahead. The bear remains active through autumn, until the bitter cold of November limits foods and makes sleep seem more favourable.

Although mountain parks and coastal salmon streams boast the highest numbers of Grizzly Bears, the status of this species throughout its range is uncertain, but we can increase the bears' chances of survival. Some of the seminal work on Grizzly Bears dates back to the foundation of conservation biology. In working with these large carnivores, pioneering biologists invented tagging, radio-tracking and other research techniques that have benefited not only Grizzly Bears, but many other carnivores as well, including the Bengal Tiger (*Panthera tigris*) and the Polar Bear.

The range of the Grizzly Bear in Canada has changed little in the last 20 years, but historically, this bear lumbered through Canada's grasslands in pursuit of American Bison.

ALSO CALLED: Brown Bear.

DESCRIPTION: The usually brownish to yellowish coat typically has white-tipped guard hairs that give it

DID YOU KNOW?

Because an adult Grizzly Bear's long foreclaws are typically blunt from digging, it cannot easily climb trees. If you think you can escape a bear by climbing a tree, however, you had better climb high and fast, because some Grizzly Bears can reach almost 4 m up a tree trunk.

a grizzled appearance (from which this animal's name is derived). Some individuals are completely black, whereas others can be nearly white. The face has a concave (dished) profile. The eyes are relatively small, and the ears are short and rounded. A large hump at the shoulder makes the forequarters appear higher than the rump in profile. The large, flattened paws have long claws, and the front claws can be nearly 10 cm long.

HABITAT: Originally, most Grizzly Bears were animals of open rangelands, where they used their long claws to dig up roots, bulbs and the occasional burrowing mammal. Although their current range is largely forested, mountain bears often forage on open slopes and on the alpine tundra.

FOOD: Although 70 to 80 percent of a Grizzly's diet is made up of plants,

including leaves, stems, flowers, roots and fruits, it eats more animals, including other mammals, fish and insects, than its Black Bear cousin. A Grizzly Bear may dig insects, ground squirrels, marmots and even mice out of the ground. In parts of British Columbia and Alaska, Grizzlies feed heavily on spawning salmon. A sow bear with cubs eagerly seeks young hoofed mammals such as Moose, Caribou and Wapiti calves, and even large adult cervids and Bighorn Sheep may be attacked and killed. Particularly after it emerges from its winter sleep, a Grizzly Bear is attracted to carrion, which it can smell up to 16 km away. This bear can eat huge meals of meat—one adult consumed an entire road-killed Wapiti in four days.

DEN: Most dens are on north- or northeast-facing slopes in areas where snowmelt does not begin until late April or early May. The den is usually in a cave or is dug into tree roots. The bear enters its den in late October or November, during a heavy snowfall that will cover its tracks. It soon falls asleep and will not eat, drink, urinate or defecate for six months.

REPRODUCTION: A sow has litters in alternate years, typically having her first after her seventh birthday. Grizzlies mate in June or July, but with delayed implantation of the embryo, the cubs are not born until sometime between January and early March, when the mother is asleep in her den. The one to four (generally two) cubs are born naked, blind and helpless. They nurse and grow while their mother continues to sleep, and they are ready to follow her when she leaves the den in April or May. By the time she emerges, she will have lost much of her body weight from her own metabolism and from having nursed her young. A sow and her cubs typically den together the following winter.

foreprint   hindprint

fast walking trail

Black Bear

Similar Species: The **Black Bear** (p. 216) is generally smaller, is tallest at the rump (it lacks the shoulder hump) and has a straight profile and shorter claws. The similar-sized **Polar Bear** (p. 224) is greyish white.

Polar Bear

Ursus maritimus

The enormous Polar Bear, a symbol of the Canadian Arctic, is a uniquely adapted bear that is ideally suited for life in frozen northern regions. The whitish, waterproof hairs are hollow and act like optical cables by transferring the sun's energy to the skin. The Polar Bear's black skin maximizes absorption of this heat energy, and when seen through a heat-sensing device, the bear appears as a "cold" spot rather than the typical mammalian "hot" spot. This bear is so well adapted to cold that, when on land during summer, it often uses its winter

RANGE: Polar Bears have a northern circumpolar distribution. They are rarely found north of 88° N latitude, and their southern limit is determined by pack and landfast ice in winter. Sometimes they are found as far south as the Pribilof Islands of Alaska in the West and Newfoundland in the East.

Total Length: 2.0–3.3 m

Shoulder Height: 122–160 cm

Tail Length: 8–13 cm

Weight: *Male:* 300–800 kg; *Female:* 150–300 kg

den again as a reprieve from the warmth of the summer (many winter dens are dug into the permafrost layer and therefore stay cool in summer). This behaviour is only observable along the shores of Hudson Bay and James Bay, the largest known region for denning Polar Bears, and the only area where this bear digs a den in land rather than ice.

Strictly speaking, Polar Bears do not hibernate, and even though they make a winter den, the time spent there is quite short and variable. Females den up for longer periods of time than males, primarily because they give birth in January. Males may den up intermittently from November to January, but Polar Bears are known to be active at any time of the year.

Unlike the other two bears in North America—the Black Bear and the Grizzly Bear—the Polar Bear is almost strictly carnivorous.

When in the water, the Polar Bear has a swimming style unlike any other North American four-legged animal. It is able to reach speeds of nearly 10 km/h, and it does so using only its forelegs. The hindlegs, apparently unnecessary for propulsion, trail behind and are probably used as a rudder.

Today, Polar Bears are an internationally protected species. Much time and research went into identifying their status and determining which areas are critical to their reproduction and survival. Because the livelihood of many northern indigenous peoples depends on these bears, many are still permitted to hunt them. The number of bears hunted each year in each region is carefully controlled. The Inuit, the Greenlanders and some indigenous peoples in northern Russia take small numbers of Polar Bears each season.

DESCRIPTION: It is a common misconception that Polar Bears are white. Typically, only the new fur after a moult is white. Most Polar Bears are yellowish or greyish. Rare individuals are almost beige. The colouration depends mainly on the season and light conditions.

DID YOU KNOW?

Many people believe that a Polar Bear will cover its black nose with a paw while stalking prey to better camouflage itself. Although a plausible and intriguing theory, it has never been shown that Polar Bears actually behave this way.

Seasonal moults change the fur colour, and exposure to sunlight oxidizes the coat and turns it yellow or tawny. Polar Bears have black lips and eyes, and under the fur, the skin is also black, as is the nose. Relative to other bears, Polar Bears have long necks and long legs. Their ears are very small and rounded, and their tails are short and triangular.

HABITAT: In their Arctic home, Polar Bears live near shorelines on the massive broken ice packs that characterize the region. Very seldom do these animals venture inland. Pregnant females visit coastline habitats when they are searching for suitable den sites. Portions of the Hudson Bay coast provide excellent opportunities for viewing these bears.

FOOD: Active at any time of the day, Polar Bears feed primarily on seals and their pups. During spring, more than half of their diet is Ringed Seal pups. Other species of seals are eaten as well. Polar Bears will scavenge whale and Walrus carcasses when available. On rare occasions, a Polar Bears will hunt a Walrus, though the two are nearly equally matched in battle and a confrontation is unlikely to end well for either animal.

DEN: Along the shores of Hudson Bay and James Bay, pregnant females and some males make dens on shore, usually within 8 km of the water. The den may be dug into a slope where snow has accumulated up to 3 m deep, but it is frequently dug into a bank or peat hummock. Den design ranges from very simple to quite complex. A simple den is a single chamber with a short tunnel, and a complex den has several chambers and connecting tunnels.

On the ice packs, the bears make a den in the snow.

REPRODUCTION: Mating occurs in April or May, but a female only mates every two or three years. The young are usually born in January. Two young is common, but only in years of favourable conditions will both survive. Cubs grow slowly and are not weaned until they are two to three years old. When the cubs are one or two years of age, the female mates again, so that by the time the young are weaned and have dispersed, she will give birth again.

hindprint

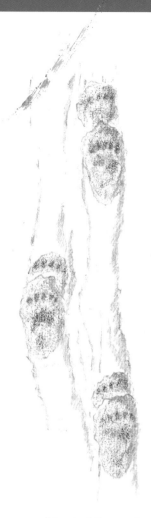

walking trail (in snow)

Similar Species: The **Grizzly Bear** (p. 220) is similar in shape and size but has a blocky head and brown fur. The **Black Bear** (p. 216) may sometimes be nearly white, but it is much smaller than the Polar Bear. Both the Grizzly Bear and wthe Black Bear have some range overlap with the Polar Bear, but it is unlikely to find them in the same habitat.

Grizzly Bear

Coyote
Canis latrans

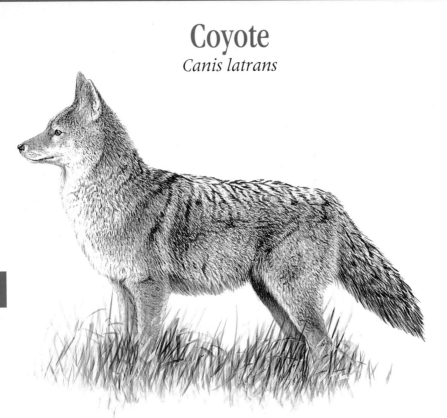

A chorus of yaps, whines, barks and howls complements the evening and morning twilight hours throughout much of Canada. Although Coyote calls are most intense during late winter and spring, corresponding to times of courtship, these excited sounds can be heard during suitable weather at any time of the day or year. Although the calls are often initiated by one animal, several family groups may join in the ruckus until a large area is alive with their energetic calls.

Two centuries ago, early explorers of this continent made frequent references in their journals to foxes and wolves, but they seldom mentioned Coyotes. Coyotes have increased their numbers across North America in the past century in response to the expansion of agriculture and forestry and the reduction of wolf populations.

RANGE: Coyotes are absent in the western third of Alaska, the tundra regions of northern Canada and the extreme southeastern U.S. Their range covers essentially the remainder of North America.

Total Length: 1.0–1.5 m

Shoulder Height: 58–66 cm

Tail Length: 30–40 cm

Weight: 8–20 kg

Despite widespread human efforts to exterminate them, they have thrived.

One of the few natural checks on Coyote abundance seems to be the Grey Wolf. As the much larger and more powerful canids of the wilderness neighbourhood, wolves typically exclude Coyotes from their territories. Prior to the 19th century, the natural condition favoured wolves, but human-induced habitat changes since then have greatly benefited Coyotes. Wolves prefer thick, unfragmented forests, whereas Coyotes are most common in open forests and grasslands. With increased fragmentation of Canada's forests, Coyote populations are on the rise. This species is now so widely distributed and comfortable with human development that almost all rural areas hold healthy populations, and even cities may have resident "urban Coyotes."

Because of their relatively small size, Coyotes typically prey on small animals such as mice, voles, ground squirrels, birds and hares, but they have also been known to kill deer, particularly fawns. Often, Coyotes form family-group packs, and they tend to hunt as a unit. When hunting, a pack may split up, with some members waiting in ambush while the others chase the prey toward them, or they may run in relays to tire their quarry—Coyotes, which are the best runners among the North American canids, typically cruise at 40 to 50 km/h.

DID YOU KNOW?

Coyotes can, and do, interbreed with domestic dogs and wolves. Hybridization with wolves is one of the biggest threats to the Grey Wolf population in Eastern Canada.

Coyotes owe their modern-day success to their varied diet, early age of first breeding, high reproductive output and flexible living requirements. They consume carrion throughout the year but also feed on such diverse offerings as eggs, mammals, birds and berries. Their variable diet and nonspecific habitat choices allow them to adapt to just about any region of North America.

DESCRIPTION: The Coyote looks like a grey, buffy or reddish grey, medium-sized dog. The underparts are light to whitish. The nose is pointed, and there is usually a grey patch between the eyes that contrasts with the rufous top of the snout. The bushy tail has a black tip. When frightened, a Coyote runs with its tail tucked between its hindlegs.

HABITAT: Coyotes are found in all terrestrial habitats in North America, except the tundra of the far north and the humid southeastern forests of the U.S. They have greatly expanded their range,

mainly because of declining Grey Wolf numbers and forestry and agriculture practices that have brought about favourable changes in habitat for Coyotes.

foreprint

hindprint

gallop group

walking trail

FOOD: Although primarily carnivorous, feeding on squirrels, mice, hares, birds, amphibians and reptiles, Coyotes will sometimes eat melons, berries and vegetation. Most farmers dislike them because Coyotes may take sheep, calves and pigs that are left exposed. They will even attack and consume domestic dogs and cats.

DEN: The den is usually a burrow in a slope, frequently a Woodchuck or American Badger hole that has been expanded to 30 cm in diameter and about 3 m in depth. Rarely, Coyotes have been known to den in a hollow tree trunk, a dense brush pile or even an abandoned car.

REPRODUCTION: A litter of 3 to 10 (usually five to seven) pups is born between late March and late May, after a gestation of about two months. The furry pups are blind at birth. Their eyes open after 14 days, and they leave the den for the first time when they are three weeks old. Young Coyotes fight with each other and establish dominance and social position at just three to four weeks of age.

Grey Wolf

Similar Species: The **Grey Wolf** (p. 232) is generally larger, with a broader snout, much bigger feet and longer legs. It carries its tail straight back when it runs. The **Red Fox** (p. 244) is generally smaller and much redder, with a white tail tip and black forelegs. The **Grey Fox** (p. 248) is smaller and has black spots on either side of the muzzle. The **Arctic Fox** (p. 236) is smaller and has different colouration.

Grey Wolf
Canis lupus

For many North Americans, the Grey Wolf represents the apex of wilderness and symbolizes the pure, yet hostile, qualities of all that remains wild. Perhaps the persecution of wolves over the last few hundred years was not really targeted at *Canis lupus,* but at the wolf-beast that lives only in the human imagination. The fear of wolves, without an understanding of their basic nature, resulted in tens of thousands of wolves being killed in vengeance for crimes that they did not commit. By studying and observing these animals, we demystify them and learn acceptance and even admiration for these remarkable creatures.

A wolf pack behaves like a "super-organism"—working cooperatively allows more animals to survive. By hunting together, pack members can catch and subdue much larger prey than if each wolf was acting alone. A pack consists of

RANGE: Much reduced from historic times, the Grey Wolf's range currently covers most of Canada and Alaska, except some of the Prairies and southern parts of eastern Canada. It extends south into Minnesota and Wisconsin and along the Rocky Mountains into Idaho, Montana and Wyoming.

Total Length: 1.4–2.0 m

Shoulder Height: 65–100 cm

Tail Length: 35–50 cm

Weight: 25–80 kg

an alpha pair (top male and female), subordinate adults, outcasts and pups and immature individuals. Usually, only the alpha animals reproduce, while the other pack members help to bring food to the pups and defend the group's territory. In most packs, the subordinate adults are nonbreeding, though they might attempt to mate in spite of the rules. The energetic pups of a wolf pack demand constant attention—they are always ready to pounce on their mother's head or tackle an unsuspecting sibling.

As the pups grow, they develop important skills that will aid them as adults. At the entrance to the den, they make their first attempts at hunting when they swat and bite at beetles and the occasional mouse or vole. These animals, however, do not react in quite the same way as larger prey does, and so the next step in the pups' apprenticeship as hunters is to watch their parents take down a big prey item such as a deer or Moose.

A wolf pack generally occupies a large territory—usually 260 to 780 km²—so individual densities are extremely low. Moreover, because of widespread extermination efforts in North America in the last century, wolves are absent over much of their historic range. Northern Canada has very high and stable numbers of wolves.

The Arctic Wolf (*C. l. arctos*), a nearly all-white subspecies of the Grey Wolf, inhabits the Arctic archipelago. Not many land mammals can survive in the cold of the Arctic as well as this wolf can. It can endure up to five months of darkness a year and can live for weeks without food.

Recent research of wolves in the Algonquin area of Ontario indicates that they may be a separate species from the Grey Wolf. If these findings are correct, the new species would be called the Eastern Timber Wolf (*C. lycaon*).

ALSO CALLED: Timber Wolf, Tundra Wolf, Arctic Wolf, Plains Wolf.

DESCRIPTION: A Grey Wolf resembles an oversized, long-legged German Shepherd with extra-large paws. Although typically

DID YOU KNOW?

Wolves are capable of many facial expressions, such as pursed lips, smile-like grins, upturned muzzles, wrinkled foreheads and angry, squinting eyes. Wolves even stick their tongues out at each other—a gesture of appeasement or submission.

thought of as being a grizzled grey colour, a wolf's coat can range from coal black to creamy white. Black wolves are most common in dense forests, whereas whitish wolves are characteristic of tundra regions. The bushy tail is carried straight behind the wolf when it runs. In social situations, the height at which the tail is held generally relates to the social status of that individual.

HABITAT: Although wolves formerly occupied numerous habitats throughout North America, they are now mostly restricted to forests, streamside woodlands and Arctic tundra.

FOOD: Grey Wolves primarily hunt cervids such as deer, Wapiti, Caribou and Moose. Although large mammals typically make up about 80 percent of the diet, wolves also prey on rabbits, mice, nestling birds and carrion when available. Arctic Wolves feed on Caribou and Muskox, as well as hares, seals, lemmings and ptarmigan.

DEN: Wolf dens are usually located on a rise of land near water. Most dens are

"Eastern Timber Wolf"

bank burrows and are often made by enlarging the den of a fox or burrowing mammal. Sometimes a rockslide, hollow log or natural cave is used. Sand or soil scratched out of the entrance by the female is usually evident as a large mound. The burrow opening is generally about 60 cm across, and the burrow extends back 2 to 10 m to a dry natal chamber with a floor of packed soil. Beds from which adults can keep watch are generally found above the entrance.

REPRODUCTION: A litter usually contains five to seven pups (with extremes of 1 to 11), which may be different colours. The newborn pups are similar to domestic dogs in their development: their eyes open at 9 to 10 days, they begin to eat solid food at three to four weeks, and they are weaned at six to eight weeks. The pups are fed regurgitated food until they begin to accompany the pack on hunts. Wolves become sexually mature a couple of months before their third birthday, but the pack hierarchy largely determines the first incidence of mating.

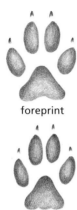

foreprint

hindprint

walking trail

trotting trail

Similar Species: The **Coyote** (p. 228) is smaller, with a much narrower and more pointed snout. A Coyote's tail is always tipped with black and held pointing downward, as opposed to horizontally in a wolf. **Foxes** (pp. 236–48) are much smaller.

Coyote

Arctic Fox
Alopex lagopus

summer coat

these traits have some level of intelligence. Cleverness has long been associated with foxes, and each species of fox has unique characteristics that epitomize the old saying, "as crafty as a fox."

The Arctic Fox exhibits a behaviour previously thought of as uniquely human. When we prepare too much for dinner or want to save summer foods for winter, we freeze them. Arctic Foxes, resourceful as they are, also freeze food for winter. In summer, food is plentiful, and they gluttonously feed on everything they can

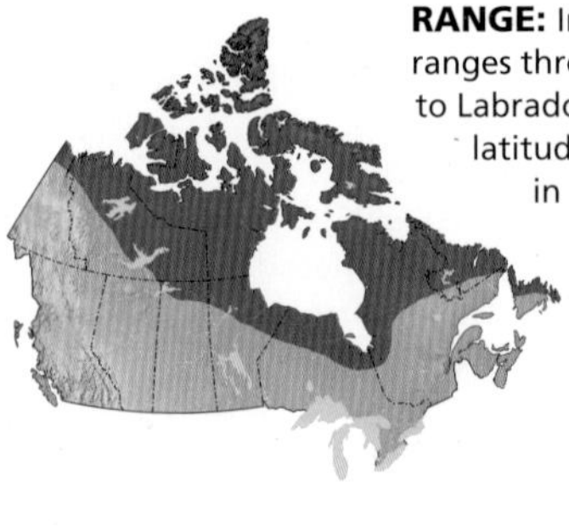

Right or wrong, humans have a strong tendency to rank animal intelligence according to how many "humanlike" behaviours the animal has. We talk about animals as "tool users" or "problem solvers" and generally assume that creatures demonstrating

RANGE: In North America, the circumpolar Arctic Fox ranges throughout the Arctic tundra, from western Alaska to Labrador. This fox has been seen as far north as 88° N latitude on the polar ice, and there are a few records in Ontario from around the 49th parallel.

Total Length: 75–91 cm

Shoulder Height: 25–30 cm

Tail Length: 27–34 cm

Weight: 1.8–4.1 kg

catch. When there is excess meat or carrion, the Arctic Fox digs a hole in the ground down to the permafrost layer and stores the food underground, where it stays frozen until the winter months.

This fox has also been observed fishing—in a rather peculiar way. When easier prey is scarce, an Arctic Fox may resort to approaching open water and tapping or stirring the surface with its paw. Fish are attracted to this movement at the surface, and when one comes within biting range, the Arctic Fox quickly nabs it with its sharply toothed, rapidly closing jaws.

The Arctic Fox typically occurs north of treeline, creatively surviving on Polar Bear or wolf kills, rodents, ground-nesting birds and their eggs, and seal pups. It also seems to follow migratory Caribou herds; even though Caribou are too large for this diminutive canid, wolves also follow the Caribou and leave many scraps on which the Arctic Fox dines. In the coldest, snowiest parts of winter, the Arctic Fox can be found a bit south of the tundra into the treeline. As soon as the snow begins to melt, however, the Arctic Fox returns to the northern tundra to feast on the bounty of the Arctic summer.

Because of its small size, this fox falls prey to several other larger carnivores. Humans are probably the most relentless pursuers of this striking canine because of its thick, luxurious coat. Its natural predators include Polar Bears, Grey Wolves and snowy owls (which can take pups). The range of the Red Fox overlaps significantly with that of the Arctic Fox, and where overlap occurs, the Red Fox is

DID YOU KNOW?

In Siberia, the Arctic Fox has been known to move as far as 1200 km south during winter.

dominant. The Red Fox does not regularly prey on the Arctic Fox, but it will harass and even kill it.

ALSO CALLED: Polar Fox, Blue Fox, White Fox.

DESCRIPTION: This cat-sized fox is the only native canid with distinct seasonal pelages. The summer coat is generally bluish grey above and pale or white below, with some white hairs on the head and shoulders, but most summer patterns are unique. The forelegs are brown and the tail is brownish above and lighter below, often with some long, white hairs. In winter, only the black nose and brownish yellow eyes contrast with the long, white coat and bushy, white tail. On rare occasions, this fox's winter pelage appears in a "blue" phase in which the entire coat is blue-black to pearl grey. The ears are short and rounded, and the soles of the feet sport abundant hair between the toe pads.

HABITAT: Although they typically occur on the treeless Arctic tundra or out on the polar ice, some Arctic Foxes may move south into the northern forests in winter.

FOOD: Arctic Foxes feed avidly on rodents, which they may dig up from

under the snow. Their diet also includes birds, eggs, young hares, insects, carrion, seal pups and the scraps left by Polar Bears or wolves.

DEN: On the tundra, Arctic Foxes make their dens under rocks or in banks or hillsides, often with multiple entrances.

foreprint

hindprint

The area around the entrance may be littered with feathers and bits of bone, and sometimes rather large pieces of bone are carried to the den. In winter, these foxes may tunnel into snowbanks.

REPRODUCTION: When there is an abundance of food, Arctic Foxes may mate as early as mid-February, but breeding is more typically delayed until late March or, if food is scarce, late April. In times of extreme hardship, Arctic Foxes might not mate at all. After a gestation of 1.5 to 2 months, the vixen usually gives birth to six to nine helpless kits in May or June. Instances where up to 22 pups are found in a den may represent multiple litters. The young are weaned at two to four weeks. If food becomes scarce, the stronger pups have been known to eat their weaker siblings. The parents continue to feed their pups until mid-August. Arctic Foxes are sexually mature by their first birthday.

loping trail

running trail

Grey Wolf

Similar Species: There is no other mammal that resembles an Arctic Fox in its white winter coat. A white **Grey Wolf** (p. 232) is much larger. The **Red Fox** (p. 244) is larger, has taller ears and is generally reddish. The **Coyote** (p. 228) is larger and greyish brown overall.

Swift Fox

Vulpes velox

There are precious few conservation success stories from the Prairies, but the reestablishment of the Swift Fox is one achievement to be celebrated. Historically, the Swift Fox was found throughout the Great Plains, ranging from the foothills of the Rockies to the Red River valley in Manitoba and as far north as the North Saskatchewan River. Swift Foxes were collateral damage in the overall attempt to eradicate native species and open the land for farming; numbers declined as a result of habitat loss, trapping and the poisoning of predator species such as Grey Wolves and Coyotes. The last wild Swift Fox on the Prairies, prior to the species' reestablishment, was seen near the southern Alberta hamlet of Manyberries in 1938.

The case for reestablishing the Swift Fox in certain areas of the Prairie provinces was strong—some relatively pristine grasslands remained, wild populations still occurred much farther south in states such as Wyoming and Colorado, and the primary causes for the species' initial decline had been eliminated. The reintroduction program began in 1983. Many of the first

RANGE: The Swift Fox's former range included prairie regions of Alberta, Saskatchewan and Manitoba south through eastern Montana and the Dakotas to northern Texas. Although its numbers are greatly reduced, the Swift Fox may still be found sprinkled throughout this range.

Total Length: 60–80 cm

Shoulder Height: 28–31 cm

Tail Length: 23–30 cm

Weight: 1.4–2.7 kg

animals that were released had been captive-raised in Alberta; later, wild-born animals were trapped and transported from the U.S. Because of the enthusiastic support of private landowner stewards and the diligent research of the Cochrane Ecological Institute, there is now a small population of about 200 Swift Foxes along the Alberta-Saskatchewan border.

Unfortunately, this cat-sized canid suffers high mortality rates. These foxes are sometimes the unintended victims of pest-control programs aimed at killing Richardson's Ground Squirrels and skunks. As well, they are preyed upon by larger grassland predators, especially raptors and Coyotes. As a result, the Swift Fox tends to den in areas with sweeping vistas, which give a denning fox ample visual warning of approaching danger. This adaptation, coupled with a progressive implementation of effective conservation strategies, will hopefully ensure that future generations of Swift Foxes will have a secure place in Canada's grassland community.

DID YOU KNOW?

The Swift Fox is sacred to nearly all Great Plains First Nations tribes, which is also an indication of their original range.

DESCRIPTION: The summer coat of this tiny fox closely matches its prairie environment—the back is mainly pale rufous or buffy grey, and the sides are a lighter yellowish buff. The long guard hairs are white- or black-tipped, giving the fox an overall grizzled appearance. Each side of the muzzle bears a distinct black spot, and the neck, backs of the ears and legs have orange highlights. The tail is large and has a noticeable black tip.

HABITAT: Historically this open-country fox inhabits moist mixed-grass, short-grass and tall-grass prairie habitats.

FOOD: Rabbits and small rodents make up the bulk of the diet, but the Swift Fox

foreprint

hindprint

also eats birds, insects, carrion and even grasses or berries.

DEN: A Swift Fox's fairly complex burrow usually has four or five entrances. The same burrow is used throughout the year, and perhaps even for a lifetime.

REPRODUCTION: Swift foxes are largely monogamous. They mate for life, are extremely selective in mate selection, are very social and not turf defensive. The kits are born in a small, bare chamber of the burrow after a gestation of about seven to eight weeks. The litter of three to six young requires great care for the first few weeks. The kits open their eyes and ears at two weeks, and they are weaned at six weeks. They disperse in autumn and are capable of breeding at one year old but in general breed at two years.

walking trail

Similar Species: The **Arctic Fox** (p. 236) is the closest to the Swift Fox in size, behaviour, vocalization and DNA but has a different range. The **Red Fox** (p. 244) is larger and has black forelegs and a white-tipped tail. The **Coyote** (p. 228) is much larger and has a tawny patch on the upper surface of its snout.

Red Fox

Red Fox
Vulpes vulpes

More than other native canid, the Red Fox has received some favourable representations in literature and modern culture. From Aesop's Fables to sexy epithets, the fox is often portrayed as a diabolically cunning, intelligent, attractive and noble animal. Many people favour having a fox nearby because of this species' skill at catching mice. In Old English, the fox had the well-deserved nickname "reynard," from the French word *renard,* which refers to someone who is unconquerable owing to his or her cleverness. The fox's intelligence, undeniable comeliness and positive impact upon most farmlands have endeared it to many people who otherwise might not appreciate wildlife.

Foxes at work in their natural habitat embody playfulness, roguishness, stealth and drama. Kits at their den wrestle and squabble in determined sibling rivalry. If its siblings are busy elsewhere, a kit may amuse itself by challenging a plaything, such as a stick or piece of old bone, to a bout of aggressive mock combat. An adult out mousing will sneak up on its rustling prey in the grass and jump stiff-legged into the air, attempting to come down directly atop the unsuspecting rodent. If the fox misses, it stomps and flattens the grass with its forepaws, biting in the air to try to catch the fleeing mouse. Usually the fox wins, but, if not, it will slip away with stately composure

RANGE: In North America, this Holarctic species occurs throughout most of Canada and the U.S., except for the High Arctic and parts of the central and western U.S. The most widely distributed carnivore in the world, it also occurs in Europe, Asia, North Africa and as an introduced species in Australia.

Total Length: 90–110 cm

Shoulder Height: 38–41 cm

Tail Length: 35–45 cm

Weight: 3.6–6.8 kg

as though the display of undignified abandon never occurred.

Oddly enough, Red Foxes exhibit both feline dexterity and a feline hunting style. They hunt using an ambush style or else creep along in a crouched position, ready to pounce on unsuspecting prey. Another undoglike characteristic is the large gland above the base of the tail, which gives off a strong musk somewhat similar to the smell of a skunk. This scent is what allows foxhounds to easily track their quarry. Foxes are territorial, and the males, like other members of the dog family, mark their territorial boundaries with urine.

The Red Fox is primarily nocturnal, and its keen senses of sight, hearing and smell enhance its elusive nature; at the first sign of an intruder, the fox will travel elsewhere. Winter may be the best time to see this canid because it is more likely to be active during the day, and its colour stands out when it is mousing in a snow-covered field.

Red Foxes have adapted well to human activity, and most of them live near farming communities and even in cities. In the northern wilderness, these midsized canids live on mice and carcasses in the shadow of the Grey Wolf.

DESCRIPTION: This small, slender, doglike fox has an exceptionally bushy, long tail.

DID YOU KNOW?

The Red Fox's signature feature, its white-tipped, bushy tail, provides balance when the fox is running or jumping, and during cold weather, a fox will wrap its tail over its face when at rest.

It usually has vivid reddish orange upperparts with a white chest and belly, but there are other colour variations: a Coyote-coloured phase; the "Cross Fox," which has darker hairs along the back and across the shoulder blades; and the "Silver Fox," which is mostly black with silver-tipped hairs. In all colour phases, the tail has a white tip and the backs of the ears and fronts of the forelegs are black.

HABITAT: The Red Fox prefers open habitats interspersed with brushy shelter year-round. It avoids extensive areas of dense, coniferous forest with heavy snowfall.

FOOD: This opportunistic feeder usually stalks its prey and then pounces on it or captures it after a short rush. In winter, small rodents, rabbits and birds make up most of the diet, but dried berries are also eaten. In more moderate seasons, invertebrates, birds, eggs, fruits and berries supplement the basic small-mammal diet.

foreprint

hindprint

DEN: The Red Fox generally dens in a burrow, which the vixen either digs herself or, more often, makes by expanding a Woodchuck hole. The den is sometimes located in a hollow log, in a brush pile or beneath an unoccupied building.

REPRODUCTION: A litter of 1 to 10 kits is born in April or May after a gestation of about 7.5 weeks. The kits weigh about 100 g at birth. Their eyes open at eight to nine days, and they are weaned when they are 8 to 10 weeks old. The parents first bring the kits dead food and later crippled animals. The male may bring several voles or perhaps a hare and some mice back to the den at the end of a single hunting trip. After the kits learn to kill, the parents start taking them on hunts. The young disperse when they are three to four months old and become sexually mature well before their first birthday.

walking trail

trotting trail

Similar Species: The larger **Coyote** (p. 228) has a dark-tipped tail and does not have black forelegs. The **Grey Fox** (p. 248) is grizzled grey over the back and has a black-tipped tail. The distinctive **Grey Wolf** (p. 232) can be more than twice the size of a Red Fox. The **Arctic Fox** (p. 236) is smaller and lacks the reddish colour.

Coyote

Grey Fox
Urocyon cinereoargenteus

Truly a crafty fox, the Grey Fox is known to elude predators by taking the most unexpected of turns—running up a tree. Unlike other canids, this fox seems comfortable in trees, and it may climb into the branches to rest and sleep. There are even rare records of this fox denning in a natural tree cavity and raising its litter in a tree den as high as 6 m off the ground. It is the only North American member of the dog family that can climb.

The Grey Fox's great rarity and nocturnal habit mean it is less frequently seen than other foxes. Furthermore, it frequents treed areas, enhancing its reputation for being elusive. This fox occurs in low numbers in the southernmost parts of Canada, from Manitoba to Québec. Foxes, on the whole, are difficult to observe, and your chances of seeing a Grey Fox are slim. If you have an optimistic nature, you can turn your eyes skyward and scan the trees, especially those with thick, heavily forked trunks or leaning branches, for this species.

After the mating season, the male stays with the female and helps raise the young. His primary role after the female gives birth is to bring her food because she must remain with the young constantly for several days. The Grey Fox often caches its food, especially large kills that cannot be consumed all at once.

RANGE: Grey Foxes have an extensive distribution in the U.S. and occur in Canada along the southern borders of Manitoba, Ontario and Québec. They range through most of the eastern states and from Texas to California and up the West Coast through most of Oregon. Some populations can be found in Colorado and Utah.

Total Length: 80–113 cm

Shoulder Height: 36–38 cm

Tail Length: 28–44 cm

Weight: 3.4–5.9 kg

Small kills may be buried right near the den during the whelping season, partly to provide the female with ready food but also to stimulate the interest of the kits. Large cache sites are made either in heaped up vegetation or in holes dug into loose dirt.

Many of the fox populations in North America suffered great losses during the peak of the fur trade. To foxes and Coyotes, humans are the worst enemy. Fortunately for the Grey Fox, its pelt is of lower quality (to humans, that is—the fox certainly appreciates it) than that of other canids because the grizzled fur is very stiff, rather than soft and long like the winter coat of a Red Fox. The Grey Fox has also suffered less persecution from farmers because it is quite shy and rarely hunts domestic animals. Unlike other canids, the Grey Fox is not inclined to take chickens, preferring to hunt the mice that abound around a henhouse. Its mousing ability is so good that it is even considered a welcome visitor to a farmyard.

DESCRIPTION: This handsome fox has an overall grizzled appearance because of the long, greyish fur over its back. Its underparts are reddish, as are the back of the head and the throat, legs and feet. Sometimes the belly is mostly white with only reddish highlights. The long tail is grey or even black on top, with a reddish underside and a black tip. The ears are pointed and are mainly grey, with patches of red on the back. A distinct black spot is present on either side of the muzzle.

HABITAT: The Grey Fox inhabits a variety of different environments, always near trees or groundcover. It prefers to forage in wooded areas rather than open environments.

FOOD: This species is more omnivorous than other foxes. It consumes a variety of small mammals such as rabbits, rodents and birds, as well as large numbers of insects and other invertebrates. Late in summer, grasshoppers, crickets and other agricultural pests constitute much of the diet. Another significant part of the diet is vegetable matter such as fruits and

grasses. Favourite items include apples and nuts.

DEN: This fox usually dens on a ridge or rocky slope or under brushy cover, but it may also den underground or in a tree. If necessary, the fox will dig a burrow itself, but it prefers to refurbish the abandoned burrow of another animal, such as a Woodchuck.

walking trail

REPRODUCTION: Grey Foxes in Canada are mostly nonbreeding dispersing animals—the foxes encountered here are mostly vagrants from breeding populations farther south. The only known breeding population in Canada is on Pelee Island in southern Ontario.

Mating occurs in January or February, and after a gestation of at least 53 days, one to seven young are born. They are born blind and almost hairless, and for the first several days, they require constant care by the mother. Their eyes open after 10 to 12 days, and they venture out of the den when they are 1.5 to 2 months old. When the kits are weaned at four months of age, they learn to hunt and accompany their parents on foraging expeditions. By the eighth month, they have dispersed to start their own dens.

foreprint

hindprint

Similar Species: The **Red Fox** (p. 244) is much redder overall and has a white-tipped tail. A Red Fox "Cross Fox" is usually darker over the shoulders and back. The **Coyote** (p. 228) is larger and lacks the black spots on either side of the muzzle.

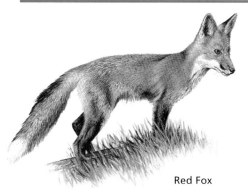

Red Fox

RODENTS

In terms of sheer numbers, rodents are the most successful group of mammals in Canada. Because we usually associate rodents with rats and mice, the group's most notorious members, many people look on all rodents as filthy vermin. Few realize that the much more endearing chipmunks, beavers and tree squirrels are also rodents.

A rodent's best-known features are its upper and lower pairs of protruding incisor teeth, which continue to grow throughout the animal's life. These four teeth have pale yellow to burnt orange enamel only on their front surfaces; the soft dentine at the rear of each tooth is worn away by the action of gnawing, so that the teeth retain knife-sharp cutting edges. Most rodents are relatively small mammals, but some species such as the American Beaver and the North American Porcupine can grow quite large.

Nutria Family (Myocastoridae)

The Nutria was introduced to North America from South America. It is the sole species in its family, and though it looks similar to a Muskrat, the two are not closely related. The Nutria has a long, rounded tail, and the webbing on each hindfoot joins only four of the five hindtoes.

Porcupine Family (Erethizontidae)

The stocky-bodied North American Porcupine has some of its hairs modified into sharply pointed quills that it uses in defence. Its sharp, curved claws and the rough soles of its feet are adapted for climbing.

Jumping Mouse Family (Zapodidae)

Jumping mice are so called because they make long leaps when they are startled. Their hindlegs are much longer than their forelegs, and the tail, which is longer than the combined length of the head and body, serves as a counterbalance during jumps. Jumping mice are almost completely nocturnal and hibernate in winter.

Mouse Family (Muridae)

This diverse group of rodents is the largest and most successful mammal family in the world. Its members include the familiar rats and mice, as well as voles and lemmings. The representatives of this family in Canada vary in size from the tiny Western Harvest Mouse to the much larger Muskrat.

Beaver Family (Castoridae)

The largest North American rodent, the American Beaver is one of two species worldwide in its family and the only representative on our continent. After humans, it is probably the animal that makes the biggest impact on the wilderness landscape of Canada.

Pocket Mouse Family (Heteromyidae)

Pocket mice belong to a group of small to medium-sized rodents that are somewhat adapted to a subterranean existence. They feed mainly on seeds, which they transport in their cheek pouches to caches in their burrows. Typically denizens of dry environments, many jumping mouse species can live for a long time without drinking water.

Pocket Gopher Family (Geomyidae)

Almost exclusively subterranean, all pocket gophers have small eyes, tiny ears, heavy claws, short, strong forelegs and a short, sparsely haired tail. Their fur-lined cheek pouches, or "pockets," are primarily used to transport food. The lower jaw is massive, and the incisor teeth are used in excavating tunnels.

Squirrel Family (Sciuridae)

This family, which includes chipmunks, tree squirrels, flying squirrels, marmots and ground squirrels, is considered the second most structurally primitive group of rodents. All its members, except the flying squirrels, are active during the day, so they are seen more frequently than other rodents.

Mountain Beaver Family (Aplodontidae)

The Mountain Beaver is the sole living member of its family, and it is usually considered the most "primitive" rodent. A Mountain Beaver resembles a small, stout marmot with a tiny tail. Despite its name, it is not related to the American Beaver (other than being a fellow rodent).

Nutria
Myocastor coypus

The Nutria was introduced to North America to be farmed for its thick fur. Most of the farmers who raised these South American rodents found the business unprofitable, and many animals were deliberately or accidentally released. Where moderate winters prevail, such as in southwestern British Columbia and the marshlands of southern Ontario and Québec, descendants of these escapees are occasionally reported.

Interestingly, some American Beavers that were imported into Argentina for fur production also escaped and are expanding their range. There appears to be no predator to control the beavers, which are responsible for widespread cutting of Argentina's riparian forests.

These two examples show the potential dangers of introducing a species into a similar habitat in a distant geographical area. Such newcomers are exempt from the intricate system of checks and balances that has evolved among native species over long periods of time, often to the detriment of the native species.

ALSO CALLED: Coypu.

DESCRIPTION: The Nutria is a greyish to brownish aquatic rodent with a long, dark brown, sparsely haired, scaly, cylindrical tail. Numerous whitish or yellowish hairs occur throughout the coat, and a distinct white patch can usually be seen

RANGE: Reports of Nutria in Canada are from the Lower Mainland and Salt Spring and Vancouver islands in British Columbia, as well as Second Marsh and other wetlands in southern Ontario. A few reports exist from Québec.

Total Length: 66–140 cm

Tail Length: 30–15 cm

Weight: 3–11 kg

on the tip of the muzzle. It has webbed hindfeet, with only four of the five hind-toes on each foot included in the web. The Nutria is very similar in appearance to the Muskrat, though it is much larger.

HABITAT: Nutrias require rivers, lakes and marshes with abundant emergent or sub-merged vegetation. They cannot live in areas where the entire water surface freezes in winter.

foreprint

hindprint

DID YOU KNOW?

In the evenings, Nutrias make loud, grunting calls and can be heard from quite a distance around their wetland homes.

FOOD: These rodents eat large amounts of wetland vegetation. Like lagomorphs, Nutrias reingest their fecal pellets to maximize nutrient absorption.

DEN: The den is dug into a bank; one or two entrances are located above the water and continue about a metre into the bank. A nest of reeds, cattails and sedges is made in a small chamber inside the den.

REPRODUCTION: Breeding takes place throughout the year, and most females produce two litters per year, starting when they are one year old. Following a 100- to 135-day gestation period, three to eight precocious young, each weighing 200 to 250 g, are born. They swim and may eat some vegetation their first day.

walking trail

Muskrat

SIMILAR SPECIES: The **Muskrat** (p. 295) is much smaller, lacks the light hairs in the coat and the white patch on the muzzle, and has a lat-erally compressed tail. The larger **American Beaver** (p. 305) has a wide, flat tail and makes unmis-takable lodges and dams.

North American Porcupine
Erethizon dorsatum

Although it lacks the charisma of the large carnivores and ungulates, the North American Porcupine's claim to fame is its unsurpassed defensive mechanism. Its formidable quills, numbering about 30,000, are actually stiff, modified hairs with shinglelike, overlapping barbs at their tips. The quills have an antibacterial coating that helps reduce the chance of infection when a porcupine stabs itself.

Contrary to popular belief, a porcupine cannot throw its quills, but if it is attacked, it will lower its head in a defensive posture and lash out with its tail. The loosely rooted quills detach easily, and they may be driven deeply into the attacker's flesh.

The barbs swell and expand with blood, making the quills even harder to extract. Quill wounds may fester, or, depending on where the quills strike, they can blind an animal, prevent it from eating or even puncture a vital organ.

The North American Porcupine is strictly vegetarian and is frequently found feeding in agricultural fields, willow-edged wetlands and forests. The tender bark of young branches seems to be a porcupine delicacy, and though you wouldn't think it from

RANGE: The North American Porcupine is widely distributed from Alaska across Canada and the northern U.S. to Pennsylvania and New England and south through most of the West into Mexico.

Total Length: 55–95 cm

Tail Length: 14–25 cm

Weight: 3.5–18 kg

the animal's size, the porcupine can move far out on thin branches with its deliberate climbing. An accomplished, if slow, climber, the porcupine uses its sharp, curved claws, the thick, bumpy soles of its feet and the quills on the underside of its tail in climbing. Despite its agility in trees, it has been known to fall on occasion, breaking bones and even stabbing itself with its own quills. In winter, this large, stocky rodent will often remain in a tree or bush for several days at a time. When it leaves a foraging site, the gnawed, cream-coloured branches are clear evidence of its activity. In summer, the porcupine can be seen in more open areas, foraging on herbaceous plants.

The North American Porcupine is primarily nocturnal in summer. It often rests by day in a hollow tree or log, a burrow, a fork in a treetop or sometimes even a rocky cave or crevice. In winter, it is not unusual to see a porcupine active by day, either in an open field or in a forest. It often chews bones or fallen antlers for calcium, and the sound of a porcupine's gnawing can sometimes be heard at a considerable distance.

Unfortunately for the porcupine, its armament is no defence against vehicles—highway collisions are a major cause of mortality—and most people only see this animal in the form of roadkill.

DESCRIPTION: This large, stout-bodied rodent has long, light-tipped guard hairs surrounding the centre of the back, where abundant, long, thick quills crisscross one another in all directions. A young porcupine is mostly black, but an adult is tinged with yellow. The upper surface of the powerful, thick tail is amply supplied with dark-tipped, white to yellowish quills. The front claws are curved and sharp. The skin on the soles of the feet is covered with tooth-like projections. There may be grey patches on the cheeks and between the eyes.

DID YOU KNOW?

The name "porcupine" comes from the Latin *porcospinus*, "spiny pig," and underwent many variations—Shakespeare used the word "porpentine"—before its current spelling was established in the 17th century.

HABITAT: Porcupines occupy a variety of habitats, including coniferous, mixed and deciduous forests and even croplands near wooded areas.

FOOD: Almost completely herbivorous, the North American Porcupine is like an arboreal counterpart of the American Beaver. It eats leaves, buds, twigs and especially young bark or the cambium layer of both broad-leaved and coniferous trees and shrubs. During spring and summer, it eats considerable amounts of herbaceous vegetation. In winter, it subsists on bark.

The porcupine typically puts on weight during spring and summer and loses it during autumn and winter. It seems to have a profound fondness for salt and will chew wooden handles, boots and other material that is salty from sweat or urine. Some carrion is consumed, likely for its sodium and mineral content.

DEN: Porcupines prefer to den in rocky caves and crevices or beneath rocks, but they sometimes move into abandoned buildings, especially in winter. They are typically solitary animals, denning alone, but may share a den during particularly cold weather. Sometimes a porcupine will sleep in a treetop for weeks, avoiding any den site, while it feeds on the tree bark.

REPRODUCTION: The North American Porcupine's impressive armament inspires many questions about how it manages to mate. The female does most of the courtship, though males may fight with one another, and she is apparently stimulated by having the male urinate on her. When the female is sufficiently aroused, she relaxes her quills and raises her tail over her back so that mating can proceed. Following mating in autumn or early winter and a gestation of 6.5 to 7 months—unusually long for a rodent—a single, precocious porcupette is born in May or June. The young Porcupine is born with quills, but they are not dangerous to the mother—the baby is born headfirst in a placental sac with its soft quills lying flat against its body. The quills harden within about an hour of birth. The porcupine has erupted incisor teeth at birth, and though it may continue to nurse for up to four months, it begins eating green vegetation in its second or third week. A porcupine becomes sexually mature when it is about 1.5 years old.

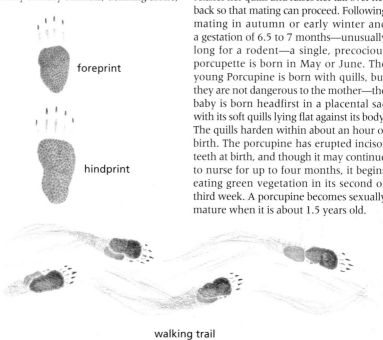

foreprint

hindprint

walking trail

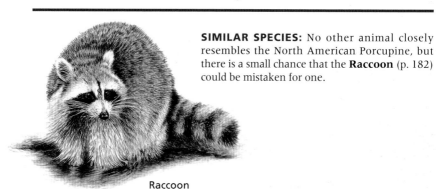

SIMILAR SPECIES: No other animal closely resembles the North American Porcupine, but there is a small chance that the **Raccoon** (p. 182) could be mistaken for one.

Raccoon

Meadow Jumping Mouse
Zapus hudsonius

HABITAT: This rodent pre- fers moist fields but also occurs in brush, marshes, brushy fields or even woods with thick vegetation.

FOOD: In spring, insects account for about half the diet. As the season progresses, the seeds of grasses and many forbs are eaten as they ripen. In summer and autumn, subterranean fungi form a significant portion of the diet.

On the rare occasions when this fascinating mouse is encountered, its method of escape reveals its true identity—startled from its sedgy home, a jumping mouse will hop away in a manner befitting a frog. Unfortunately, this rodent's speed and abundant hideouts prevent extended observation.

Jumping mice are dormant up to seven months of the year. True hibernators, their metabolism slows to the barest minimum, and they survive solely on the body's fat stores. Adults are underground by the end of August, and only those few juveniles that are below their minimum hibernation weight are active until mid-September.

DEN: The Meadow Jumping Mouse hibernates in a nest of finely shredded vegetation in a burrow or other protected site. Its summer nest is built on the ground or in a small shrub.

REPRODUCTION: Breeding typically takes place in April or May, depending on the latitude. The female bears two to nine young after a 19-day gestation. The eyes open after two to five days, and nursing continues for a month, after which the young leave the nest. Some females may have a second litter.

DESCRIPTION: The brownish back has a dark dorsal stripe. The sides are yellow- ish, and the belly is whitish. The long, naked tail is dark above and pale below, and there is no white tip. The hindfeet are greatly elongated.

SIMILAR SPECIES: The **Woodland Jumping Mouse** (p. 263) in eastern Canada has a distinctive, white-tipped tail. The **Western Jumping Mouse** (p. 261) has black hairs in the dorsal stripe. The **Pacific Jumping Mouse** (p. 262) has a different range.

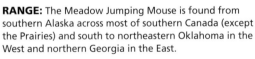

RANGE: The Meadow Jumping Mouse is found from southern Alaska across most of southern Canada (except the Prairies) and south to northeastern Oklahoma in the West and northern Georgia in the East.

Total Length: 19–22 cm

Tail Length: 11–14 cm

Weight: 15–25 g

Western Jumping Mouse
Zapus princeps

In the field, jumping mice are virtually indistinguishable from each other, but no other rodent can easily be confused with jumping mice—their hopping escapes and supremely long tails are sufficiently distinctive for even a novice naturalist to identify one as a jumping mouse, even if the exact species remains elusive.

Jumping mice are usually active within their soggy domains after dark and only during seasons filled with the distracting whine of mosquitoes.

DESCRIPTION: A broad, dark, dorsal stripe extends from the nose to the rump. It is primarily clay-coloured, with some blackish hairs. The mouse's sides are yellowish olive, often with some orangey hairs. The belly is a clear, creamy white. The flanks and cheeks are golden yellow. The naked tail is olive brown above and whitish below. The hindfeet are greatly elongated.

HABITAT: This jumping mouse prefers areas of tall grass, often near streams, that may have brush or trees. In the mountains, it ranges from valley floors up to treeline, and even into tundra sedge meadows. It appears to swim well, diving as deep as 1 m.

FOOD: In spring and summer, this mouse eats berries, vegetation, insects and a few other invertebrates. In autumn, grass seeds and the fruits of forbs are taken more frequently. Subterranean fungi are also favoured.

DEN: Hibernation takes place in a nest of finely shredded vegetation, 30 to 60 cm below the surface in a burrow that is 1 to 3 m long. The summer breeding nest is typically built aboveground among interwoven, broad-leaved grasses or in sphagnum moss in a depression.

REPRODUCTION: Breeding takes place within a week after the female emerges from hibernation. Following an 18-day gestation period, four to eight young are born in late June or early July. The eyes open after two to five days, and the young nurse for one month, after which they leave the nest. Some females have two, or even three, litters per year.

SIMILAR SPECIES: The **Meadow Jumping Mouse** (p. 260) lacks black hairs in the dorsal stripe. The **Pacific Jumping Mouse** (p. 262) is usually distinctly tricoloured. The **Woodland Jumping Mouse** (p. 263) in found only in Eastern Canada.

RANGE: This western species is found from the southern Yukon southeast to North Dakota and south to central California and northern New Mexico.

Total Length: 20–26 cm

Tail Length: 12–15 cm

Weight: 19–33 g

Pacific Jumping Mouse
Zapus trinotatus

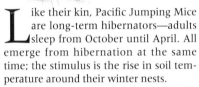

L ike their kin, Pacific Jumping Mice are long-term hibernators—adults sleep from October until April. All emerge from hibernation at the same time; the stimulus is the rise in soil temperature around their winter nests.

These mice jump by pushing off with their hindfeet and landing on their forefeet. They can leap well over a metre almost straight up when trying to escape a threat. Their long tails are critical for balance when jumping and are also used for noise-making when fighting. By vibrating their tails against the ground, jumping mice can produce a distinctive drumming sound.

DESCRIPTION: A dark brown dorsal stripe, flecked heavily with black, extends down the back. The sides are ochre to golden, and the underparts are creamy white. The forelegs are short, and the hindlegs are extremely long. The thin, bicoloured, scaly tail is equal to or longer than the length of the body and is sparsely haired. The ears are dark with light edges, and the whiskers are abundant.

HABITAT: These mice prefer areas with dense plant cover, such as streamsides, thickets, moist fields and some woodlands. In mountains, they range from valley floors to above treeline in wet alpine sedge meadows.

FOOD: Underground fungi, grass seeds, berries and tender vegetation are staples. In spring, many insects and other invertebrates are eaten.

DEN: A hibernation nest of finely shredded vegetation is located 30 to 60 cm underground in a burrow that is 1 to 3 m long. The spherical breeding nest is built aboveground among interwoven, broad-leaved grasses or in sphagnum moss in a depression.

REPRODUCTION: Breeding takes place between mid-May and mid-June. Following a gestation of 18 to 23 days, four to eight naked, hairless, blind young are born. The young nurse for a month and grow slowly. Females occasionally have a second litter.

SIMILAR SPECIES: The **Western Jumping Mouse** (p. 261) is usually less distinctly tricoloured. The **Meadow Jumping Mouse** (p. 260) and **Woodland Jumping Mouse** (p. 263) have different ranges.

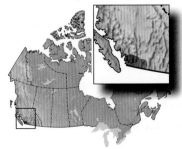

RANGE: This jumping mouse is found from extreme southwestern British Columbia (not including Vancouver Island) south to north-coastal California.

Total Length: 21–25 cm

Tail Length: 11–15 cm

Weight: 20–30 g

Woodland Jumping Mouse
Napaeozapus insignis

The remarkable Woodland Jumping Mouse is capable of making leaps up to 1.8 m in length, an astonishing 18 times the length of its own body. That's roughly equivalent to a human covering 30 m in one leap. Even a baby mouse seems to be born with the desire to jump. It requires only about 12 days after birth to stand on its feet, albeit shakily, and try out its hindlegs. By day 19, it regularly attempts jumps of about 2 or 3 cm but usually falls on its side and has to wriggle to get back up. A quick learner, by day 28, a young jumping mouse can easily leap at least six times the length of its body.

DESCRIPTION: This jumping mouse is distinctly tricoloured, with a blackish to brownish back, orange sides and white underparts. The long tail is dark on top and light below, with a distinctive white tip. The hindfeet are conspicuously large, and the forefeet are relatively small.

HABITAT: These mice are found in the dense foliage of moist, forested regions, in both coniferous and hardwood forests.

FOOD: Primary foods include seeds, fungi, a variety of plant material and insects. These mice hibernate but do not store food. To survive winter, they put on a layer of fat before dormancy.

DEN: These mice dig their own burrows or use those of other small mammals.

Their grass-and-leaf nests are found either underground or in brushy debris aboveground.

REPRODUCTION: After 29 days of gestation, the young are born in the nest in spring. Females have one litter per year of two to seven young. The young are altricial and require at least 34 days to gain the appearance of an adult.

SIMILAR SPECIES: The **Meadow Jumping Mouse** (p. 260) is slightly smaller and lacks the white-tipped tail. The **Western Jumping Mouse** (p. 261) and **Pacific Jumping Mouse** (p. 262) have different ranges. *Peromyscus* **mice** (pp. 265–67) have shorter tails and hindfeet.

RANGE: The Woodland Jumping Mouse occurs in southeastern Canada from Manitoba to Labrador and in the northeastern U.S. Southerly populations are found only in the Allegheny Mountains of the eastern U.S.

Total Length: 20–26 cm

Tail Length: 12–16 cm

Weight: 17–26 g

Western Harvest Mouse
Reithrodontomys megalotis

shrub borders and grasslands, especially where there is good cover overhead.

A leading candidate for the title of Canada's smallest mouse, the Western Harvest Mouse is most active during the two hours after sunset. Its activity may continue almost until dawn, particularly on dark, moonless nights. It often uses vole runways through thick grass to reach foraging areas, and it is named for its habit of collecting grass cuttings in mounds along trail networks. The Western Harvest Mouse does not store food in any great quantities, however, which is understandable for an animal that usually lives for less than a year.

DESCRIPTION: This native mouse is small and slim, with a small head and pointed nose. It has a conspicuous, long, sparsely haired tail and large, naked ears. The bicoloured tail is greyish above and lighter below. The upperparts are tawny, and the underparts are greyish white or sometimes pale cinnamon.

HABITAT: This species occurs in both arid and moist places, but it tends to favour

FOOD: The Western Harvest Mouse eats lots of green vegetation in spring and early summer. During most of the year, however, seeds and insects dominate the diet.

DEN: This mouse builds its ball-shaped nest, which is about 7.5 cm in diameter, either on the ground, low in a shrub or among weeds. The nest is made of dry grasses and is lined with soft material such as cattail fluff. One nest may house several mice and have multiple entrances.

REPRODUCTION: Reproduction may occur at any time of year if conditions are favourable. The average litter of three is born after a 23- to 24-day gestation. Hair is visible by five days, the eyes open after 10 to 12 days, and the young are weaned at 19 days. A female becomes sexually mature at four to five months of age and may have multiple litters per year.

SIMILAR SPECIES: The **House Mouse** (p. 275) is generally larger and has a hairless tail. The **Deer Mouse** (p. 267) has much whiter underparts.

RANGE: This mouse ranges from extreme southern Alberta, Saskatchewan and British Columbia south nearly to the Yucatán Peninsula of Mexico. It does not inhabit the roughest parts of the Rocky Mountains.

Total Length: 11–14 cm

Tail Length: 5.0–6.5 cm

Weight: 9–19 g

Keen's Mouse
Peromyscus keeni

One of the major identifying features that distinguishes the Keen's Mouse from the Deer Mouse is its long tail. Long tails appear to benefit animals that climb into shrubs and bushes, and the Keen's Mouse makes good use of its lengthy appendage when it harvests berries.

This mouse was named in honour of the Reverend John Henry Keen, who contributed greatly to our knowledge of the natural history of Haida Gwaii (Queen Charlotte Islands).

DESCRIPTION: The Keen's Mouse is greyish or slightly brown above and bright white below. The large ears extend well above the fur on the head, and the beady, black eyes protrude. The tail is distinctly bicoloured, slate grey above and white below. The feet and heels are white. On many islands, this species may have markings on the chest.

HABITAT: This mouse inhabits coastal rainforests dominated by western hemlock, Sitka spruce and red alder. The usually heavy underbrush of these areas includes blueberry, salmonberry and devil's club.

FOOD: The Keen's Mouse has a diverse, omnivorous diet that changes seasonally and includes seeds, assorted vegetation and insects. The primary food is seeds and berries, with an emphasis on spruce and hemlock seeds. This species appears to eat fewer invertebrates than the Deer Mouse does.

DEN: Nests are located in burrows, logs, buildings, tree cavities, among rocks or in other sheltered areas. The nest is a sphere of grass and fine, dry vegetation and is about 10 cm in diameter.

REPRODUCTION: The breeding season extends from May to September. Following a gestation of 23 to 25 days, two to seven young are born. The pink, naked, blind young weigh less than 2 g at birth and are weaned about three to four weeks later. During summer, the female breeds immediately after giving birth, so the weanlings are evicted from the nest to make room for the new litter.

SIMILAR SPECIES: The **Deer Mouse** (p. 267) has a slightly shorter tail.

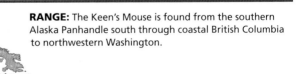

RANGE: The Keen's Mouse is found from the southern Alaska Panhandle south through coastal British Columbia to northwestern Washington.

Total Length: 18–23 cm

Tail Length: 9–11 cm

Weight: 10–30 g

White-footed Mouse
Peromyscus leucopus

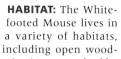

HABITAT: The White-footed Mouse lives in a variety of habitats, including open woodlands, riparian areas, shrubby areas and some agricultural lands bordering wooded areas. It is not considered detrimental to agriculture but is known to occupy buildings.

The White-footed Mouse is nearly impossible to distinguish from the Deer Mouse without a specimen in hand. When measurements are obtained, this mouse has a slightly shorter tail than the Deer Mouse. Nevertheless, confusion between the two species is common, even with a specimen and a mammal key.

This mouse demonstrates a strong swimming ability, and dispersing individuals are frequently found colonizing islands of lakes. Like the Deer Mouse, the White-footed Mouse is able to live in a wider variety of habitats then most other *Peromyscus* species. The only major habitat requirement for this excellent climber is some form of canopy or shrub cover.

DESCRIPTION: Colouration varies, but most individuals are pale to dark reddish brown above and white below. The bicoloured tail is brownish or greyish above and white below. The large ears extend well above the fur on the head, and the beady, black eyes protrude. The feet and heels are white.

FOOD: The diverse, omnivorous diet changes seasonally. Insects, seeds, vegetation, berries and nuts are the primary foods.

DEN: Nests are located in logs, tree cavities, buildings or other sheltered areas. The nest is a sphere of grass and fine, dry vegetation and is about 10 cm in diameter. This mouse may refurbish and "cap" an abandoned bird's nest.

REPRODUCTION: The breeding season extends from March to October, and females have multiple litters per season. Following a gestation of at least 22 days, four to six young are born. The pink, naked, blind young weigh less than 2 g at birth and are weaned about three to four weeks later.

SIMILAR SPECIES: The **Deer Mouse** (p. 267) is difficult to distinguish from the White-footed Mouse in the field; a specimen and key are needed. The **House Mouse** (p. 275) lacks the distinctive, bright white belly.

RANGE: The White-footed Mouse is found in the eastern U.S., north to southern Ontario, Québec and Nova Scotia, southwest as far as Arizona and northwest as far as extreme southern Alberta and Saskatchewan.

Total Length: 15–20 cm

Tail Length: 6.0–9.5 cm

Weight: 15–25 g

Deer Mouse
Peromyscus maniculatus

Upon first seeing a Deer Mouse, many people are surprised by its cute appearance. The large, protruding, coal black eyes give it a justifiably inquisitive look, while its dainty nose and long whiskers continually twitch, sensing the changing odours in the wind.

Wherever there is groundcover, from thick grass to deadfall, Deer Mice scurry about with great liveliness. These small mice are abundant over much of their range, and they may well be the most numerous mammals in much of Canada. When you walk through forested wilderness areas, they are in your company, even if their presence remains hidden.

Although Deer Mice most frequently forage along the ground, they regularly climb into trees and shrubs to reach food. During winter, they are the most common of the small rodents to travel above the snow. In doing so, however, they are vulnerable to nighttime predators. The tiny skulls of these rodents are among the most common remains in the regurgitated pellets of owls, a testament to the Deer Mouse's ubiquity and importance in the food web.

This mouse, which is named for the similarity of its colouring to that of the White-tailed Deer, commonly occupies farm buildings, garages and storage sheds, often alongside the House Mouse.

DESCRIPTION: The Deer Mouse has protruding black eyes, large ears, a pointed nose, long whiskers and a sharply bicoloured tail, with a dark top and a light underside. In contrast to these constant characteristics, the colour of the adult's upperparts is quite variable and can be yellowish buff, tawny brown, greyish brown or blackish brown. The upperparts, however, are always set off sharply from the bright white underparts and feet. A juvenile has uniformly grey upperparts.

HABITAT: These ubiquitous mice occupy a variety of habitats, including grasslands, mossy depressions, brushy areas, tundra and heavily wooded regions. They also greatly favour human-built structures—our warm, food-laden homes are palatial residences to Deer Mice.

FOOD: Deer Mice use their internal cheek pouches to transport large quantities of

RANGE: The Deer Mouse is the most widespread mouse in North America. Its range extends from Labrador west almost to Alaska and south through most of North America to south-central Mexico.

Total Length: 14–21 cm

Tail Length: 5–10 cm

Weight: 18–35 g

seeds or fruits from grasses, chokecherries, buckwheat and other plants to their burrows for later consumption. They also eat insects, nestling birds and eggs.

DEN: As the habitat of this mouse changes, so does its den type. In meadows, it nests in a small burrow or makes a grassy nest on raised ground, whereas in wooded areas, it makes a nest in a hollow log or under debris. Nests can also be made in rock crevices and certainly in human-made structures. This mouse is not meticulous about keeping its nest clean—every few weeks it abandons its nest and starts fresh.

REPRODUCTION: Breeding takes place between March and October, and gestation lasts for three to four weeks. Females have multiple litters in one season. There can be up to nine (usually four or five) helpless young, each weighing about 2 g at birth. They open their eyes between days 12 and 17 and venture out of the nest about four days later. At three to five weeks of age, the young are completely weaned and are soon on their own. Females are sexually mature in about 35 days and males in about 45 days.

running group

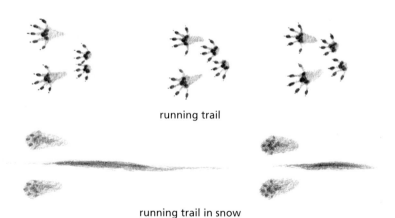

running trail

running trail in snow

SIMILAR SPECIES: In the field, the **White-footed Mouse** (p. 266) is difficult to distinguish from the Deer Mouse; a specimen and key are needed. It also has a smaller range. The **House Mouse** (p. 275) lacks the distinctive, bright white belly. **Jumping mice** (pp. 260–63) have much longer tails.

White-footed Mouse

Northern Grasshopper Mouse
Onychomys leucogaster

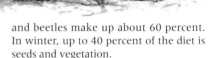

The Northern Grasshopper Mouse is the bulldog of mice. Its stocky body befits its predatory nature—up to 90 percent of its diet consists of animal matter, primarily grasshoppers and other insects, but also includes prey as large as other mice and voles. Studies of this mouse indicate that it has characteristics normally associated with carnivores—it exhibits social bonds and elaborate courtship, and it has long claws, enhanced jaw muscles and teeth suitable for shearing animal matter. Unlike other mice, it even "howls."

DESCRIPTION: The back is grey to yellowish buff, and the entire belly is white. The nose is pointed, the dark eyes protrude noticeably, and the ears are large. The short tail, which is less than twice the length of the hindfoot, is thick, sharply bicoloured (darker above, white below) and has a white tip.

HABITAT: This mouse occurs in a wide variety of open habitats with sandy or gravelly soils, from grasslands to sandy brushlands, but it avoids alkali flats, marshy areas and rocky sites. Although it does not occur in the mountains, its range approaches the foothills in a few places.

FOOD: Only a little more than 10 percent of the summer diet consists of vegetation, mostly seeds, while grasshoppers, crickets and beetles make up about 60 percent. In winter, up to 40 percent of the diet is seeds and vegetation.

DEN: The nest burrow is U-shaped, about 4 cm in diameter and located 15 cm below the surface. The entrance is plugged by day to retain moisture. This mouse may also nest under vegetation.

REPRODUCTION: Females may bear two or three litters between March and August. A female's first pregnancy lasts about one month, but subsequent litters are typically born 32 to 38 days after mating. The three or four naked, blind young weigh 1.7 to 2.7 g each at birth. The incisors begin to erupt at nine days, the eyes open at two to three weeks and weaning follows by day 24.

SIMILAR SPECIES: The **Deer Mouse** (p. 267) has similar colouration but lacks the burly proportions and has a longer, thinner tail.

RANGE: The Northern Grasshopper Mouse ranges through much of the Prairies and Great Plains, the Rocky Mountains and the Great Basin, from southern Alberta, Saskatchewan and Manitoba south into northern Mexico.

Total Length: 13–19 cm

Tail Length: 3–4 cm

Weight: 20–52 g

Bushy-tailed Woodrat
Neotoma cinerea

are likely attracted to males who have secure nests, and several females may be found nesting with a single male.

While most people have heard of "packrats," few people realize that this nickname applies to woodrats. The Bushy-tailed Woodrat is a fine example of a packrat—it is widely known for its habit of collecting all manner of objects into a heap. This rodent is also sometimes called a "trade rat," because it is nearly always carrying something in its teeth, only to drop that item to pick up something else instead. Thus, camping gear, false teeth, tools or jewellery may disappear from a campsite, with a stick, bone or pine cone kindly left in their place.

Bushy-tailed Woodrats tend to nest in rocky areas, and because the nests are large and messy, woodrat homes are easier to find than the residents. The places in which woodrats can build their nests are limited, and rival males fight fiercely over prime locations. Female woodrats

The Bushy-tailed Woodrat likely has the longest whiskers proportionally of any rodent in the country. Extending well over the width of the animal's body on either side, a woodrat's whiskers serve it well as it feels its way around in the darkness of caves, mines and the night. This woodrat is most active after dark, so a late-night prowl with flashlight in hand may catch the reflective glare of woodrat eyes as the animal forages or investigates its territory.

DESCRIPTION: This woodrat has large, protruding, black eyes, big, fur-covered ears and long, abundant whiskers. The back is grey, pale pinkish or grizzled brown. The belly is white. The long, soft, dense, buffy fur is underlain by a short, soft undercoat. The long, bushy, almost squirrel-like tail is grey above and white below. There are distinct juvenile and

RANGE: The Bushy-tailed Woodrat is the most northerly woodrat. Its range extends from the southern Yukon southeast to western North Dakota and south to central California and northern New Mexico.

Total Length: 28–46 cm

Tail Length: 11–22 cm

Weight: 80–510 g

subadult pelages—a juvenile's back is grey and it has short tail hairs, whereas a subadult has brown hues in its back and its tail has bushy guard hairs. The tawny adult pelage develops in autumn.

HABITAT: The Bushy-tailed Woodrat's domain usually includes rocks and shrubs or abandoned buildings, mine shafts or caves. It has a greater elevational range than other woodrats and can be found from grasslands to the alpine.

FOOD: Leaves of shrubs are probably the most important component of the diet, but conifer needles and seeds, juniper berries, mushrooms, fruits, grasses, rootstocks and bulbs are all eaten or stored for later consumption. To provide adequate winter supplies, a woodrat must gather food into several caches. One of the largest caches known weighed 50 kg.

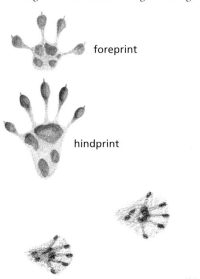

foreprint

hindprint

walking trail

DID YOU KNOW?

When a very old woodrat nest in an old cabin near the Banff Springs Hotel was torn apart some years ago, a collection of silverware dating back to the earliest days of the hotel was found.

DEN: Large numbers of sticks, plus a wide variety of bark, dung and other materials, are piled in talus or a rock cleft near the nest site. There are often no inner passages or chambers in this accumulation. Instead, a lined, ball- or cup-shaped nest is built of fibrous material and is situated nearby, usually more than 3 m above the ground, either in a narrow crevice, in the fork of a tree, on a shelf or sometimes in a stove in an abandoned cabin.

REPRODUCTION: Mating usually takes place between March and June. Following a 27- to 32-day gestation, three or four helpless young are born. They are 12 to 18 g at birth, and their growth is rapid. Special teeth help them hold onto their mother's nipples almost continuously. Their incisors erupt at 12 to 15 days, and their eyes open on day 14 or 15. The young first leave the nest at about 22 days and are weaned at 26 to 30 days. They reach sexual maturity the spring following their birth. Some females bear two litters in a season.

SIMILAR SPECIES: The **Norway Rat** (p. 272) is a similar size but does not have a bushy tail. It tends to live near areas of human activity. The **American Pika** (p. 373) has a shorter muzzle and no visible tail.

Norway Rat

Norway Rat
Rattus norvegicus

It is said that absence makes the heart grow fonder, but it's a sure bet that no one misses rats. The cold, northern regions of Canada are inhospitable to the Norway Rat, but this rodent is common in developed areas and farmland to the south. Everywhere that Norway Rats occur, they are subject to public scorn and intense pest control measures. This species is not native to our continent.

Norway Rats were introduced to North America in about 1775, and they have established colonies in most cities and towns south of the boreal forest. These pests feed on a wide variety of stored grain, garbage and carrion, gnaw holes in walls and contaminate stored hay with urine and feces. They have also been implicated in the transfer of diseases to both livestock and humans.

The geography and climate of much of Canada help limit the spread of rats through the country. Rats can disperse 5 to 8 km in summer, but if they are unable to find shelter in a building or garbage dump, winter temperatures of –18°C will prove fatal. The greatest influx of rats in most regions comes courtesy of modern transportation. Rats hitchhiking on trucks and trains are of concern because they often get deposited in warm buildings in cities and towns in previously rat-free areas. Rats are even transported in this way into Alberta, a province that fights an ongoing battle to maintain its rat-free status.

Perhaps more so than any other creature, the Norway Rat is viewed with disgust by most people. As one of the world's most studied and manipulated animals, however, much of our biomedical and psychological knowledge can be directly attributed to experiments involving the Norway Rat—a rather significant contribution for such a hated pest.

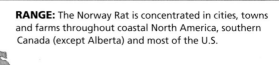

RANGE: The Norway Rat is concentrated in cities, towns and farms throughout coastal North America, southern Canada (except Alberta) and most of the U.S.

Total Length: 33–46 cm

Tail Length: 12–22 cm

Weight: 200–480 g

ALSO CALLED: Common Rat, Brown Rat, Water Rat, Sewer Rat.

DESCRIPTION: The back is grizzled brown, reddish brown or black. The paler belly is greyish to yellowish white. The long, round, tapered tail is darker above, lighter below and is sparsely haired and scaly. The prominent ears are covered with short, fine hairs. Occasionally, someone releases an albino, white or piebald Norway Rat that was kept in captivity.

HABITAT: Norway Rats nearly always live in proximity to human habitation, and they can only live away from human structures in the warmest parts of the country. In areas where they are found away from humans, they prefer thick vegetation with abundant cover. Abandoned buildings in the wilderness are more frequently occupied by native rodents than by Norway Rats.

DID YOU KNOW?

Some historians attribute the end of the Black Death epidemics in Europe to the southward invasion of the Norway Rat and its displacement of the less-aggressive, plague-carrying Black Rat, which was much more apt to inhabit human homes.

FOOD: This rat eats a wide variety of grains, insects, garbage and carrion. It may even kill the young of chickens, ducks and other small animals. Green legume fruits are also popular items, and some shoots and grasses are consumed.

DEN: A cavity scratched beneath a fallen board or a space beneath an abandoned building may hold a bulky nest of grasses, leaves and often paper or chewed rags. Although Norway Rats are able to, they seldom dig long burrows.

REPRODUCTION: After a gestation of 21 to 22 days, 6 to 22 pink, blind babies are born. The eyes open after 10 days. The young are sexually mature in about three months. Norway Rats seem to breed mainly in the warmer months of the year, but in some large cities, they may breed year-round.

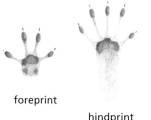

foreprint

hindprint

walking trail

Black Rat

SIMILAR SPECIES: The **Black Rat** (p. 274) has a slightly smaller body and a proportionally longer tail. The **Muskrat** (p. 295) is larger and has a laterally compressed tail.

Black Rat
Rattus rattus

The Black Rat is a more common stowaway on ships than the Norway Rat, so it is continually reintroduced at seaports such as Vancouver and Victoria. It lives in a wild state on Vancouver Island and Haida Gwaii (Queen Charlotte Islands), where it is abundant. Although this rat can be pitch black, as its name suggests, it can also be brownish with lighter underparts.

DESCRIPTION: The upperparts are brownish or sooty grey, contrasting with the greyish to whitish underparts. The ears are large and prominent. The tail, which is longer than the body, is scaly, sparsely haired and uniformly dark.

HABITAT: This introduced rat is typically found in and around human structures.

Where it overlaps with the more aggressive Norway Rat, the Black Rat typically occupies the upper levels of a building. In the wild, it occurs in second-growth forests and along forest edges.

FOOD: This rat eats a wide variety of grains, vegetation, seeds, nuts, insects, garbage and carrion.

DEN: The Black Rat will nest in a human-made structure or in a hollow log or stump. It occasionally nests in the crotch of a tree, where it builds a home of twigs, leaves and bark.

REPRODUCTION: This rat can breed year-round, but in Canada, it is more likely to breed during the warmer months. The gestation period is about 21 days, and the litter size is two to eight young.

SIMILAR SPECIES: The **Norway Rat** (p. 272) is generally more robust and has a proportionally shorter tail.

RANGE: An invasive species worldwide, in North America, the Black Rat is primarily found around human habitation in the southern and coastal U.S. It ranges north along the Atlantic Coast to Maine and along the Pacific Coast to British Columbia.

Total Length: 33–46 cm

Tail Length: 16–25 cm

Weight: 120–340 g

House Mouse
Mus musculus

Thanks to its habit of catching rides with humans, first aboard ships and now in train cars, trucks and containers, the House Mouse is found in most countries of the world. In fact, its dispersal closely mirrors the agricultural development of our species. As humans began growing crops on the great sweeping plains of Central Asia, this mouse, native to that region, began profiting from our storage of surplus grains and our concurrent switch from a nomadic to a relatively sedentary lifestyle. It is not native to North America.

Within the span of a few hundred human generations, farmed grains began to find their way into Europe and Africa for trade. Along with these grain shipments, stowaway House Mice spread to every corner of the globe. Even in the parts of Canada where the climate is harsh, they are found wherever humans provide free room and board. Unlike many of the introduced animals in the country, however, these mice seem to have had a minimal negative impact on native animal populations.

Most people who have spent some time on farms or in warehouses, unkempt places and university labs are familiar with the House Mouse. The white mice commonly used as laboratory animals are an albino strain of this species.

RANGE: The House Mouse is widespread in North America, inhabiting nearly every city, hamlet or farm from the Pacific to the Atlantic and north to the tundra.

Total Length: 13–20 cm

Tail Length: 6–10 cm

Weight: 14–25 g

DESCRIPTION: The back is yellowish brown, grey or nearly black, the sides may have a slight yellow wash, and the underparts are light grey. The nose is pointed and surrounded by abundant whiskers. The ears are large and almost hairless, and the black eyes protrude. The long, tapered tail is hairless, grey and slightly paler below than above. The brownish feet tend to be whitish at the tips.

HABITAT: This introduced mouse inhabits houses, outbuildings, barns, granaries, haystacks and trash piles. It cannot tolerate temperatures below −10°C around its nest and seems to be unable to survive winters in the northern forests without access to heated buildings or haystacks. In summer, it may disperse slightly more than 3 km from its winter refuge into fields and grasslands, only to succumb the following winter. Deer Mice are far more likely to invade wilderness cabins than House Mice are.

FOOD: Seeds, stems and leaves constitute the bulk of the diet, but insects, carrion and human food, including meat and milk, are eagerly consumed.

DID YOU KNOW?

The word "mouse" is probably derived from the Sanskrit *mus*—also the source, via Latin, of the genus name—which itself came from the word *musha*, meaning "thief."

DEN: The nest is constructed of shredded paper and rags, or sometimes fur and vegetation, combined into a 10-cm ball beneath a board, inside a wall, in a pile of rags or in a haystack. These mice may be found at any level in a building. They sometimes dig short tunnels but generally do not use them as nest sites.

REPRODUCTION: If abundant resources are available, as in a haystack, breeding may occur throughout the year, but populations away from human habitations seem to breed only during the warmer months. Gestation is usually three weeks but may be extended to one month if the female is lactating when she conceives. The litter usually consists of four to eight helpless, pink, jellybean-shaped young. Their fur begins to grow in two to three days, the eyes open at 12 to 15 days, and they are weaned at 16 to 17 days. The young become sexually mature at six to eight weeks of age.

SIMILAR SPECIES: The **Deer Mouse** (p. 267) and the **White-footed Mouse** (p. 266) have bright white underparts, dark dorsal stripes and istinctively bicoloured tails. **Voles** (pp. 277–94) have shorter, furred tails.

Deer Mouse

Southern Red-backed Vole
Clethrionomys gapperi

This attractive little vole is active both day and night and can be heard rustling in the leaf litter of forests throughout much of Canada. It is almost never seen, however, because it scurries along on its short legs through almost invisible runways on the forest floor.

The Southern Red-backed Vole is a classic example of a subnivean wanderer, a small mammal that lives out cold winters between the snowpack and the frozen ground. The snow's insulating qualities create a layer at ground level within which the temperature is nearly constant. This vole does not cache food; instead, it forages widely under the snow for vegetation or any other digestible items.

DESCRIPTION: The reddish dorsal stripe makes this one of the easiest voles to recognize. On rare occasions, the dorsal stripe is a rich brownish black or even slate brown. The sides are greyish buff, and the underparts and feet are greyish white. Compared with most voles, the black eyes seem small and the nose looks slightly more pointed. The short, slender, scantily haired tail is grey below and brown above. The rounded ears project somewhat above the thick fur.

HABITAT: This vole is found in a variety of habitats including damp coniferous forests, bogs, swampy land and sometimes drier forests.

FOOD: Green vegetation, grasses, berries, lichens, seeds and fungi form the bulk of the diet.

DEN: Summer nests are made in shallow burrows, rotten logs or rock crevices and are lined with fine materials such as dry grasses, mosses and lichens. Winter nests are subnivean.

REPRODUCTION: Mating occurs between April and October. Following a gestation of 17 to 19 days, two to eight (usually four to seven) pink, helpless young are born. By two weeks, they are well furred and their eyes are open. Once the young are weaned at about 17 days, they are no longer permitted in the vicinity of the nest. This vole reaches sexual maturity at two to three months of age.

SIMILAR SPECIES: The **Northern Red-backed Vole** (p. 278) has a tail that is reddish above and tawny below. Other **voles** (pp. 279–94) lack the reddish dorsal stripe.

RANGE: This vole is widespread across most of the southern half of Canada and south through the Cascade and the Rocky mountains, as far as northern New Mexico and through the Appalachians to North Carolina.

Total Length: 12–16 cm

Tail Length: 3–6 cm

Weight: 12–43 g

Northern Red-backed Vole
Clethrionomys rutilus

The rounded ears extend only slightly above the long hair on the head. The sides, flanks and cheeks are yellowish orange, and the underparts are creamy to greyish white. About 10 percent have a brownish grey instead of a reddish back.

HABITAT: These voles prefer the short birch, willow and alder thickets of the northern boreal forest. Although their homes are often in damp, mossy areas, they avoid standing water.

The bright colour, long, lax fur, protruding, beady, black eyes and short tail of the Northern Red-backed Vole combine to make a very handsome little creature. This vole seldom bites when handled, but it may not be so gentle when it comes to its relatives—the southern limit of this vole's range closely matches the northern limit of the Southern Red-backed Vole's range, suggesting strong competition between the two.

There is some evidence that these voles store food for winter. They do not hibernate, instead remaining active beneath the snow in runways cut in the moss. This species is undoubtedly an important prey item for owls, as well as for American Martens and other weasels.

DESCRIPTION: A broad, reddish stripe extends from the forehead to the rump. The tail is reddish above and tawny below.

FOOD: Most of the diet consists of berries, buds, leaves and twigs. This vole does not eat mosses or lichens.

DEN: The nest of moss and shredded vegetation is located beneath rocks, in a rotten tree or in a short burrow.

REPRODUCTION: Breeding usually begins in May. Following a 17- to 19-day gestation, 4 to 10 helpless, blind, jellybean-shaped young are born, each weighing about 2 g. They nurse persistently, and after their eyes open on day 10 or 11, they leave the nest. Some females breed in their first summer.

SIMILAR SPECIES: The **Southern Red-backed Vole** (p. 277) has a tail that is grey below and brown above. Other **voles** (pp. 279–94) lack the reddish dorsal stripe.

RANGE: The Northern Red-backed Vole occurs across Alaska, south into northern British Columbia and east to the western shore of Hudson Bay.

Total Length: 12–14 cm

Tail Length: 2–4 cm

Weight: 17–30 g

Western Heather Vole
Phenacomys intermedius

The Western Heather Vole generally occupies the alpine tundra, but it may descend to the same northern woodlands as the Northern Red-backed Vole, and skulls of both species are frequently found in the same owl pellets.

This vole's common name refers to its preference for high-elevation environments where heathers are common. As well, it may feed on the inner bark of heathers. It consumes a high percentage of bark seasonally, and it has a cecum (a functional appendix) with modified, 1-cm-long intestinal villi that assist in digesting this fibrous and lignin-rich food.

eats the inner bark of various shrubs from the heather family.

DESCRIPTION: The short, thin, bicoloured tail is slate grey above, sometimes with a few white hairs, and white below. The tops of the feet are silvery grey, and the belly hairs have pale tips, giving the underparts a light grey hue. Dorsal colours vary, but the most common is a grizzled buffy brown. The roundish ears scarcely extend above the fur. There is tawny or orangey hair inside the front of the ears.

HABITAT: This vole seems to prefer open areas in a variety of habitats in the mountains, including alpine tundra and coniferous forests.

FOOD: The Western Heather Vole feeds primarily on green vegetation, grasses, lichens, berries, seeds and fungi. It also

DEN: The summer nest is made in a burrow up to 20 cm deep and is lined with fine, dry grasses and lichens. In winter, the nest is built on the ground in a snow-covered runway.

REPRODUCTION: Mating occurs between April and October. Following a gestation of about three weeks, one to eight (usually four or five) pink, helpless young are born. By two weeks of age, they are well furred, and their eyes are open. Females become sexually mature at four to six weeks of age and may breed in their first year.

SIMILAR SPECIES: The **Eastern Heather Vole** (p. 280) has a different range. The **Southern Red-backed Vole** (p. 277) has a distinctive, reddish dorsal stripe and a longer tail. The larger **Meadow Vole** (p. 288) has blackish fur on the tops of its feet.

RANGE: The Western Heather Vole occurs from northwestern British Columbia south through the western mountains to central California and northern New Mexico.

Total Length: 11–16 cm

Tail Length: 2–4 cm

Weight: 25–50 g

Eastern Heather Vole
Phenacomys ungava

HABITAT: These voles occur in a variety of habitats, primarily open coniferous forests and shrubby areas on forest edges. Birch and willow thickets often attract this species.

FOOD: The green foliage of shrubs and forbs forms the bulk of the diet in summer. The bark and buds of shrubs form most of the winter diet.

Stories abound among naturalists of how gentle this little vole is when captured. Among the variety of explanations for this tranquillity, the most unusual is that diet may play an important role in temperament. These voles feed heavily on willow bark, and certain compounds in the bark and leaves of willows are known to have calming and even analgesic properties. (Willows are the source of salicylic acid, the natural precursor to aspirin.) Other animals, such as the ptarmigan, also feed on willow bark, and they, too, have gentle demeanours.

DESCRIPTION: This vole is greyish brown over its back, with light grey to steel grey underparts. The tops of its feet are silvery grey. Generally, there are a few orange hairs at the base of the ears. The eyes are small, the ears are short and rounded, and the face is yellowish. The fur is long and silky, and the tail is white beneath and grey above.

DEN: The nest is made of heather twigs and lichens gathered into a ball 15 cm in diameter and located in a sheltered place. The natal den is in a burrow in a rocky area, under a log, beneath a root or stump or at the base of a shrub.

REPRODUCTION: Females mate from May to August. After a gestation of 19 to 24 days, two to eight altricial young are born. Their eyes open at 14 days, whereupon they are weaned and begin eating vegetation. Females can mate as young as six weeks of age. Males mate after their first winter.

SIMILAR SPECIES: The **Western Heather Vole** (p. 279) has a different range. The **Southern Red-backed Vole** (p. 277) has a distinct reddish dorsal stripe and a longer tail. The larger **Meadow Vole** (p. 288) has blackish fur on the tops of its feet. Other voles may be difficult to distinguish without a specimen and a key in hand.

RANGE: The range of the Eastern Heather Vole closely corresponds to the range of the boreal forest from the Yukon down the eastern edge of the Rockies and across to Labrador.

Total Length: 11–16 cm

Tail Length: 2–4 cm

Weight: 25–50 g

Rock Vole
Microtus chrotorrhinus

A common—and usually true—assertion is that voles make up a large percentage of the mammal population in nearly any natural community. Alas, the Rock Vole seems to have difficulty achieving ubiquitous status. This vole has been identified in Pleistocene deposits, and in each case of fossil documentation, it was the least common microtine rodent. Even today, when biologists trap microtines, the Rock Vole is present only in one-third the numbers of the other vole species. This low proportion of individuals seems to be consistent over time, even in the most favourable habitats. No one knows what causes this natural check on Rock Vole populations.

ALSO CALLED: Yellow-nosed Vole.

DESCRIPTION: This medium-sized vole is tawny or brown above and grey below. The relatively long tail is similarly coloured. The nose is usually yellowish, orangey or slightly pink—a helpful field mark.

HABITAT: The Rock Vole inhabits rocky areas in moist deciduous woodlands and some disturbed areas. It requires water near its home.

FOOD: This vole feeds heavily on bunchberries, other fruit, grasses, leaves and fungi. It also takes insects and their larvae.

DEN: These voles form extensive surface runways and subterranean burrows. Nests are either inside a shallow burrow or in a rock crevice. Separate burrow chambers or designated surface areas are used as latrines.

REPRODUCTION: Mating occurs several times between early spring and late autumn. Females may have two or three litters per year, with one to seven young per litter. The young are altricial at birth.

SIMILAR SPECIES: The **Meadow Vole** (p. 288) has a shorter tail. Other voles with overlapping ranges lack the yellowish orange nose.

RANGE: Rock Voles are found from south-central Ontario and extreme northeastern Minnesota east to Labrador and south to New York and parts of New England. An isolated population occurs in the Great Smoky Mountains.

Total Length: 14–19 cm

Tail Length: 4.2–6.4 cm

Weight: 30–48 g

Long-tailed Vole
Microtus longicaudus

mountain slopes, coniferous forests, alpine tundra and among alders or willows near water. It does not follow well-defined trails and ranges widely at night.

FOOD: Summer foods consist of green leaves, grasses and berries. In winter, this vole eats the bark of heathers, willows and other trees.

Long-tailed Voles inhabit much of Western Canada, though with an unusual distribution pattern. These voles are among the alpine elite, thriving above treeline in the mountain parks, but they also live among their flatland kin on grassland plateaus. In both communities, Long-tailed Voles choose to live in wet meadows with stunted thickets. They are the only voles on most central and northern islands on the West Coast.

DESCRIPTION: The upperparts vary in colour, ranging from grizzled greyish to dark grey-brown, but the black tips on the guard hairs may give this vole a dark appearance. The sides are paler than the back, and the underparts are paler still. The tail is about 6 cm long and indistinctly bicoloured. The tops of the feet are grey.

HABITAT: This vole lives in a variety of habitats, including moist, grassy areas,

DEN: The simple burrows made by this vole under logs or rocks are often poorly developed. The nest chamber is lined with fine, dry grasses, mosses or leaves. Winter nests are subnivean.

REPRODUCTION: Mating is presumed to occur from May to October. Females often have two litters of two to eight (usually four to six) young per year. Gestation is about three weeks. The young are helpless at birth, but at about the same time their eyes open at two weeks of age, they are weaned and leave the nest. Some young females have their first litter when they are only six weeks old.

SIMILAR SPECIES: The **Meadow Vole** (p. 288) has a shorter tail and dark feet. Most other voles with overlapping ranges have shorter tails.

RANGE: The Long-tailed Vole ranges south from eastern Alaska and the Yukon along the Rocky Mountains to New Mexico and Arizona. From this eastern limit, its range extends west to the Pacific as far south as California.

Total Length: 17–23 cm

Tail Length: 6–7 cm

Weight: 35–57 g

Singing Vole
Microtus miurus

True to its name, the Singing Vole actually sings. Or, at least, it makes a metallic, trilling sound that is unusual enough among voles as to warrant being called singing. While other voles, such as the Meadow Vole, may make faint, squeaky sounds, the Singing Vole's vocalization is unique. The sound is believed to be a warning call or perhaps territorial.

Another characteristic of this vole is its rather large "haystacks." Many voles collect piles of neatly cut grasses and vegetation along their runways, but this vole has been known to make haystacks as large as 30 L in volume.

ALSO CALLED: Alaska Vole, Alaska Haymouse.

DESCRIPTION: This small, short-tailed vole is greyish brown with slightly lighter buffy grey sides and underparts. Its short tail is similarly coloured. The distinct ears are rounded. The feet have noticeably long claws.

HABITAT: This vole lives almost exclusively in Arctic tundra, alpine and subalpine habitats. It prefers areas with rocky cover and well-drained soil.

FOOD: Singing Voles probably feed on a variety of available vegetation, especially willows, lupines and sedges.

DEN: This vole excavates burrows under rocks in well-drained soil. It also forms a network of worn runways. It nests in a specific chamber within its burrow system.

REPRODUCTION: Little is known about the mating behaviour of Singing Voles, but presumably mating occurs in spring. Gestation is probably about 22 days, and litter size ranges from two to nine.

SIMILAR SPECIES: Other *Microtus* voles in the same region, such as the **Yellow-cheeked** (p. 293), **Meadow** (p. 288) and **Tundra** (p. 286) **voles**, all have longer tails.

RANGE: This vole has a horseshoe-shaped range that extends across northern Alaska, into much of the Yukon and some of the Northwest Territories, and then back into southern Alaska.

Total Length: 12.5–16.8 cm

Tail Length: 2.0–3.6 cm

Weight: 23–60 g

Montane Vole
Microtus montanus

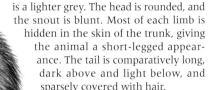

is a lighter grey. The head is rounded, and the snout is blunt. Most of each limb is hidden in the skin of the trunk, giving the animal a short-legged appearance. The tail is comparatively long, dark above and light below, and sparsely covered with hair.

HABITAT: This vole is found in mountain meadows, valleys and arid sagebrush communities.

FOOD: Green shoots form the majority of the diet when they are available. At other times of the year, seeds or even bark may be eaten.

In south-central British Columbia, the Montane Vole is one of the most abundant small mammals, so if you see a vole in that region, there is a good chance it is a Montane Vole. More importantly, the high numbers of this vole indicate its ecological importance—it is a steady food supply for larger creatures, such as owls, raptors, weasels, Coyotes and more. At its highest density, this vole is reported to reach numbers of more than 1000 per hectare, though its population cycles and numbers can sometimes be as low as about 200 per hectare.

DESCRIPTION: This small, thickset mammal has short ears that are largely hidden in the fur. The dark eyes protrude. The back is brown to black, and the belly

DEN: The winter nest is often located aboveground along a well-used runway. This vole's runways and nests are easiest to observe soon after the snow melts in spring. The summer nest is in a short burrow, fallen log or at the base of a shrub.

REPRODUCTION: Montane Voles may have several litters per year, but reproduction usually takes place between spring and autumn. Gestation is about 21 days, after which six to eight young are born.

SIMILAR SPECIES: Other voles may not be readily distinguishable. The **Long-tailed Vole** (p. 282) has a longer tail. The **Meadow Vole** (p. 288) tends to be darker.

RANGE: The Montane Vole occurs from southern British Columbia to Montana, south to New Mexico and California.

Total Length: 13–18 cm

Tail Length: 3–6 cm

Weight: 15–50 g

Prairie Vole
Microtus ochrogaster

Though the Prairie Vole prefers grassland habitats in Manitoba and Saskatchewan, it seems to favour aspen parkland areas in Alberta. It seems to be less antagonistic toward other members of its own species than other voles, and small groups have even been noted, typically during autumn and winter. These aggregations may be family groups using the same runways, tunnels and nests within a total home range that is often no larger than a tennis court.

Native peoples of the Prairies habitually gathered the Prairie Vole's autumn caches. These larders would often yield several pounds of desirable grass seeds, ground beans and Jerusalem artichoke rhizomes.

DESCRIPTION: The upperparts are dull grey with cinnamon highlights. Many of the long, lax dorsal hairs have light tips that produce a salt-and-pepper effect. The underparts are light grey or sometimes silvery grey. The legs are short, and the short, rounded ears scarcely protrude from the fur. The tail, which is about twice as long as the hindfoot, is indistinctly bicoloured.

HABITAT: The preferred habitats are arid grassland regions and sagebrush flats, though it may inhabit some moist grassland areas.

FOOD: The new green shoots of grasses and the flowers and leaves of forbs comprise the bulk of the spring and summer diet. In autumn and winter, this vole relies on ripened fruits, seeds, bulbs, roots, corms and the inner bark of shrubs.

DEN: This vole digs shallow but extensive burrows in damp soils following rains. The tunnels lead to a grass-lined nest chamber, which is the hub of the network of burrows and runways.

REPRODUCTION: Mating occurs mainly between April and September. After a gestation of about three weeks, a litter of three or four helpless young is born. The young grow extremely rapidly and are weaned and out of the nest before they are three weeks old.

SIMILAR SPECIES: The **Meadow Vole** (p. 288) has a slightly longer tail and greyish underparts.

RANGE: This open-country vole occurs from east-central Alberta southeast to Oklahoma, Tennessee and West Virginia.

Total Length: 12–17 cm

Tail Length: 2.5–4.1 cm

Weight: 20–58 g

Tundra Vole
Microtus oeconomus

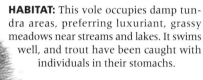

HABITAT: This vole occupies damp tundra areas, preferring luxuriant, grassy meadows near streams and lakes. It swims well, and trout have been caught with individuals in their stomachs.

FOOD: Sedge and grass clippings make up most of the summer diet. Grass seeds and the rhizomes of licorice root and knotweed are stored in autumn for winter use.

DEN: This vole digs a shallow burrow in loose soil above the permafrost; a pile of dirt may be apparent at the entrance. The bulky nest, located inside a hummock, is about 15 cm in diameter and is made of dried vegetation and lined with grasses. A series of runways with neat latrines in side branches is often visible.

REPRODUCTION: The female bears two or three litters between May and early September. Usually there are 5 to 11 young in a litter. The gestation period and maturity rates are not known for this species but are probably similar to those of other similar-sized voles.

SIMILAR SPECIES: The **Meadow Vole** (p. 288) usually lacks the yellowish tinge to the fur. **Lemmings** (pp. 297–304) have much shorter tails. This vole is often found in association with the **Long-tailed Vole** (p. 282).

The Tundra Vole prefers wet tundra areas in northwestern parts of the country. Few other vole species are as well suited to the short summers and long winters of this region, so the aptly named Tundra Vole is most commonly seen in these northern climes. It appears to be dominant within this habitat, even compared to the larger Brown Lemming. Like other small rodents of the tundra, it is an important prey species for predators such as raptors and foxes.

DESCRIPTION: The medium-sized Tundra Vole has a short tail and short ears. The tail is dark brown above, pale grey beneath and well furred. The long, lax fur is grizzled brown on the back and buffy grey beneath. The flanks and rump may be slightly yellowish brown. The winter hair is longer and silkier than the summer coat.

RANGE: The Tundra Vole is found throughout Alaska, the western Yukon, some of the western Northwest Territories and Nunavut, and just slightly into northwestern British Columbia.

Total Length: 15–19 cm

Tail Length: 4–5 cm

Weight: 25–70 g

Creeping Vole
Microtus oregoni

Decades of walking through prime Creeping Vole habitat will, on very rare occasions, produce encounters with this secretive animal. "Creeping" probably refers to this vole's preference for underground burrows and runways. This kind of concealment means that seeing one in the wild is very unlikely. The Creeping Vole is also the smallest vole in its range, making it that much more difficult to spot.

DESCRIPTION: This small, slender vole appears to have very plush fur, because the guard hairs are about as long as the underfur. It is dull or sooty brown above and grey below. The feet and tail are dark grey. The eyes and ears are small.

HABITAT: This species prefers grassy openings in moist coniferous forests at sea level and in the mountains. It makes burrows in crumbly woodland soils and mossy bogs. It is also found in brushlands beside cultivated fields.

FOOD: The Creeping Vole eats a wide variety of plant stems and roots. Potatoes and fallen apples are favourites when available, and moles are often blamed for the vole's garden pilfering.

DEN: The nest is a mass of grasses or fine vegetation located beneath a log or rotten stump. Numerous tunnels just below the soil surface extend out from the nest. Most of the time this vole lives a subterranean existence and often uses burrows of the Coast Mole.

REPRODUCTION: Mating begins in March, and after a gestation of 23 to 24 days, one to five naked, pink, blind young are born. A female may have up to five litters in a summer, but three are more common. The young are weaned at 13 days and then disperse. The females are sexually mature at about 35 days and males at about 45 days. Very few Creeping Voles live for a whole year.

SIMILAR SPECIES: The **Long-tailed Vole** (p. 282) and the **Townsend's Vole** (p. 292) have longer tails. The **Meadow Vole** (p. 288) is larger.

RANGE: The Creeping Vole is a species of the Pacific Northwest, ranging from southwestern British Columbia to northern California.

Total Length: 12–15 cm

Tail Length: 3–4 cm

Weight: 16–23 g

Meadow Vole
Microtus pennsylvanicus

The Meadow Vole is well adapted to the winters of Canada. It is subnivean in winter, meaning that it remains active beneath the snow but above the ground. When the snow melts in spring, an elaborate network of Meadow Vole activity is exposed. Runways, chambers and nests, previously insulated from winter's cold by deep snows, await the growth of spring vegetation to conceal them once again.

These voles are an important prey species—many of them die in their first months, and very few live as long as a year. With two main reproductive cycles per year, it is unlikely that many voles get to experience all the seasons.

DESCRIPTION: The upperparts vary from brown to blackish, and the underparts are grey. The beady eyes are small and black. The rounded ears are mostly hidden in the long fur of the rounded head. The tops of the feet are blackish brown, and the tail is about twice as long as the hindfoot.

HABITAT: The Meadow Vole can be found in a variety of habitats, provided that grasses are present. Grasslands, pastures, marshy areas, open woodlands and tundra are all potential homes for this vole.

FOOD: The green parts of sedges, grasses and some forbs make up the bulk of the spring and summer diet. In winter, seeds, some bark and insects are eaten. Other foods include grains, roots and bulbs.

DEN: The summer nest is made in a shallow burrow and lined with fine materials, such as dry grasses, mosses and lichens. The winter nest is subnivean.

REPRODUCTION: Spring mating occurs between late March and the end of April. Gestation is about 20 days, and the average litter size is four to eight young. From birth, the helpless young nurse almost constantly to support their rapid growth. Their eyes open in 9 to 12 days, and they are weaned at 12 to 13 days. At least one more litter is born, usually in autumn.

SIMILAR SPECIES: The Meadow Vole is difficult to distinguish from the other voles found in Canada, but no other vole has such dark feet.

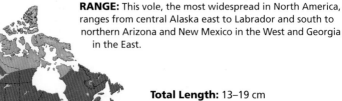

RANGE: This vole, the most widespread in North America, ranges from central Alaska east to Labrador and south to northern Arizona and New Mexico in the West and Georgia in the East.

Total Length: 13–19 cm

Tail Length: 3–5 cm

Weight: 30–64 g

Woodland Vole
Microtus pinetorum

The Woodland Vole can be found only in a few isolated areas in the Carolinian Forest region. Even though its range is small, it is very numerous where it occurs in southern Ontario and Québec. Because of its limited distribution and its tendency to live a somewhat subterranean existence, you are not likely to see one of these handsome voles. If you do see one, it will look very similar to the much more common Southern Red-backed Vole, even though they are from two different genera. The two voles, however, are not likely to have any range overlap, so identification should be easy.

DESCRIPTION: This vole is quite small and has a short tail. It is faintly tricoloured, with a brownish back, reddish sides and greyish belly. Its short ears and small eyes are largely hidden in the fur. It has long whiskers, a rounded head and a blunt snout. The body appears cylindrical, with short limbs. Some individuals are greyish overall.

HABITAT: This vole is found mainly in deciduous wooded or thick, shrubby areas, especially where sandy soils are available for burrowing. Sometimes individuals are found in heavily grassed areas.

FOOD: The Woodland Vole feeds mainly on forbs, grasses, roots, seeds, fruit, bark and sometimes insects and fungi. The choice of food varies depending on seasonal availability. It stores large amounts of food in underground burrows.

DEN: This vole digs burrows, and its nest chamber is located in the burrow and typically has multiple exits. The nest is lined with dried grasses, leaves and roots.

REPRODUCTION: The Woodland Vole has one to four litters per year, and reproduction takes place from May to October. Gestation is about 21 days, and the average litter size is two or three young.

SIMILAR SPECIES: The **Southern Red-backed Vole** (p. 277) has a longer tail, is reddish on the back rather than on the sides and has a different range. The **Meadow Vole** (p. 288) lacks the reddish colour.

RANGE: The Woodland Vole is found throughout the eastern U.S., just into southwestern Ontario in the Carolinian Forest region and into southern Québec.

Total Length: 11–15 cm

Tail Length: 1.2–2.9 cm

Weight: 19–39 g

Water Vole
Microtus richardsoni

If you linger next to streams when hiking high-elevation trails in the mountain parks, you may have an opportunity to become familiar with the Water Vole. This large vole is like a small alpine Muskrat in many ways because it dives and forages with ease along the icy snowmelt creeks. Unfortunately, it is almost exclusively nocturnal, so you are more likely to see its distinctive signs than the animal itself.

The Water Vole's diagnostic, well-worn runways crisscross the margins of alpine streams, connecting the burrows and foraging areas of small colonies. These damp pathways often run under the mat of roots and plant debris on the ground surface.

Vegetation cuttings often line the paths. This semi-aquatic vole's burrows may be in such close proximity to the water that each rainfall brings the potential of flooding.

Water Voles appear to abandon these tunnel networks through the winter months, remaining adequately protected from the winter chill by the snows that deeply cover and insulate their habitat. Like other voles, they are heavily depredated, and few of them survive more than one winter.

DESCRIPTION: This large vole is brownish black above, with paler grey sides. The belly is grey with a brownish white wash.

RANGE: The Water Vole occurs in two distinct populations, each associated with mountains. The western population extends along the Cascade Mountains from central British Columbia to southern Oregon. The eastern population occurs in the Rocky Mountains from central Alberta to south-central Utah.

Total Length: 19–
Tail Length: 5–10 cm
Weight: 30–120 g

The fur is thick, with an abundant, water-repelling undercoat. The extremely long hindfeet aid in swimming. The tail is indistinctly bicoloured, blackish above and dark grey below. The ears are rounded and scarcely extend above the thick fur. The eyes are small, black and protruding.

HABITAT: True to its name, this vole lives primarily along alpine and subalpine streams and lakes. It favours clear, swift, gravel-bottomed streams bordered with mixed stands of low willows and dense herbage.

FOOD: In summer, Water Voles feed on the culms of various sedges and grasses, plus the leaves, stems, roots and flowers of forbs. Winter foods include the bark of willows and bog birch, various roots and rhizomes and the fruits and seeds of available green vegetation.

DEN: This vole digs extensive burrow systems, with tunnels up to 10 cm in diameter, through moist soil at the edges of streams and waterbodies. The nest chamber, which is about 10 cm high and 15 cm long, is lined with mosses and dry grasses or leaves. It is often situated under a rise, log or stump. A Water Vole will excavate and re-excavate its burrow system throughout summer. The winter nest is located farther from the water in a snow-covered runway.

REPRODUCTION: Water Voles probably breed periodically from May through September, with usually two or more litters of 2 to 10 young born each year. Gestation is at least 22 days. The young are helpless at birth but grow rapidly, reaching maturity quickly. They may even breed during their birth year.

SIMILAR SPECIES: The Water Vole's large hindfoot, which is more than 2.5 cm long, and its generally large size distinguish it from all other voles in its range.

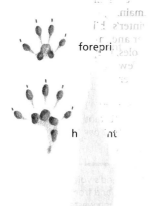

foreprint

hindprint

walking trail

Townsend's Vole
Microtus townsendii

HABITAT: This vole is found in moist fields and sedge meadows on the Fraser River delta and in similar habitats on Vancouver Island, where it also occurs in subalpine and alpine meadows.

FOOD: Townsend's Voles prefer tender marsh and grassland vegetation. They may also consume the bark of shrubs, some stems and roots of conifers and starchy roots.

The Townsend's Vole is one of largest voles in North America. As with many vole species, its populations periodically erupt and then abruptly crash. The mechanism that triggers these fluctuations is unknown.

This species' runways are used by generations of voles and may be up to 5 cm deep. In these well-used networks, the intersections often serve as latrines. In extreme cases, the pile of droppings may be 18 cm long by 8 cm wide and may create a ramp 13 cm high—an obstacle the vole simply scurries over as it travels the runways.

DESCRIPTION: This large, dark brown vole has broad ears that extend noticeably above the fur. It has a long, blackish brown tail and similarly coloured feet with brown claws. The protruding, black eyes measure more than 4 mm in diameter.

DEN: During rainy seasons, the nest is built on or above the soil surface, often on a hummock. During dry periods, a subterranean burrow system and an underground nest are maintained. The nest is made of dry grasses.

REPRODUCTION: Breeding occurs from early February until October. Following a gestation of 21 to 24 days, one to nine young are born in the grassy nest. They are weaned and leave the nest at 15 to 17 days. Young females born early will mate and bear litters their first summer. Males mature later.

SIMILAR SPECIES: The **Creeping Vole** (p. 287) is smaller. The **Long-tailed Vole** (p. 282) does not have a uniformly dark tail. The **Meadow Vole** (p. 288) has a shorter tail.

RANGE: The Townsend's Vole occurs throughout Vancouver Island and from the Lower Mainland south through western Washington and Oregon into northern California.

Total Length: 17–23 cm

Tail Length: 5–7 cm

Weight: 47–82 g

Yellow-cheeked Vole
Microtus xanthognathus

With its short ears and tail, the Yellow-cheeked Vole is well adapted to the cold taiga where it lives. When encountered, this little rodent is very active and inquisitive, chirping in the manner of a ground squirrel at the approach of an intruder. Although records exist from the northern parts of the Prairies, the current range of this vole is not well documented.

Skulls of this resolutely boreal species have been discovered in Virginia and Pennsylvania. These 11,000-year-old remains reinforce the notion that many of the species we now attribute to northern Canada were found much farther south during periods of extensive glaciation.

ALSO CALLED: Chestnut-cheeked Vole, Taiga Vole.

DESCRIPTION: The back is a dark greyish brown with intermixed coarse, black hairs. The short tail is indistinctly bicoloured, blackish on top and dusky grey below. The breast and the tops of the feet are sooty. The belly is dusky grey. A bright rusty yellowish patch appears on the nose and in front of each ear. Sometimes the cheeks and the fur around the eyes are yellowish.

HABITAT: This vole seems to favour forests in post-fire successional stages with heavy moss groundcover. It may occupy upland, grassy slopes in aspen-spruce and shrubby evergreen communities.

FOOD: In summer, this species feeds on horsetails, berries, some grasses, willows and lichens. The winter diet is from stores of the rhizomes of horsetails and fireweed.

DEN: These voles dig deep, extensive burrows and form loose colonies. Only one pair of voles occupies each burrow. The burrows have large spoil piles at the entrances. The nest chamber is lined with dry grasses, lichens or mosses.

REPRODUCTION: Mating occurs anytime from May to September, and females may produce two litters during this time. The litter of 6 to 13 helpless young is born after a gestation of about three weeks. The young grow rapidly but do not reach sexual maturity in the summer of their birth.

SIMILAR SPECIES: The **Water Vole** (p. 290) has much longer hindfeet.

RANGE: This vole's former range was from central Alaska and the Yukon east to Hudson Bay in northern Manitoba and south to central Alberta. There are, however, no recent records from Alberta or Manitoba—it appears to have abandoned the southern and eastern portions of its former range.

Total Length: 19–23 cm

Tail Length: 4.5–5.3 mm

Weight: 52–170 g

Sagebrush Vole
Lemmiscus curtatus

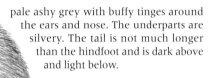

pale ashy grey with buffy tinges around the ears and nose. The underparts are silvery. The tail is not much longer than the hindfoot and is dark above and light below.

HABITAT: In keeping with its name, the Sagebrush Vole thrives in arid grassland regions and sagebrush flats.

FOOD: The spring and summer diet includes a variety of plant and some insect material, especially the new green shoots of grasses and the flowers and leaves of forbs. In autumn and winter, the diet switches to ripened fruits, seeds, bulbs, roots, corms and the inner bark of shrubs.

The Sagebrush Vole, the only member of its genus, is rather chunky and looks more like a small lemming than a vole. It tends to live in small colonies in the arid regions of southern Alberta and Saskatchewan. The arid, short-grass prairie seems to offer little concealment for even this small rodent, but its pale colours help it blend in with the surroundings. Years of very low vole densities may account for the difficulty in spotting one, but infrequent population irruptions can result in a 10-fold increase in numbers.

The prairie rattlesnake seems to have little trouble finding this vole. A recent study found that in some areas, this rodent makes up about one-fifth of the snake's diet.

DESCRIPTION: This small, stout vole has short ears and legs and long, lax hair. It is

DEN: Shallow but extensive burrows lead to grass-lined nest chambers.

REPRODUCTION: Mating occurs mainly between April and September. After a gestation period of about 25 days, a litter of 1 to 13 helpless young is born. The young grow extremely rapidly, and they are weaned and out of the nest within three weeks.

SIMILAR SPECIES: The **Long-tailed** (p. 282), **Meadow** (p. 288) and **Prairie** (p. 285) **voles** all have longer tails.

RANGE: The U-shaped range, which skirts much of western Montana and northern Idaho, extends from southern Alberta and Saskatchewan east to North Dakota, south to northern Colorado, southwest to southern Nevada and north to central Washington.

Total Length: 11–14 cm

Tail Length: 1.8–2.7 cm

Weight: 21–39 g

Muskrat
Ondatra zibethicus

Although some Muskrats spend their winters in ice-free climates, the ones found in much of Canada are restricted to living beneath the ice of their ponds and in their burrows. When the snow and ice melt, these large rodents can be seen out on dry land. In early spring, many first-year animals, now sexually mature, venture from their birth ponds to establish their own territories. These dispersing Muskrats are commonly seen travelling over land. The journey ends tragically for many—their numbers are all too easily tallied on May roadkill surveys.

The Muskrat is not a "mini-beaver," nor is it a close relative of that large rodent; rather, it is a highly specialized aquatic vole that shares many features with the American Beaver as a result of their similar environments. Like a beaver, a Muskrat can close its lips behind its large, orange incisors, which allows it to chew underwater without getting water or mud in its mouth. Its eyes are placed high on its head, and it can often be seen swimming with its head and sometimes its tail above water. The Muskrat dives with ease—it can remain submerged for more than 15 minutes and can swim the length of a football field before surfacing.

Muskrats lead busy lives. They continually gnaw cattails and bulrushes, whether they are eating the tender shoots or gathering the coarse vegetation to build homes. Muskrat homes are of tremendous importance not only to these aquatic rodents, but also to geese and ducks, which make use of Muskrat homes as nesting platforms.

RANGE: This wide-ranging rodent occurs from the southern limit of the Arctic tundra across nearly all of Canada and the lower 48 states, except most of Florida, Texas and California.

Total Length: 46–61 cm

Tail Length: 20–28 cm

Weight: 0.8–1.6 kg

Both sexes have perineal scent glands that enlarge and produce a distinctly musky discharge during the breeding season. Although this scent is by no means unique to the Muskrat, its potency is sufficiently notable to have influenced this animal's common name.

DESCRIPTION: The coat generally consists of long, shiny, tawny to nearly black guard hairs overlying a brownish grey undercoat. The flanks and sides are lighter than the back. The underparts are grey, with some tawny guard hairs. The long tail is black, nearly hairless, scaly and laterally compressed, with a dorsal and ventral keel. The legs are short. The hindfeet are large, partially webbed and have an outer fringe of stiff hairs. The tops of the feet are covered with short, dark grey hair. The claws are long and strong.

HABITAT: Muskrats occupy sloughs, lakes, ponds, marshes and streams with cattails, rushes and open water. They often occupy American Beaver ponds, and the two species coexist harmoniously.

FOOD: The summer diet includes a variety of emergent herbaceous plants. Cattails, rushes, sedges, irises, water lilies and pondweeds are staples, but a few frogs,

> **DID YOU KNOW?**
>
> Muskrats are highly regarded by First Nations peoples. In one story, it was Muskrat who brought some mud from the bottom of the flooded world to the water's surface. This mud was spread over Turtle's back, thus creating dry land for the human race.

turtles, clams, snails, crayfish and an occasional fish may be eaten. In winter, Muskrats feed on submerged vegetation.

DEN: Muskrat houses are built entirely of herbaceous vegetation, without the branches or mud of beaver lodges. The dome-shaped piles of cattails and rushes have an underwater entrance. Muskrats may also dig bank burrows, which are 5 to 15 m long and have entrances that are below the usual water level.

REPRODUCTION: Breeding takes place between March and September. Females produce two or sometimes three litters per year, depending on location. Gestation lasts 25 to 30 days, after which six to seven young are born. Their eyes open at 14 to 16 days, the young are weaned at about four weeks, and they are independent at one month. Both males and females are sexually mature the spring after their birth.

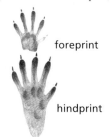

foreprint

hindprint

walking trail

SIMILAR SPECIES: The **American Beaver** (p. 305) is larger and has a broad, flat tail. Typically only its head is visible above water when it swims. The introduced **Nutria** (p. 254) is larger and has a cylindrical tail. The **Norway Rat** (p. 272) is smaller and has a cylindrical tail. The similar-sized **Woodchuck** (p. 328) has a furred tail and is not associated with water.

American Beaver

Brown Lemming
Lemmus trimucronatus

The Brown Lemming is a colourful Arctic furball that tolerates some of the most inhospitable environments in North America. Every part of its body is covered with a long coat—most appropriate for an animal that typically lives on the tundra—and this coat also provides the lemming with excellent buoyancy when it swims. The soles and toes of its feet are covered in stiff bristles, an adaptation that helps with burrowing.

DESCRIPTION: The body, ears, feet, head and stubby tail are all completely covered with long, lax fur. In summer, the lower back is chestnut coloured, grading to grizzled grey over the head and shoulders. The rump is a lighter brown, and the cheeks and sides are tawny. The underparts are primarily light grey. In autumn, the lemming moults into a longer, greyer coat. The strong, curved claws aid in digging elaborate winter runways.

HABITAT: Bogs, alpine meadows, tundra and even spruce woods may support large lemming colonies.

FOOD: Grasses, sedges and other monocots form the bulk of the diet. In times of scarce vegetation, any emergent plant is eaten down to the ground.

DEN: Summer nests are located 5 to 30 cm underground in tunnels. The nests are made of dry grasses and fur, with nearby chambers for wastes. Winter nests are subnivean.

REPRODUCTION: Breeding occurs anytime from spring through autumn, and sometimes in winter. Gestation lasts about three weeks, after which four to nine young are born. Lemmings resemble pink gummy bears at birth. Their growth is rapid—at seven days they are furred, the ears open at eight to nine days, and the eyes open at 10 to 12 days. The young are weaned at 16 to 21 days and are probably sexually mature soon thereafter.

SIMILAR SPECIES: The **Northern Bog Lemming** (p. 298) is generally smaller, with a bicoloured tail and grooved upper incisors. **Voles** (pp. 277–94) are differently coloured and typically have bicoloured tails.

RANGE: The Brown Lemming primarily inhabits Alaska, the Yukon, the Northwest Territories and Nunavut, ranging south into the Rockies of northern British Columbia and Alberta.

Total Length: 10–17 cm

Tail Length: 1–3 cm

Weight: 50–110 g

Northern Bog Lemming
Synaptomys borealis

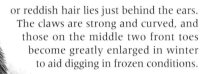

The Northern Bog Lemming is found across much of Canada. Its favoured habitat is cool sphagnum bogs, but black spruce forests and tundra sedge meadows can also host populations. This bog lemming can be found at sea level in suitable habitat along the Hudson Bay coast.

Although these animals are rarely seen, their workings are easy enough to identify. The mossy runways are frequently marked by evenly clipped grasses stacked in neat piles, like harvested trees awaiting logging trucks along haul roads. Another sign of recent Northern Bog Lemming activity is their greenish fecal pellets in little piles along the runways.

DESCRIPTION: The ears of this stout lemming scarcely project above the fur of the head. The entire body is covered in thick fur. The sides and back are usually chestnut or dark brown, and the underparts are usually greyish. A little patch of tawny or reddish hair lies just behind the ears. The claws are strong and curved, and those on the middle two front toes become greatly enlarged in winter to aid digging in frozen conditions.

HABITAT: This lemming thrives in wet tundra conditions such as tundra bogs, meadows and even spruce woods.

FOOD: The diet is mostly grasses, sedges and similar plants. If this vegetation is scarce, other emergent plants are eaten.

DEN: In summer, a nest of dry grasses and fur is built in a tunnel about 15 cm underground. There are nearby chambers for wastes. In winter, the nest is subnivean.

REPRODUCTION: Little is known about the reproduction of this lemming, but it is thought to breed between spring and autumn, with a gestation of about three weeks. The litter contains two to six helpless young. Growth is rapid—the young are furred by one week, weaned by three weeks, and leave to start their own families soon thereafter.

SIMILAR SPECIES: The **Brown Lemming** (p. 297) is larger. The **Southern Bog Lemming** (p. 299) lacks the reddish spot at the base of the ears. Most voles have longer tails.

RANGE: The Northern Bog Lemming is found across most of Alaska and Canada south of the Arctic tundra. It occurs as far south as northern parts of Washington, Idaho and western Montana.

Total Length: 11–14 cm

Tail Length: 1.7–2.7 cm

Weight: 23–35 g

Southern Bog Lemming
Synaptomys cooperi

The Southern Bog Lemming is the most southerly of all lemmings. While most inhabit Arctic or near-Arctic regions, this lemming prefers grassy meadows and open forests in southeastern Canada and the central and eastern U.S. Unlike its northern relative, the Northern Bog Lemming, it does not inhabit bogs.

In suitable habitat, this lemming's presence is indicated by well-used runways and neatly clipped piles of grass at intervals along the paths. As well, little green fecal pellets hint at recent activity.

Where the range of this lemming and the Meadow Vole overlap, the vole easily outcompetes the lemming. This bog lemming does not appear to have a high density anywhere in Canada.

DESCRIPTION: The coat is brownish above and silvery or greyish below. The very short tail is faintly bicoloured, brown above and greyish below. In summer, the nape of the neck and shoulders may be rich brown. The strong, curved claws aid in digging winter runways.

HABITAT: The Southern Bog Lemming is found primarily in open forests, grassy meadows and shrub or sedge areas.

FOOD: Grasses, clovers and sedges compose the majority of the diet, though fungi, algae, roots, bark and mosses are occasionally eaten. In times of scarce vegetation, any emergent plant is eaten down to the surface.

DEN: This lemming digs its own burrows but may use the burrows of other small mammals. The nest can be above or below ground. An aboveground nest is usually located in a sheltered area and looks like a spherical, grassy clump. An underground nest is usually less than 15 cm below the surface in a chamber off the main burrow.

REPRODUCTION: Breeding occurs from spring through autumn and sometimes in winter in southern populations. Gestation is about 25 days, after which one to eight young are born. The young are weaned at 16 to 21 days and are probably sexually mature soon thereafter.

SIMILAR SPECIES: Most voles in this range have longer tails. The **Northern Bog Lemming** (p. 298) has a reddish spot at the base of the ears.

RANGE: The Southern Bog Lemming can be found from southeastern Manitoba to eastern Québec and south to Kansas in the West and Virginia in the East.

Total Length: 12–15 cm

Tail Length: 1.3–2.4 cm

Weight: 21–50 g

Northern Collared Lemming
Dicrostonyx groenlandicus

densely furred, as is the short, stubby tail. The short ears are entirely concealed in the fur. In winter, the coat is all white and the foreclaws appear forked because of the development of a second snow claw.

HABITAT: This lemming is known to inhabit high, dry areas of the tundra.

FOOD: Primary foods in summer include grasses, sedges, berries, flowers and other vegetation. Winter foods are primarily woody material from willows and birches.

The unusual Northern Collared Lemming inhabits tundra regions of the North. Unlike most small, drab brown rodents, this lemming has a variably coloured summer coat and becomes entirely white in winter.

Some populations of collared lemmings exhibit the "boom and bust" characteristic that makes lemmings famous. In some areas, the population increases suddenly and dramatically, only to crash immediately afterward. Although the mechanism for this is not well understood, food availability and predation are likely factors.

ALSO CALLED: Northern Varying Lemming, Northern Hoofed Lemming.

DESCRIPTION: The summer coat is brownish or greyish, often with alternating speckles of darker grey and buffy brown. There is a distinct, dark stripe down the back. The "collar" is tawny or reddish brown. The upper incisors are not grooved. The soles of the feet are

DEN: The Northern Collared Lemming makes runways, either through vegetation or under the snow, and also burrows into the ground above the level of the permafrost. The grassy nest is located in a chamber of the underground burrow and may sometimes be above ground, hidden in vegetation.

REPRODUCTION: Mating occurs from January to September. Gestation is about 20 days. Females may have two or three litters per year, each with up to 11 young. The young are weaned when they are 15 to 20 days old.

SIMILAR SPECIES: Other *Dicrostonyx* **lemmings** (pp. 301–04) have different ranges. The **Brown Lemming** (p. 297) is slightly larger, does not turn white in winter and inhabits moist areas.

RANGE: The Northern Collared Lemming is found in most of Alaska and across the northern tundra of Canada to Hudson Bay.

Total Length: 11–18 cm

Tail Length: 10–20 mm

Weight: 30–112 g

Ungava Collared Lemming
Dicrostonyx hudsonius

The Ungava Collared Lemming, once common during the Pleistocene, is now found only in northern Québec and Labrador. Every three or four years, its population cycles, and at population peaks, individuals can be found dispersing to new areas. Members of the *Dicrostonyx* genus are the only rodents that become completely white in winter.

These lemmings are an important prey species on the tundra and are eaten by Arctic Foxes, Least Weasels, Short-tailed Weasels, parasitic jaegers and snowy owls.

ALSO CALLED: Labrador Collared Lemming, Varying Lemming, Hoofed Lemming.

DESCRIPTION: The summer coat is brownish or greyish, often with a yellowish stripe down the sides. There is a distinct, dark stripe down the back. The "collar" is tawny or cinnamon-coloured. This lemming often has reddish patches behind the ears. Its upper incisors are not grooved. The soles of the feet are densely furred, as is the short, stubby tail. The short ears are entirely concealed in fur. In winter, the coat is all white and the foreclaws appear forked because of the development of a second snow claw.

HABITAT: This lemming prefers well-drained slopes and high, rocky meadows of the tundra.

FOOD: As with other lemmings, the main foods in summer are grasses and fresh vegetation including buds and blossoms. Berries may also be eaten. The winter foods are mainly willow and birch.

DEN: A grassy nest is made in a burrow underground above the permafrost. Burrows are also excavated in the snow just above the ground. In summer, runways through the vegetation can be seen.

REPRODUCTION: The breeding season is from March to September, and gestation is 22 to 26 days. Females will have up to three litters per year, each containing up to seven young. Female young are sexually mature within one month.

SIMILAR SPECIES: Other *Dicrostonyx* lemmings (pp. 300, 302–04) have distinctly different ranges. The **Brown Lemming** (p. 297) lacks the dark dorsal stripe and is found in moist areas.

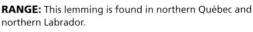

RANGE: This lemming is found in northern Québec and northern Labrador.

Total Length: 12.5–16.5 cm

Tail Length: 9–17 mm

Weight: 35–85 g

Victoria Collared Lemming
Dicrostonyx kilangmiutak

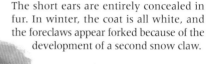

The short ears are entirely concealed in fur. In winter, the coat is all white, and the foreclaws appear forked because of the development of a second snow claw.

HABITAT: These lemmings prefer dry meadows and rocky slopes in the tundra. They avoid wet areas except in times of population booms.

Virtually identical to the Northern Collared Lemming, the Victoria Collared Lemming has only recently been given separate species status. Little is known about it, other than what is known of collared lemmings in general. Genetic studies focusing on the status of the Victoria Collared Lemming used individuals collected from Pearce Point in Nunavut. Like all collared lemmings, they turn white in winter. The shortening day length also triggers the growth of a second claw on each of their front toes. These double claws are found on all collared lemmings in winter and are used to help dig through snow.

ALSO CALLED: Kilangmiutak Collared Lemming.

DESCRIPTION: The summer coat is brownish or greyish, often with alternating speckles of darker grey and buffy brown. There is a distinct, dark stripe down the back. The "collar" is tawny or reddish brown. The upper incisors are not grooved. The soles of the feet are densely furred, as is the short, stubby tail.

FOOD: The primary summer foods include grasses, green vegetation, flowers and berries. Winter foods are basically woody material such as the bark and roots of willows.

DEN: In addition to runways through vegetation, this lemming makes subnivean burrows in winter, as well as shallow underground burrows. The grassy nest may be hidden in vegetation in summer or located in an underground burrow.

REPRODUCTION: Little is known about the reproduction of this species, but it is probably similar to other collared lemmings. Females likely have two to three litters per year, sometime between March and September.

SIMILAR SPECIES: Other *Dicrostonyx* **lemmings** (pp. 300–01, 303–04) have different ranges. The **Brown Lemming** (p. 297) is usually slightly larger, does not turn white in winter and inhabits moist areas.

RANGE: Victoria Collared Lemmings are found in a small region of mainland Arctic as well as Banks, Victoria and King Williams islands.

Total Length: 11–18 cm

Tail Length: 10–20 mm

Weight: 30–112 g

Ogilvie Mountain Collared Lemming
Dicrostonyx nunatakensis

Whether or not this lemming deserves the status of a unique species is a topic of considerable debate among mammalogists. In fact, several *Dicrostonyx* species have alternated between separate species and subspecies status (usually of *D. groendlandicus*) for many years. At this time, *D. nunatakensis* is distinct, though little else is known about its life history.

As with other collared lemmings, the shortening day length is responsible for this species' coat changing colour, as well as other changes. As daylight wanes, the lemming's metabolism slows, allowing the creature to gain a lot of weight before the onset of winter. In some cases, it will even double its weight.

DESCRIPTION: The summer coat is brownish or greyish, often with alternating speckles of darker grey and buffy brown. There is a distinct, dark stripe down the back. The "collar" is tawny or reddish brown. The upper incisors are not grooved. The soles of the feet are densely furred, as is the short, stubby tail. The short ears are entirely concealed in fur. In winter, the coat is all white, and the foreclaws appear forked because of the development of a second snow claw.

HABITAT: This lemming prefers rocky alpine tundra.

FOOD: This species likely feeds on grass and vegetation in summer and woody material from willows, birches and other shrubs in winter.

DEN: This lemming digs shallow burrows above the permafrost and also makes tunnels in the snow just above the ground. The grassy nest is located in one of the underground burrows. In summer, runways are evident, and a grassy nest may be hidden in the vegetation.

REPRODUCTION: Lemmings exhibit high fecundity; females usually have up to three litters per year, usually from March to September. Young females are sexually mature by the time they are one month old.

SIMILAR SPECIES: Other *Dicrostonyx* **lemmings** (pp. 300–02, 304) have different ranges. The **Brown Lemming** (p. 297) is slightly larger, lacks the dark dorsal stripe and is found in moist areas.

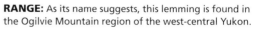

RANGE: As its name suggests, this lemming is found in the Ogilvie Mountain region of the west-central Yukon.

Total Length: 11–18 cm

Tail Length: 10–20 mm

Weight: 30–112 g

Richardson's Collared Lemming
Dicrostonyx richardsoni

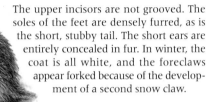

The upper incisors are not grooved. The soles of the feet are densely furred, as is the short, stubby tail. The short ears are entirely concealed in fur. In winter, the coat is all white, and the foreclaws appear forked because of the development of a second snow claw.

HABITAT: This lemming lives on rocky hillsides and in meadows on the tundra.

The population "boom and bust" that is typical of lemmings has been well documented in the Richardson's Collared Lemming. Although populations are believed to cycle over three to four years, researchers are not able to detect the pattern in some. In one study in Manitoba, however, the population "boom" was quite evident. In a three-year time span, the Richardson's Collared Lemmings in the study area increased in number from just one mature individual per 15 hectares to 325 per 15 hectares. After a boom, individuals either disperse or are caught by predators.

ALSO CALLED: Labrador Collared Lemming.

DESCRIPTION: The summer coat varies from reddish to greyish brown, often with reddish underparts. There is a distinct, dark stripe down the back. The "collar" is tawny or cinnamon-coloured.

FOOD: Primary foods in summer include green vegetation, grasses, flowers and berries. Winter foods are limited to woody material from willows and other shrubs.

DEN: Burrows are subnivean and underground above the permafrost. Runways in vegetation are common. The grassy nest is found in one of the underground burrows or is sometimes hidden in vegetation in summer.

REPRODUCTION: Females typically have two to three litters per year, usually between March and September. Litter size is four to eight young. Females are sexually mature after about one month and males soon after.

SIMILAR SPECIES: Other *Dicrostonyx* **lemmings** (pp. 300–03) have different ranges. The **Brown Lemming** (p. 297) is slightly larger, does not turn white in winter and inhabits moist areas.

RANGE: The Richardson's Collared Lemming is found in a small region directly west of Hudson Bay.

Total Length: 11.5–15 cm

Tail Length: 9–17 mm

Weight: 35–90 g

American Beaver
Castor canadensis

The American Beaver is truly an iconic North American mammal. Its highly valued pelt motivated the earliest explorers to discover the riches of the wilderness, and, even today, the beaver serves as an international symbol for wild places. Quite surprisingly to many Canadians, foreign tourists often hold great hopes of seeing this aquatic specialist during their visits. Fortunately, the beaver can be found in wet areas throughout the country south of treeline, where its engineering marvels can be studied in awe-inspiring detail.

One of the few mammals besides humans that significantly alters its habitat to suit its needs, the American Beaver often sets back ecological succession and brings about changes in vegetation and animal life. The deep pools that the beaver's dams create allow it to remain active beneath the ice in winter, at a cost of vast amounts of labour—a single beaver may cut down hundreds of trees each year to ensure its survival. A researcher using images from Google Earth recently discovered the largest beaver dam in the world in Wood Buffalo National Park in Alberta. The dam is maintained by many beavers of at least two families and stretches across 850 m of wetland.

RANGE: American Beavers can be found wherever there is water, from the northern limit of deciduous trees south to northern Mexico. They are absent only from the Great Basin, deserts of the southwestern U.S. and extensive grassland areas devoid of trees.

Total Length: 90–120 cm

Tail Length: 28–53 cm

Weight: 16–31 kg

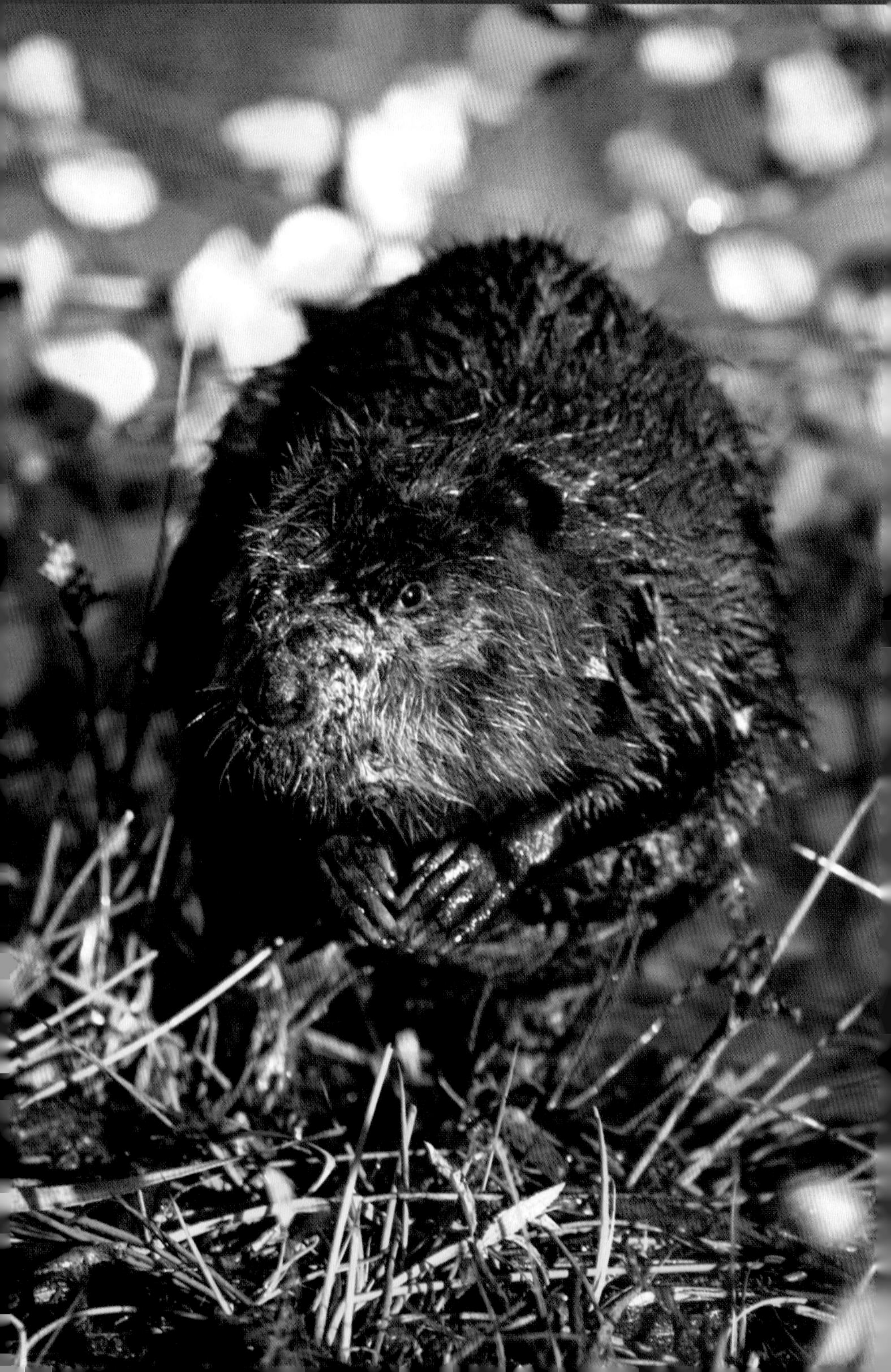

Beavers live in groups that generally consist of a pair of mated adults, their yearlings and a set of young kits. This family group usually occupies a tightly monitored habitat that consists of several dams, terrestrial runways and a lodge. In most cases, the lodge is ingeniously built of branches and mud. Some beavers, especially adult males, tunnel into the banks of rivers, lakes or ponds for their den sites. In areas where trees do not commonly grow or currents are swift, females may also occupy bank dens.

Although not a fast mover, the beaver is extremely strong. It is not unusual for this solidly built rodent to handle and drag a 9-kg piece of wood with its jaws. The flat, scaly tail, for which the beaver is so well known, increases the animal's stability as it cuts a tree; the tail is also slapped on the water or ground to communicate alarm.

The beaver is well adapted to its aquatic lifestyle. It has valves that allow it to close its ears and nostrils when it is submerged, and clear membranes slide over the eyes. Because the lips can form a seal over the incisors, the beaver can chew while it is submerged without having water or mud enter its mouth. In addition to its waterproof fur, the beaver has a thin layer of fat to protect it from cold water, and the oily secretion it continually grooms into its coat keeps its skin dry.

An impressive and industrious animal, the beaver shapes the physical settings of many wilderness areas. Although most tree cutting and dam building occur at dusk or at night, you may see the beaver during the day—sometimes working, but usually sunning itself.

DESCRIPTION: The chunky, dark brown beaver is the second largest rodent in the world, taking a backseat only to the Capybara (*Hydrochoerus hydrochaeris*) of South America. It has a broad, flat, scaly tail, short legs, a short neck and a broad head, with short

DID YOU KNOW?

Beavers are not bothered by lice or ticks, but there is a tiny, flat beetle that lives in a beaver's fur and nowhere else. This beetle feeds on beaver dander, and its meanderings probably tickle sometimes, because beavers often scratch themselves when they are out of water.

ears and massive, protruding, orange-faced incisors. The underparts are paler than the back and lack the reddish brown hue. The nail on the next-to-outside toe of each webbed hindfoot is split horizontally, allowing it to be used as a comb for grooming the fur. The forefeet are not webbed.

HABITAT: Beavers occupy freshwater environments wherever there is suitable woody vegetation. They are sometimes even found feeding on dwarf willows above treeline.

FOOD: Bark and cambium, particularly that of aspen, willow, alder and birch, is favoured, but aquatic vegetation is eaten in summer. Beavers sometimes come ashore to eat grains or grasses.

DEN: Beaver lodges are cone-shaped piles of mud and sticks. These remarkable rodents first construct a great mound of material and then chew an underwater

access tunnel into the centre and hollow out a den. The lodge is typically located away from shore in still water; in flowing water, it is generally on a bank. Access to the lodge is from about 1 m below the water's surface. A low shelf near the two or three plunge holes in the den allows much of the water to drain from the beavers before they enter the den chamber. Beavers often pile more sticks and mud on the outside of the lodge each year, and shreds of bark accumulate on the den floor. Some "bank beavers" do not live in a lodge, but dig bank burrows at the water's edge. These burrows—the entrances to which are below water—may be as long as 50 m, but most are much shorter.

REPRODUCTION: Most mating takes place in January or February, but occasionally as much as two months later. After a gestation of four months, a litter of usually four kits is born. A second litter may be born in some years. At birth, the 340- to 650-g kits are fully furred, their incisors are erupted, and their eyes are nearly open. They can swim when they are only four days old. The kits begin to gnaw before they are one month old, and weaning takes place at two to three months of age. Beavers become sexually mature when they are about two years old, at which time they often disperse from the colony.

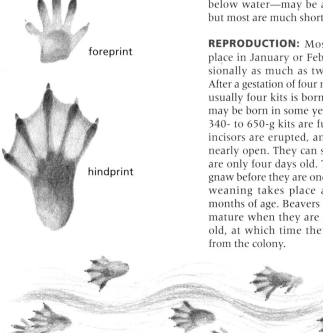

foreprint

hindprint

walking trail

SIMILAR SPECIES: The **Muskrat** (p. 295) is much smaller, and its long tail is laterally compressed rather than paddle-shaped. The introduced **Nutria** (p. 254) has a blockier head and a cylindrical tail. The **Northern River Otter** (p. 174) has a long, round, tapered, fur-covered tail, a streamlined body and a small head.

Muskrat

Olive-backed Pocket Mouse
Perognathus fasciatus

Unlike most hibernating rodents, pocket mice do not build up a store of fat; instead, they pack their burrows with vast numbers of seeds. When outside food supplies dwindle because of cold winter temperatures or during periods of extreme summer heat, pocket mice retreat to their burrows and enter torpor, a state of dormancy that is not as deep as hibernation. They rouse periodically to urinate and feed on their stored seeds, but they consume less than half the amount of food that they would typically eat. Pocket mice do not need to drink because their metabolism generates water through the digestion of lipids in the seeds.

DESCRIPTION: This tiny mouse has a dominant buffy back colour modified by blackish or olive-coloured hairs. The dark hairs end abruptly, and the back contrasts with the buffy sides and the white or buffy feet and underparts. There is a buffy spot behind each ear. The long, thin tail is uniformly coloured.

HABITAT: This pocket mouse is restricted to grasslands on light, sandy soils.

FOOD: Seeds with a higher than average oil content, including thistle, knotweed, lamb's-quarters, blue-eyed grass and foxtail, seem to be the mainstay of the diet. Some green vegetation and a few insects are also consumed. Grasshopper eggs may be stored in the burrow.

DEN: Summer tunnels, about 2 cm in diameter, form a network of storage and refuge burrows 30 to 46 cm deep. Tunnel entrances are plugged with soil by day. The summer nesting chamber is bare. Winter burrows may penetrate 2 m beneath the surface and contain a grass nest. The first 1 m of the burrow is plugged with soil. The entire system may have a diameter of 10 m.

REPRODUCTION: Breeding occurs in spring. After four weeks of gestation, a litter of usually four to six young is born. Females have two litters per season. Young pocket mice become sexually mature the spring after their birth.

SIMILAR SPECIES: The **Great Basin Pocket Mouse** (p. 310) is found only in British Columbia. The **Ord's Kangaroo Rat** (p. 311) is larger, has an extremely long tail and rests with its body approaching a more upright stance.

RANGE: The range of this pocket mouse extends from southeastern Alberta to western Manitoba and south to New Mexico.

Total Length: 10–15 cm

Tail Length: 5–7 cm

Weight: 7–14 g

Great Basin Pocket Mouse
Perognathus parvus

HABITAT: These pocket mice live in sandy soils in arid and semi-arid areas where the dominant plant cover is sagebrush and other small shrubs.

FOOD: Grass seeds form the bulk of the diet, but this mouse also eats the grain and seedlings of winter wheat, and, in spring, caterpillars and adult insects supplement a diet that includes the seeds of thistle, wild mustards, bitterbrush and pigweed.

For Great Basin Pocket Mice, a bright future is one that is invested in seeds. This animal collects massive quantities of seeds, and then burrows 1 to 2 m into the ground to spend the winter. The number of seeds stored by an individual is even more phenomenal when you consider that each seed is handled individually and that most are smaller than the head of a pin. Equally outstanding is the fact that in some parts of its range, there is considerable competition between this pocket mouse and ants for the seeds.

DESCRIPTION: The back is a glossy, yellowish buff, with many black-tipped hairs that overlay the fur, giving it a peppered appearance. A narrow, buffy line separates the back colour from the uniform white or buffy white underparts and feet. The tail is generally more than half the animal's total length and is darker above and lighter below. The hindfoot is 2.2 to 2.5 cm long.

DEN: The extensive burrow system has shallow summer tunnels and deep winter burrows. Tunnel entrances are plugged with soil by day and throughout winter. The summer nesting chamber is bare, but a grass nest is built for winter.

REPRODUCTION: Mating occurs in April. About four to six helpless young are born after a gestation period of 21 to 28 days. The young are weaned at 25 days, when they weigh about 7 g. A few females have a second litter in late summer.

SIMILAR SPECIES: The **Olive-backed Pocket Mouse** (p. 309) and the **Ord's Kangaroo Rat** (p. 311) are found on the Prairies.

RANGE: The Great Basin Pocket Mouse is found in semi-desert areas from the interior of British Columbia south to California, Nevada and northern Arizona and east as far as southwestern Wyoming.

Total Length: 15–20 cm

Tail Length: 8–11 cm

Weight: 14–28 g

Ord's Kangaroo Rat
Dipodomys ordii

Finding sand dunes within this rodent's range is challenging in itself; locating an Ord's Kangaroo Rat requires luck and knowledge. This nocturnal hopper is best seen on moonless or overcast nights. If you shine a flashlight across the dunes or drive slowly along sandy road cuts, this sand-dweller might be revealed. By day, the kangaroo rat retires to its plugged sand burrow.

Much of this species' food is taken from the sand. The kangaroo rat forages slowly, sifting out seeds with its sharp foreclaws. Seeds that are to be eaten immediately are first husked, but those to be stored are left intact. The kangaroo rat transports the food to its burrow in spacious cheek pouches, which can be inverted for cleaning and combing with the foreclaws. The sand is also critical to this mammal's cleanliness—dust bathing is an important part of the kangaroo rat's grooming routine.

The Ord's Kangaroo Rat can live its entire life without drinking free water. It can survive on the metabolic water produced through the breakdown of the oils and fats in the seeds it eats. This water is used in digestion, reproduction and milk production. Despite this ability, in the wild, kangaroo rats may lap droplets of dew and sometimes eat a bit of green vegetation.

DESCRIPTION: The back is yellowish buff with a few black hairs down the centre, the sides are clear buff, and the belly is white. The eyes are large, luminous and protruding. There is a white spot above the eye and behind the brownish black ear. A black patch on the side of the nose, above the white lip, marks the base of the whiskers. There is a diagonal, white line across the hip. The extremely long hindfeet are white on top and brownish black on their hairy soles. The greatly reduced forelegs are held up and are often not visible in profile as the animal sits hunched over, supporting itself on its hindfeet. The tail, which is at least as long as the body, is tufted at the tip, and has white sides and brownish black upper and lower surfaces.

RANGE: The widely distributed Ord's Kangaroo Rat occurs in disjunct populations from southeastern Alberta and southwestern Saskatchewan south through the Great Plains and western Texas into Mexico, and from eastern Oregon south through the Great Basin and Arizona.

Total Length: 23–28 cm

Tail Length: 14–16 cm

Weight: 44–96 g

HABITAT: This kangaroo rat occupies active sand dunes and sand flats as well as sandy grasslands and semi-desert sagebrush flats. Disturbed, vegetation-free areas, produced by drifting sand seem particularly attractive.

FOOD: Seeds make up more than three-quarters of the year-round diet. Insects such as ants, butterfly pupae, adult beetles and larval antlions account for one-fifth of the diet in spring. Grasshoppers and roots are eaten in summer.

DEN: Burrows, which are usually located in the sides of sand dunes, dry, eroded channels or road slopes, are about 7.5 cm in diameter. The entrances are plugged

during the day. The tunnels branch frequently, with some branches used for food storage and at least one as a nesting chamber. Most of the burrow system is within 30 cm of the surface. The burrow is actively defended against other kangaroo rats. The Canadian population is the only one known to use torpor to save energy in the winter.

REPRODUCTION: Breeding occurs from early spring to midsummer, and females may have up to four litters per year. After a 29- to 30-day gestation period, a litter of usually three to five young is born in a nest built just beforehand. The helpless newborns, each weighing about 6 g, are groomed by the mother. At two weeks, their eyes open, and at three weeks, they have functioning cheek pouches. After the young reach adult size at five to six weeks of age, they disperse to develop their own burrows. The mortality rate in the first year may be as high as 80 percent.

slow hop group

fast hopping trail

SIMILAR SPECIES: No other rodent has the extremely long tail and hindlegs, big head and stocky body of the Ord's Kangaroo Rat. The **Olive-backed Pocket Mouse** (p. 309) is smaller, with a shorter tail, and it usually holds its body more horizontally when it rests on its hindfeet.

Olive-backed Pocket Mouse

Northern Pocket Gopher
Thomomys talpoides

The Northern Pocket Gopher is one of nature's rototillers. This ground-dwelling rodent continually tunnels through dark, rich soils, and one individual is capable of turning over 15,000 kg of soil every year. Evidence of pocket gopher activity is commonplace on the land in the form of freshly churned earth neatly piled in mounds, or "gopher cores," without visible entrances. In many agricultural areas, the Northern Pocket Gopher is the most controlled "nuisance" mammal because of these mounds, which can damage machinery and cover vegetation.

Pocket gopher "pushups" hide the access holes to a system of burrows. From the rodent's viewpoint, the surface provides a space to dump the dirt from tunnel excavation. When the ground is covered by snow, pocket gophers still bring waste soil up to the surface and pack it into snow tunnels. When the snow melts, these soil cores, or "crotovinas," are left exposed.

The Northern Pocket Gopher is extremely well suited to an underground existence. It has small eyes, which it rarely needs in its darkened world, reduced external ears that do not interfere with tunnelling, short, lax fur that does not impede backward or forward movement in the tunnels, and a short, sparsely haired tail that serves as a tactile organ when the animal runs through tunnels in reverse.

To dig its elaborate burrows, the pocket gopher has heavy, stout claws on short, strong forelegs and a massive lower jaw armed with long incisors. Once the soil has been loosened with tooth and claw, it is pushed back under the body, initially with the forefeet, and then farther with the hindfeet. When sufficient soil has accumulated behind the animal, the gopher turns, guides the mound with its forefeet and head and pushes with its

RANGE: This pocket gopher occupies most of the southern Prairies, northern Great Plains and western mountains from Manitoba to British Columbia and south to Nebraska and northern Arizona. It occurs as far east as western Minnesota and as far west as the Cascades and the Sierra Nevada.

Total Length: 19–25 cm

Tail Length: 4–8 cm

Weight: 75–150 g

hindlegs until the soil has been deposited in a side tunnel or on the surface.

Pocket gophers are named for their large, externally opening, fur-lined cheek pouches. As in the related pocket mice and kangaroo rats, these "pockets" are used to transport food, but they have no direct opening to the animal's mouth. Many people incorrectly call pocket gophers "moles," but true moles look more like shrews with oversized forelegs.

DESCRIPTION: This squat, bullet-headed rodent has visible incisors, long foreclaws and a thick, nearly hairless tail. A row of stiff hairs surrounds the naked soles of the forefeet. The upperparts, which are slightly darker than the underparts, often match the soil colour—individuals may be black, dark grey, brown or even light grey.

HABITAT: This adaptable animal avoids only dense forests, wet or waterlogged soils, very shallow, rocky soils or areas exposed to strong winter freezing of the soil.

FOOD: Succulent underground plant parts are the staple diet, but pocket gophers emerge from their burrows on summer nights to collect green vegetation.

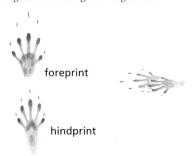

foreprint

hindprint

walking trail

DEN: The burrow system may spread 45 to 150 m laterally and extend 5 cm to 3 m deep. Tunnels are about 5 cm in diameter. Some lateral tunnels serve as food storage and others function as latrines. Several nesting chambers, 20 to 25 cm in diameter and filled with fine grasses, are located below the frost line. Soil from tunnelling is spread fanwise to one side of the burrow entrance, then the burrow is plugged from below. Only a single gopher occupies a burrow system, except during the breeding season, when a male may share a female's burrow for a time.

REPRODUCTION: Breeding occurs once per year, in April or May. Following a 19- to 20-day gestation, three to six young are born in a grass-lined nest. Weaning takes place at about 40 days. When the young weigh about 40 g, they leave to either occupy a vacant burrow system or begin digging one of their own. They are sexually mature the following spring.

SIMILAR SPECIES: The **Plains Pocket Gopher** (p. 315) tends to be browner and lives in moister areas with meadows and sandy soils. **Voles** (pp. 277–94) do not have the large, external cheek pouches, nearly hairless tail or long front claws of pocket gophers, though their colour patterns are similar.

Plains Pocket Gopher

Plains Pocket Gopher
Geomys bursarius

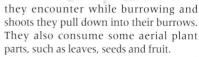

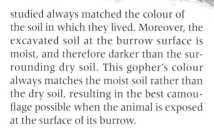

For most small mammals, a brown or grey colour is ideal for blending into their earthy homes. Studies of the Plains Pocket Gopher, however, show that its camouflage is surprisingly precise. The individuals studied always matched the colour of the soil in which they lived. Moreover, the excavated soil at the burrow surface is moist, and therefore darker than the surrounding dry soil. This gopher's colour always matches the moist soil rather than the dry soil, resulting in the best camouflage possible when the animal is exposed at the surface of its burrow.

DESCRIPTION: The coat is dark or greyish brown above and slightly paler below. The short tail is tapered and sparsely furred, usually tawny or grey. The ears are rounded and inconspicuous. This creature looks disproportionate, with an overly large head and forefeet.

HABITAT: These rodents live in sandy loam soils, favouring grassy meadows and cultivated land.

FOOD: Pocket gophers feed on vegetation of all sorts, especially roots and tubers they encounter while burrowing and shoots they pull down into their burrows. They also consume some aerial plant parts, such as leaves, seeds and fruit.

DEN: Like other pocket gophers, this species spends about 70 percent of its life underground. It digs nest chambers, special waste tunnels, deep, lateral tunnels for nests and food storage, and shallow tunnels for foraging routes.

REPRODUCTION: Plains Pocket Gophers may have delayed fertilization or delayed implantation, because gestation is reported to vary between 18 and 51 days. Females give birth to two to six young.

SIMILAR SPECIES: The **Northern Pocket Gopher** (p. 313) is generally not as brown, and where both species occur, it is found in drier, stony, upland sites.

RANGE: The Plains Pocket Gopher is found only in the Roseau River area south of Winnipeg.

Total Length: 24–31 cm

Tail Length: 6–9 cm

Weight: 120–250 g

Yellow-pine Chipmunk
Neotamias amoenus

The sound of scurrying among fallen leaves, a flash of movement and sharp, high-pitched chip calls will direct your attention to the nervous behaviour of a Yellow-pine Chipmunk. Using fallen logs as runways and the leaf litter as its pantry, this busy animal inhabits much of central and southern British Columbia and the mountains of Alberta. It is often the most commonly seen chipmunk in mountain parks.

The word "chipmunk" is thought to have been derived from the Algonquian word for "headfirst," which is the manner in which chipmunks descend trees, but contrary to cartoon-inspired myths, chipmunks spend very little time in high trees. They prefer the ground, where they bury food and dig golf ball–sized entrance holes to their networks of tunnels. Chipmunk burrows are known for their well-hidden entrances, which never have piles of dirt to give away their locations.

In certain heavily visited parks and on some golf courses, Yellow-pine Chipmunks that have grown accustomed to human handouts can be very easy to approach. These exchanges contrast dramatically with the typically brief sightings of wilder chipmunks, which usually scamper away at the first sight of humans. Normally, chipmunks rely on

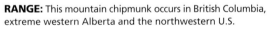

RANGE: This mountain chipmunk occurs in British Columbia, extreme western Alberta and the northwestern U.S.

Total Length: 20–24 cm

Tail Length: 8–11 cm

Weight: 46–85 g

their nervous instincts to survive in their predator-filled world.

DESCRIPTION: This chipmunk is usually brightly coloured, from tawny to pinkish cinnamon, but it is the most variably coloured of the chipmunks. There are three dark and two light stripes on the face, and five dark and four light stripes on the back. The light stripes are white or greyish. The dark stripes are nearly black, and the central three extend all the way to the rump. The sides of the body and the underside of the tail are greyish yellow. Females tend to be larger than males.

HABITAT: The Yellow-pine Chipmunk inhabits a wide variety of habitats, including open coniferous forests, sagebrush flats, rocky outcroppings and pastures with small shrubs. It may be seen on ranches and farms well away from mountains or forests, attracted by livestock feed.

FOOD: This chipmunk loves to dine on ripe berries such as chokecherries, pincherries, strawberries, raspberries or blueberries. Other staples in the diet include nuts, seeds, grasses, mushrooms and even insects. It may be an important predator of eggs and nestling birds during the nesting season. A chipmunk may be attracted to animal feed, and sometimes one can be seen filling its cheek pouches from a pile of oats shared by a horse.

DID YOU KNOW?

During summer, a chipmunk's body temperature is 35° to 42°C. During winter, when it is hibernating in its burrow, its body temperature drops to 5° to 7°C.

DEN: The Yellow-pine Chipmunk usually lives in a burrow that has a concealed entrance. It can sometimes be found in a tree cavity but seldom builds a tree nest.

hibernating adult

REPRODUCTION: The young are born in May or June, after spring mating and about one month of gestation. Usually five or six young are born in a grass-lined chamber in the burrow. They are blind and hairless at birth, but their growth is rapid, and they are usually weaned in about six weeks.

SIMILAR SPECIES: It is very difficult to distinguish between the chipmunks; range differences often help. The underside of the **Red-tailed Chipmunk**'s (p. 320) tail is brick red. The **Least Chipmunk** (p. 318) tends to have duller colours. **Townsend's Chipmunk** (p. 321) has a light-tipped tail and less-contrasted stripes. The **Golden-mantled Ground Squirrel** (p. 336) and **Cascade Golden-mantled Ground Squirrel** (p. 341) are larger and do not have facial stripes.

Red-tailed Chipmunk

Least Chipmunk
Neotamias minimus

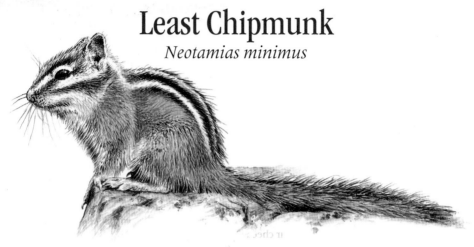

The Least Chipmunk, the smallest of Canada's chipmunks, has the largest range. From the central Yukon to western Québec, this little rodent at home in a variety of different habitats.

The five dark and four pale stripes down the back of a Least Chipmunk have a slightly grey tone to them, which distinguishes this species from the common Yellow-pine Chipmunk. The Least Chipmunk undergoes two full moults each year, so its appearance changes seasonally, if only slightly—the summer coat is bright, and the winter coat looks dusty.

Late in summer, Least Chipmunks start to dig their winter burrows, which consist of one main passageway and one sleeping chamber. By October or November, their energetic food gathering stops, and they start to spend more time in their burrows. Adults usually bed down first; youngsters must feed as late into the year as possible to catch up to the size and weight of the adults.

When Least Chipmunks hibernate, their periods of dormancy, which last from a couple of days to weeks at a time, are broken by wakeful periods during which they eat from their well-packed food cache. The frequency of the wakeful periods depends on the latitude and climate of their home. Like many chipmunks, Least Chipmunks do not put on a layer of fat before hibernating, so they must eat frequently to survive winter.

DESCRIPTION: This tiny chipmunk has three dark and two light stripes on its face and five dark and four light stripes on its body. One of the dark facial stripes runs through the eye. The central dark stripe goes from the head to the base of the tail, but the other dark stripes end at the hips. The overall colour is greyer and paler than

RANGE: The extensive range of this species spreads from the central Yukon southeast to western Québec and south to northern California, New Mexico and North Dakota.

Total Length: 18–24 cm

Tail Length: 7.5–11 cm

Weight: 35–72 g

other chipmunks, and the underside of the tail is yellower. The coat of this chipmunk changes seasonally—in summer, its coat is new and bright, and in winter, the coat is duller, as if the chipmunk had rolled in dust to mute its colours. The tail is quite long and is more than 40 percent of the animal's total length.

HABITAT: Least Chipmunks can live in a variety of habitats, including open areas of boreal and mountain forests, tundra and sagebrush plains. They prefer trees or shrubs nearby for cover.

FOOD: These chipmunks prefer seeds and nuts, but they also consume insects and fruit when available. They use the   k

> **DID YOU KNOW?**
>
> In one study, a researcher discovered a Least Chipmunk's food cache that contained 478 nuts and over 2000 cherry pits.

pouches to carry food to caches that support them throughout winter. As is true for most chipmunks, they will eat baby voles, mice and birds when encountered.

DEN: The majority of Least Chipmunks den in underground burrows, which have concealed entrances, but some individuals live in tree cavities or even make spherical leaf-and-twig nests among the branches in the manner of tree squirrels.

ODUCTION: Breeding occurs about tw   eeks after the chipmunks emerge fro.   berration in spring. After about a one   nth gestation, a litter of two to seven (ally four to six) helpless young is born   a grass-lined nest chamber. The you.   eigh about 2 g each. They develop ra   ly and are weaned at about 36 days old. the mother may later transfer them to   tree cavity or tree nest. The young are probably sexually mature at about 10 months of age.

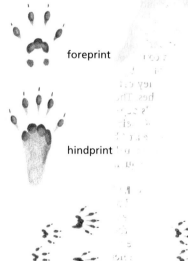

foreprint

hindprint

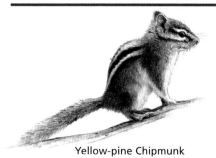

running trail

SIMILAR SPECIES: The **Yellow-pine Chipmunk** (p. 316) has brighter colours. The underside of a **Red-tailed Chipmunk**'s (p. 320) tail is brick red. The larger **Eastern Chipmunk** (p. 322) has a shorter tail, lacks the stripe through the eye and has shorter dorsal stripes.

Yellow-pine Chipmunk

Red-tailed Chipmunk
Neotamias ruficaudus

on the back are brownish. The rump is greyish. In keeping with its name, the tail is rufous above and brick red below, bordered with black and pale pinkish orange.

HABITAT: This chipmunk inhabits coniferous mountain forests and boulder-covered slopes below treeline.

FOOD: Although conifer seeds, nuts, some berries and insects form most of the diet, it is not uncommon for these chipmunks to feed on eggs, fledgling birds, young mice or even carrion.

True to its name, the Red-tailed Chipmunk has a distinctly rust-coloured tail. White underparts, cinnamon-coloured feet and a grey rump also help distinguish this species.

Red-tailed Chipmunks are more arboreal than other chipmunks. Skillful climbers, they prefer to live in dense coniferous forests, nesting either in trees or on the ground. A typical tree nest looks like a shaggy ball of grass and leaves, 6 to 18 m off the ground. Treetop architects, these chipmunks design their tree nests to be hollow and roomy inside and about 30 to 40 cm in diameter.

DESCRIPTION: This large chipmunk has three dark and two light stripes on the face, and five dark and four light stripes on the back. The inner three dark stripes on the back are black, whereas the dark facial stripes and the outermost dark stripes

DEN: As with all chipmunks, the Red-tailed Chipmunk usually spends winter in a burrow. A mother often bears her young in a tree nest or cavity. This chipmunk makes spherical tree nests with the skill of a tree squirrel.

REPRODUCTION: Breeding occurs in spring, and after a one-month gestation, a litter of usually four to six young is born in May or June. The young are born blind and hairless. They grow rapidly, and they are usually weaned in about six weeks. Age of sexual maturity varies considerably but is not earlier than 10 months of age.

SIMILAR SPECIES: The **Least Chipmunk** (p. 318) is smaller, and both it and the **Yellow-pine Chipmunk** (p. 316) have a greyish yellow, not brick red, underside of the tail.

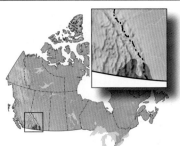

RANGE: The small range of this chipmunk includes only southeastern British Columbia, the extreme southwestern corner of Alberta, northeastern Washington, northern Idaho and western Montana.

Total Length: 21–25 cm

Tail Length: 9–12 cm

Weight: 53–74 g

Townsend's Chipmunk
Neotamias townsendii

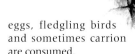

The telltale feature of a Townsend's Chipmunk is its upright tail. Whether it's darting across a hot stretch of beach sand or jumping through a fern-shaded coastal forest, this chipmunk's tail advertises its whereabouts.

Few chipmunks outweigh this distinctly dark-coated species. The female is the largest, and a newborn Townsend's Chipmunk is the biggest chipmunk baby.

An excellent climber and avid explorer, this robust chipmunk may travel more than 1 km in search of diverse food items, and with cheek pouches that can hold more than 100 oat seeds, it is able to transport large quantities of food to its larder.

DESCRIPTION: The stripes over the face and back are indistinct and of low contrast—the dark stripes are dark brown, never black, and the light stripes are ochre or tawny, never white. The underparts range from tawny to nearly white, and the tail is greyish above and reddish below, with a light or white tip.

HABITAT: This chipmunk occurs in areas of dense cover such as driftwood beaches, dense hardwood forests and moist, fern-filled, coniferous forests. It is found at elevations up to 2000 m.

FOOD: A wide variety of foods including roots, grass and conifer seeds, hazelnuts, berries, dandelion flowers, fungi, insects, eggs, fledgling birds and sometimes carrion are consumed.

DEN: The burrow is usually located at the base of a tree or stump or in a crevice among rocks. It descends about 30 cm, levels off for 1 m, then terminates in a chamber 10 to 13 cm in diameter and filled with a nest of shredded vegetation.

REPRODUCTION: These chipmunks emerge from hibernation in late April to early May and mate within a week. Following a 30-day gestation, two to seven blind, hairless babies are born. They are weaned five weeks later, and by the end of August, they disperse. Females have a single litter each year, and the young are sexually mature after their first winter.

SIMILAR SPECIES: The **Yellow-pine Chipmunk** (p. 316) is smaller, lacks the light tail tip and usually shows much more contrast in its dorsal stripes.

RANGE: The Townsend's Chipmunk is found from extreme southwestern British Columbia south through western Washington and most of western Oregon.

Total Length: 22–36 cm

Tail Length: 9–15 cm

Weight: 50–115 g

Eastern Chipmunk
Tamias striatus

These common chipmunks of Eastern Canada are quite similar to their western counterparts. Chipmunk nomenclature has been extremely variable over the last few decades. At one time, Eastern Chipmunks were the sole members of the genus *Tamias*, and the western species were classed as *Eutamius* species. Later, all the chipmunks were grouped together in the genus *Tamias*. Now, again, the Eastern Chipmunk is the only *Tamias* species, and all others are in the genus *Neotamias*.

An Eastern Chipmunk nests in underground burrows and uses a tricky technique when creating its home. To excavate the main passageway and sleeping chamber of an underground nest, the chipmunk starts a "work hole." All the dirt excavated is strewn on the ground outside this work hole. When the burrow is finished, the work hole is closed and a main entrance is opened elsewhere. This way, the main entrance has no mound of dirt to advertise its whereabouts and is therefore known only to the chipmunk that built it.

With an insatiable drive, the Eastern Chipmunk rapidly gathers food for winter storage. When collecting berries, it runs along the twigs and quickly snips the stems with its sharp teeth. So speedy is its work that the ground beneath the berry bush seems bombarded with a storm of red pellets. Once enough berries have been cut, the chipmunk scampers to the ground and carries the berries to a nearby cache. Most seeds, berries and nuts are collected in this manner. Throughout winter, the chipmunk remains in its burrow, eating the collected food items during wakeful periods of its hibernation.

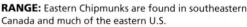

RANGE: Eastern Chipmunks are found in southeastern Canada and much of the eastern U.S.

Total Length: 23–30 cm

Tail Length: 7.2–10 cm

Weight: 66–139 g

DESCRIPTION: This robust chipmunk differs from western species in its stripe pattern. The dark dorsal strip is quite narrow and runs from the back of the head to the rump. On either side of the dorsal stripe is a pale stripe that is either greyish or the same colour as the body. After this pale stripe, there is a dark stripe on each side, followed by a creamy yellow stripe and then another dark stripe. The outermost dark stripes are much shorter than the others. The body colouration varies regionally, ranging from light oaky brown to dark walnut. This chipmunk's underside is quite pale, nearly white. There are two pale facial stripes—one above and one below the eye—but overall, the facial stripes are indistinct. The tail is relatively short, brown on top and edged with black. The ears are prominent and rounded.

HABITAT: Eastern Chipmunks inhabit both urban and wilderness areas. They live as comfortably in backyards and city parks as they do in open deciduous woodlands, forest edges, brushy areas and rocky outcroppings.

FOOD: Devoted gathers, these chipmunks continuously run between food sources and their storage chambers to store as much food as possible. Favourite items include berries, nuts, seeds and mushrooms. Perishable foods such as slugs, insects and snails are usually eaten right away. Like other chipmunks, Eastern Chipmunks will feed on carrion when the opportunity arises.

DEN: Primarily ground dwellers, Eastern Chipmunks excavate simple to complex

DID YOU KNOW?

Although this chipmunk is essentially a ground- and shrub-dwelling species, it may also climb high into large oak trees in autumn to get at the ripe acorns.

burrows. Simple burrows consist of one main passageway with a single storage chamber and one den chamber. Usually there is a secret hidden entrance. More complex burrows may have multiple passageways and entrances, several storage chambers and various smaller galleries or pockets in which debris or food is stored. Occasionally, a female with her young may be found in a simple nest inside a hollow tree.

REPRODUCTION: Mating occurs in early spring, and the female gives birth to three to five young sometime in May. They are born either in a special nest chamber of the burrow or in a safe nest inside a hollow tree. The young are altricial and require great care for several weeks. By the time they are one month old, they resemble small adults.

burrow cross-section

SIMILAR SPECIES: The smaller **Least Chipmunk** (p. 318) has more distinct stripes over the face, including a dark stripe through the eye, and its dorsal stripes continue to the base of the tail. No other chipmunks occur in the same range as the Eastern Chipmunk.

Least Chipmunk

Hoary Marmot
Marmota caligata

These stocky sentinels of alpine vistas pose majestically on boulders, gazing for untold hours at the surrounding mountain scenery. They customarily emerge from their burrows soon after sunrise but remain hidden on windy days and during snow, rain or hailstorms.

Hoary Marmots occupy exclusively high-elevation environs, where long summer days allow rapid plant growth during a growing season that often lasts only 60 days per year. Despite the shortened summer, these creatures seldom seem hurried, spending most of their time staring off into the distance, perhaps on the lookout for predators or perhaps in simple appreciation of the spellbinding landscape.

Where they are frequently exposed to humans, Hoary Marmots become surprisingly tolerant of our activities. In the backcountry, the presence of an intruder in an alpine cirque or on a talus slope is greeted by a shrill and resounding whistle, from which the Hoary Marmot's old nickname, "Whistler," is derived. When alarmed, marmots travel surprisingly gracefully over rocks and through boulder fields, quickly finding one of their many escape tunnels.

Being chunky is most fashionable in Hoary Marmot circles. Although at first glance their alpine surroundings may appear to hold few dietary possibilities, these areas are, in fact, rich in marmot foods. These rodents consume great quantities of green vegetation throughout summer, putting on thick layers of fat as their metabolism slows. They rely on this fat during their eight- to nine-month

RANGE: The Hoary Marmot occurs from northern Alaska south through the mountains to southern Montana and central Idaho.

Total Length: 70–80 cm

Tail Length: 18–24 cm

Weight: 5–7 kg

hibernation. A considerable portion of stored fat remains when the marmots emerge from hibernation, but they need it for mating and for other activities they perform before the green vegetation reappears.

ALSO CALLED: Whistler Rockchuck, Mountain Marmot.

DESCRIPTION: The head is grey and white with contrasting black markings. The cheeks are grey. A black band across the bridge of the nose separates the white nose patches from the white patches below the eyes. The ears are short and black. The underparts and feet are grey. A black stripe extends from behind each ear toward the shoulder. The shoulders and upper back are grizzled grey, changing to buffy brown on the lower back and rump, where black-tipped guard hairs top the underfur. The bushy, brown tail is so dark it often appears black. This marmot often fails to groom its lower back, tail and hindquarters, so the fur there appears matted and rumpled.

HABITAT: Hoary Marmots require large talus boulders or fractured rock outcrops

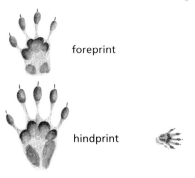

foreprint

hindprint

near abundant vegetation in moist surroundings. They most commonly occur in alpine tundra and high subalpine areas.

FOOD: Copious quantities of many tundra plants are consumed so avidly that the vegetation near the burrows is often lawnlike from frequent clipping. Grasses, sedges and broad-leaved herbs are all eaten.

DEN: Burrows run about 2 m into the slopes, where they may end beneath a large rock as a cave up to 1 m in diameter. The nest chamber is often filled with soft grasses.

REPRODUCTION: A litter of three to five young is born in mid to late May, after a gestation of about 30 days. The fully furred young first emerge from the burrows in about the third week of July, when they weigh about 30 g. They are weaned soon after emerging and grow rapidly until they enter hibernation in September. Sexual maturity is achieved during their third spring.

walking trail

Woodchuck

SIMILAR SPECIES: The **Woodchuck** (p. 328) is uniformly brownish. The **Yellow-bellied Marmot** (p. 326) has a bright, buffy yellow belly and a grizzled brown back. Where the range of the Hoary Marmot overlaps with either of these species, the Hoary Marmot occurs at higher elevations.

Yellow-bellied Marmot
Marmota flaviventris

When architects designed the Head-Smashed-In Buffalo Jump Interpretive Centre near Fort McLeod, Alberta, they likely did not have Yellow-bellied Marmots in mind, but in building this unobtrusive prairie monument, they created a marmot motel. It is quite common to see these chunky rodents happily poised upon the building protuberances that rise from the hillside.

The Yellow-bellied Marmot, though well established in British Columbia, is a newcomer to the prairie wildlife scene. The first official record of this species dates back to 1952, and few verbal accounts predate this record by more than a decade. This marmot is commonly encountered in the U.S. Rockies in northern Montana, and it may have spilled north into the Waterton Lakes area, where it was once (but is no longer) found. Other individuals probably entered Alberta via the Sweetgrass Hills of Montana, which lie just south of the Milk River, a region where these animals are now common. A mountain dweller through most of its range, north of Montana's Flint Creek Range, the Yellow-bellied Marmot becomes a lowland animal—the Hoary Marmot occupies the mountains. The possibility of human assistance in the Yellow-bellied Marmot's travels

RANGE: Yellow-bellied Marmots are found from central British Columbia and extreme southern Alberta and Saskatchewan, south into central California and northern New Mexico.

Total Length: 47–67 cm

Tail Length: 13–19 cm

Weight: 2.8–5.2 kg

should not be discounted because its present-day concentrations appear discontinuous with typical marmot habitat. It is possible that the animals that now pose atop the Head-Smashed-In Interpretive Centre were moved to that spot just a few short marmot generations ago.

Marmot colonies have a strict social order, and whenever members are eating or wrestling, at least one marmot plays watchdog. This sentinel is responsible for warning the others if danger approaches. The alarm call is a loud whistle, which may vary in duration and intensity depending on the nature of the threat. Short, steady notes probably translate as "Heads up, pay attention," whereas loud, shrill notes convey the message "Into your burrows, now!" Different urgent warnings are reserved for immediate dangers, such as a circling eagle or an approaching fox.

ALSO CALLED: Yellow-footed Marmot, Rockchuck, Mountain Marmot.

DESCRIPTION: The back is tawny or yellow-brown and grizzled by the light tips of the guard hairs. The feet and legs are blackish brown. The head has whitish grey patches across the top of the nose, from below the ear to the shoulder and from the nose and chin toward the throat, which leaves a darker brown patch surrounding the ear, eye and upper cheek on each side of the face. The ears are short and rounded. The whiskers are dark and prominent. The dark, grizzled, bushy tail is often arched behind the animal and flagged from side to side. The bright buffy yellow sides of the neck, maxilla, belly and hips are responsible for this marmot's name.

DID YOU KNOW?

Yellow-bellied Marmots frequently bask in the morning sun, probably to warm up. At about midday they retire to their cool burrows, but in late afternoon, they reemerge to feed. They seem to have poor control of their body temperature—in summer, it may range from 34° to 40°C.

HABITAT: Within its range, this species may be found from valley bottoms to alpine tundra but never in dense forests.

FOOD: Abundant herbaceous or grassy vegetation must be available within a short distance of the den. This marmot occasionally feeds on road-killed carrion, and there have been reports of the cannibalization of young.

DEN: Each adult maintains its own burrow, and marmots with the highest social status are nearest the colony centre. A burrow is typically 20 to 35 cm in diameter. It slants down for 50 to 100 cm, and then extends another 3 to 4.5 m to end in a bulky nest lined with grasses beneath or among large rocks.

REPRODUCTION: A litter of three to eight young is born in June after a 30-day gestation. Naked and blind at birth, they first emerge from the burrow at three to four weeks of age. Well-fed females become sexually mature before their first birthday. Males and females born at higher altitudes usually do not get a chance to breed until they are at least two years old.

Hoary Marmot

SIMILAR SPECIES: The **Hoary Marmot** (p. 324) has grey cheeks and a grey belly, and it occupies higher elevations. The **Woodchuck** (p. 328) is uniformly brownish, without the yellowish belly, and occurs farther north.

Woodchuck
Marmota monax

For much of the year, Woodchucks are tucked quietly away more than 2 m underground, relying on their lethargic metabolism during hibernation to keep them alive. They lie motionless, breathing an average of once every six minutes and maintaining life's requirements with a metabolic pilot light fed by a trickle of fatty reserves. In late April (never as early as Groundhog Day in early February), Woodchucks awaken from their catatonic slumbers to breed and to forage on the delectable new, green shoots that emerge with the warmer weather.

Woodchucks are found across much of Canada but are much less common in northern areas. They make their burrows in rock piles, under old buildings and along rivers. In general, they are more solitary in nature than other kinds of marmots, and they are rarely seen far from their protective burrows, valuing security over the temptations of foraging. For feeding, Woodchucks tend to favour the early evening or the period shortly after dawn, but they can be seen at any time of day. In most cases, they are wary and can usually outrun intruders in an all-out sprint back to their burrows. A shrill whistle of alarm typically accompanies their disappearance.

Historically, the Woodchuck lived in forested areas. This mammal can still be found in open woodlands but now lives in great numbers on cultivated land—it is among the few mammals to

RANGE: The Woodchuck occurs from central Alaska east to Labrador and south to northern Idaho in the West and eastern Kansas, northern Alabama and Virginia in the East.

Total Length: 46–66 cm

Tail Length: 11–16 cm

Weight: 1.8–5.4 kg

have prospered from human activity. Unabashed about pilfering, a Woodchuck that lives near humans often grazes in sweet alfalfa crops to help expand its waistline. The luckiest Woodchucks find their way into people's backyards, where they stuff themselves on tasty apples, carrots, peas, strawberries and other garden delights.

ALSO CALLED: Groundhog, Marmot.

DESCRIPTION: This short-legged, stout-bodied, ground-dwelling marmot is brownish, with an overall grizzled appearance. It has a prominent, slightly flattened, bushy tail and small ears. The feet and tail are dark or even black. Some individuals have whitish or tawny patches around the mouth.

HABITAT: Woodchucks favour pastures, meadows and open woodlands. They are sometimes found in ravines and other natural areas within city limits.

FOOD: In wild areas, this ground dweller follows the standard marmot diet of grass, leaves, seeds and berries, which it supplements with bark and sometimes a bit of carrion. The Woodchuck loves garden vegetables, and if it makes its way into an urban area, it will happily dine on fresh backyard produce. It may even make a temporary home in a backyard with a tempting garden.

DEN: The Woodchuck uses its powerful digging claws to excavate a burrow in an area with good drainage. The main burrow is 3 to 15 m long and ends

> **DID YOU KNOW?**
>
> Woodchucks are superb diggers and are responsible for turning over large amounts of earth each year. As they burrow, they periodically turn around and bulldoze loose dirt out of the tunnel with their suitably stubby heads.

in a comfortable, grass-lined nest chamber. Plunge holes, without dirt piles, often lead directly to the chamber. A separate, smaller chamber is used for wastes.

REPRODUCTION: Mating occurs in spring, within a week after the female emerges from hibernation. Some evidence suggests that a male may enter a female's den and mate with her before she arouses from her winter sleep. After a gestation of about one month, one to eight (usually three to five) young are born. The helpless newborns weigh only about 26 g. In four weeks, their eyes open, and they look like proper Woodchucks after five weeks. The young are weaned at about 1.5 months. Their growth accelerates once they begin eating plants, and they continue growing throughout summer to put on enough fat for winter hibernation and early spring activity.

burrow cross-section

SIMILAR SPECIES: The **Hoary Marmot** (p. 324) has grey cheeks and a grey belly, and it occupies higher elevations. The **Yellow-bellied Marmot** (p. 326) has a bright, yellowish belly and occurs farther south. The similar-sized **Muskrat** (p. 295) has a naked tail and is almost always associated with water.

Hoary Marmot

Vancouver Island Marmot
Marmota vancouverensis

The rarest marmot of them all, the Vancouver Island Marmot wears a lustrous brown coat and sports flashy white markings on its nose, chin and underside. Likely numbering fewer than 100 in the wild, this marmot's range is restricted to a small area of high-elevation habitat on Vancouver Island. Its unique, solid-coloured coat may change throughout summer from deep ebony to light walnut brown, depending on how intensely an individual sun-bathes. The annual moult for this marmot, completed in July, renews the intense dark brown of its coat.

Voracious eaters, Vancouver Island Marmots double or even triple their weight from May to September, gorging themselves on lush and abundant summer foodstuffs. Coupled with the high food intake is a lowering of their metabolic rate, resulting in terrific fat gain. Marmots do not store food in their burrows, so they must put on a layer of fat in summer to last them through their cold winter hibernation.

Colonies of these shy marmots have well-structured hierarchies. A male dominates the colony and usually has a couple of females subordinate to him. Vancouver Island Marmots are highly sociable, and they enjoy play-fighting and nuzzling to strengthen familial bonds. The strength of these ties may be the reason that young marmots do not

RANGE: This marmot is found only on Vancouver Island. Many colonies have disappeared, and the species is declining; most known colonies now occur within an area of approximately 80 km².

Total Length: 63–72 cm

Tail Length: 20–30 cm

Weight: 3–6 kg

leave the colony until as late as their third year, rather than their second year, as is usual for other marmot species.

The territory of a colony is well defined. Using scent glands on their cheeks, Vancouver Island Marmots mark large rocks at the limits of their home ranges. Antagonistic behaviour is rare among marmots, but trespassers are greeted with intense growls and hisses. Despite the colony's patriarchy, females are the more aggressive sex. These marmots can be fearsome fighters, and both males and females will vigorously defend themselves when faced with a predator such as a golden eagle, Grey Wolf or Mountain Lion. Unfortunately, their efforts are not enough, as overpredation is one of the causes of their population decline.

In addition to increased predation, habitat disturbance may have contributed to a 50 to 60 percent decline in this marmot's population during the past 10 years. Efforts to provide protection for this species have included protecting their habitat; however, because most colonies occur on privately owned land, widespread protection is not practical. Captive breeding and reintroduction programs have been initiated in an attempt to increase the population, and the captive population is currently about 200 individuals.

DESCRIPTION: This marmot's colour ranges from dark chocolate brown to rusty or walnut brown. There is a grey to whitish patch around the nose and lips and often a similarly coloured patch on the chest. The tail is moderately bushy. This species is less vocal than other marmots but will emit five short whistles when an avian predator is seen and longer whistles for a terrestrial predator.

HABITAT: Vancouver Island Marmots live on south- and west-facing slopes of the mountains on southeastern Vancouver Island, in alpine or subalpine meadows. Rocky outcrops with good vegetation are favoured.

DID YOU KNOW?

With so few individuals remaining, the Vancouver Island Marmot is among the world's most endangered mammals.

FOOD: Marmots are strongly herbivorous, and the Vancouver Island Marmot's foods include blueberries, bluebells, ferns, asters, lupines, pearly everlasting and grasslike plants. Logging several years ago created a vegetation mix attractive to the marmots but seems to have increased their susceptibility to predation over time.

DEN: The den is located in a burrow under rocks. The burrow is typically about 1 m deep and 4 m long and has multiple entrances. Marmots spend most of their time in their burrows; they come above ground for only a few hours in the morning and again in the late afternoon to sun themselves or feed. In autumn, each marmot selects a burrow for hibernation, always in a place that will be completely covered by the snowpack. A grass-and-mud plug is made to seal the entrance. A marmot will use a good hibernation burrow for many winters.

REPRODUCTION: Mating occurs within three weeks of the marmots emerging from hibernation in May. Gestation is approximately one month, but pups do not emerge above ground until they are about one month old. The young always hibernate with their mother for their first winter, and most return to hibernate with her for their second winter, too. These marmots reach sexual maturity at three or four years of age. Females have one litter of two to five young every two years, and males may mate with more than one female each year.

SIMILAR SPECIES: No other marmot occurs on Vancouver Island, and other similar-sized mammals are readily distinguishable.

Columbian Ground Squirrel
Spermophilus columbianus

From montane valleys to alpine meadows, the Columbian Ground Squirrel is a common sight in mountainous regions of British Columbia and Alberta. Nearly every meadow has a thriving population of this large rodent living among the grasses. At heavily visited day-use areas and campgrounds, colonies of this ground squirrel attract a great deal of tourist attention.

Columbian Ground Squirrels are robust, sleek and colourful animals that chirp loudly, often at the first sight of anything unusual. The chirp coincides with a flick of the tail and, in the event of real danger, is followed by a sharp trill and a split-second plunge down a burrow. The ground squirrels give different alarms for avian versus terrestrial predators, and for squirrel intruders from outside the colony. Making sense of the repertoire of different ground squirrel sounds may require more effort than most people are prepared to devote.

Colony members interact freely and nonaggressively with one another in most instances, sniffing and kissing their neighbours upon each greeting. The dominant male has his burrow near the centre of the colony and maintains his central location throughout the breeding season. Ground squirrels from outside the colony are typically attacked by one or several members of the community and are quickly driven away.

Dispersing individuals forced to emigrate from their home colony are exceedingly vulnerable to predation. Away from the sanctuary of communal life, these large

RANGE: The Columbian Ground Squirrel is restricted to the mountains from east-central British Columbia and western Alberta south into northeastern Oregon and western Montana.

Total Length: 33–40 cm

Tail Length: 8.3–12 cm

Weight: 440–570 g

rodents are a much-valued food item for other mammals and birds. Several hawk and falcon species, for example, seasonally focus their hunting efforts on Columbian Ground Squirrels.

DESCRIPTION: The entire back is cinnamon buff, but because the dorsal guard hairs have black tips, a dappled, black-and-buffy effect results. The top of the head and the nape and sides of the neck are rich grey with black overtones. There is a buffy eye ring. The nose and face are a rich tawny colour, sometimes fading to buffy ochre on the forefeet, but more frequently the tawny colour continues over the forefeet, underparts and hindfeet. The base of the tail is sometimes tawny or, more rarely, rufous. The moderately bushy tail is brown, overlain with hairs having black subterminal bands and buffy white tips.

HABITAT: This wide-ranging ground squirrel may occupy montane valleys, forest edges, open woodlands, alpine tundra and even open prairies. Although it is primarily an animal of meadows and grassy areas, some individuals may learn to climb trees.

FOOD: All parts of both broad-leaved and grassy plants are consumed. Carrion is eaten when it is found, and there are several reports of adults, especially males, cannibalizing the young. Insects and other invertebrates are also eaten. Individual squirrels ordinarily store only seeds and bulbs in their burrows.

DEN: The colony develops its burrow system on well-drained soils, preferably loams, on north- or east-facing slopes in

> **DID YOU KNOW?**
>
> Columbian Ground Squirrels have been known to hibernate for up to 220 days. During hibernation, the squirrels wake at least once every 19 days to urinate and sometimes defecate, and to eat stored food.

the mountains. The tunnels are 8 to 11 cm in diameter and descend 1 to 2 m. Each colony member's burrow has from 2 to 35 entrances and may spread to a diameter of more than 25 m. The central nest chamber, up to 75 cm in diameter, is filled with insulating vegetation. Around the colony, several other burrows may serve as temporary refuges, each up to 1.5 m long and with a single entrance.

REPRODUCTION: Mating occurs in the female's burrow soon after she emerges from hibernation. After a gestation of 23 to 24 days, she delivers a litter of two to seven young. The upper incisors erupt by day 19, the eyes open at about day 20, and the young are weaned at about one month of age. All Columbian Ground Squirrels are sexually mature after two hibernation periods, though some females may mate after their first winter.

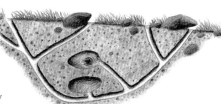

burrow cross-section

Golden-mantled Ground Squirrel

SIMILAR SPECIES: The **Golden-mantled Ground Squirrel** (p. 336) has stripes. The **Arctic Ground Squirrel** (p. 338) is found farther north. The **Yellow-bellied Marmot** (p. 326) is much larger and has dark facial markings.

Franklin's Ground Squirrel
Spermophilus franklinii

Although these ground squirrels are common in the right habitat, they are visible for only a short time each year. They live in edge communities where long grass or brushy areas meet woodlands and spend much of their time underground—adults that have emerged from hibernation in mid-April are typically back underground in August or early September, which amounts to an astonishing 7 to 8.5 months of hibernation. Only the young of the year are still aboveground in September because they need more time than adults do to attain their minimum hibernation weight. If you find this animal's burrows and wait patiently nearby, you may be rewarded with a good sighting as the ground squirrel runs off to a nearby food source.

With its large, bushy, grey tail and its affinity for shrubs and even open forests, the Franklin's Ground Squirrel superficially resembles a tree squirrel. In fact, the resemblance is more than superficial; this ground squirrel is often found within sprinting distance of a tree, and it is an active climber. It tends to prefer deciduous forests, but in some parts of central Canada, this squirrel is found where conifers prevail. In Ontario, it could easily be mistaken for an Eastern Grey Squirrel.

The name of this ground squirrel honours the famed English Arctic explorer Sir John Franklin, who led three

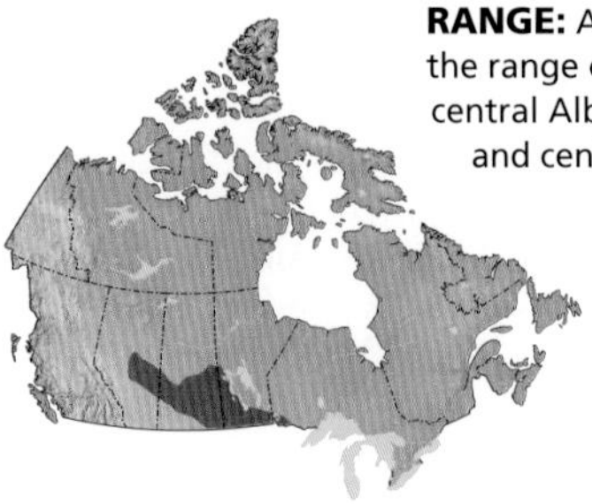

RANGE: Associated with the Prairies and Great Plains, the range of the Franklin's Ground Squirrel extends from central Alberta and Saskatchewan southeast to Kansas and central Illinois.

Total Length: 33–43 cm

Tail Length: 12–16 cm

Weight: 360–700 g

expeditions to attempt to locate the Northwest Passage. (He finally lost his life, and the lives of his crew, in a failed attempt in 1846.) The first scientific collection of the Franklin's Ground Squirrel was made by Sir John Richardson, a medical doctor who also served aboard Franklin's expeditions as a naturalist, and for whom another squirrel, the Richardson's Ground Squirrel, is named.

ALSO CALLED: Whistling Ground Squirrel, Grey Gopher.

DESCRIPTION: This ground squirrel's most noticeable feature is its tail, which makes up about one-third of the animal's total length and is almost as bushy as a tree squirrel's. The tawny to olive tail is sprinkled with black and white hairs and has a white border. The overall body colour is grey, darker on the top of the head and lighter around the snout and on the sides of the face. A whitish ring surrounds the eye. The back is brownish grey, with indistinct light dapples and brownish transverse barring that becomes more pronounced on the rump. The underparts are grey or buffy coloured, and the feet are grizzled grey.

HABITAT: The Franklin's Ground Squirrel typically inhabits tall- and mid-grass prairies and brushy regions that border open woodlands or in grassy forest meadows, on aspen bluffs or even along the edges of dense coniferous forests.

FOOD: Grasses, green vegetation, berries and seeds make up about two-thirds of

DID YOU KNOW?

During the courtship season, rival males fight violently. The combatants bite one another, particularly on the genitalia, while the females await the outcome. The victorious male generally pursues the females through dense brush.

the diet; the remaining third is animal matter. These ground squirrels are effective predators that may take mice, young birds, eggs, frogs, toads, other ground squirrels and even small rabbits and ducks. They also feed on carrion.

DEN: This species is mainly solitary but may gather in small, loose colonies in good habitat with abundant food sources. The burrow is usually well concealed and may descend 1 to 2 m underground. The details of the burrow system are not well known, but presumably the design is similar to that of other ground squirrels, with multiple entrances and one main nest chamber. The female lines her nest with grasses. The burrow entrance usually lacks a spoil pile of dirt.

REPRODUCTION: Mating occurs in the first week of May. After a gestation of about four weeks, a litter of 2 to 13 (usually seven to nine) young is born. The newborns resemble pink gummy bears and are completely reliant on their mother. Their eyes open at 20 to 27 days, and by day 29 or 30, they are weaned and foraging by themselves. They reach adult size by mid-September.

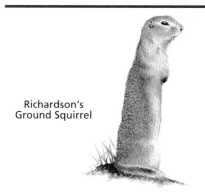

Richardson's
Ground Squirrel

SIMILAR SPECIES: The **Richardson's Ground Squirrel** (p. 339) lacks the bushy tail and prefers short-grass habitat.

Golden-mantled Ground Squirrel
Spermophilus lateralis

and disappear among the boulders. If you can imitate their high-pitched cries, the ground squirrels may approach, suddenly appearing perched on a rock surprisingly close by. At close range, you can often see their bulging cheek pouches crammed with seeds and other food, ready to be stored in their burrows.

Although these ground squirrels, which are often common around campsites and picnic areas, frequently mooch handouts from visitors, feeding them (or any other wildlife) is illegal in national parks. Handouts often lead to extreme obesity in animals, which is unhealthy, and dependency on humans. Perhaps when visitors to the parks become better informed, they will resist the temptation to feed these and other "friendly" animals, instead satisfying their nurturing instincts with detailed observations and quiet awe.

In spite of this squirrel's companionable association with human visitors in mountain parks, it is frequently the victim of mistaken identity. Misled by its long, white and black side stripes, onlookers often think it is a chipmunk. Closer inspection, however, reveals a distinct difference—the stripes stop short at this ground squirrel's neck, whereas all chipmunks have stripes running through their cheeks.

Golden-mantled Ground Squirrels are attractive, stocky, brown-eyed charmers. Bold, buffy white rings frame their endearing eyes, which seem to give expression to their antics. On talus slopes, these squirrels are found alongside pikas, and both of these small mammals continually appear

DESCRIPTION: The head and front of the shoulders are rich chestnut brown. The buffy white eye ring is broken toward the ear. Two black stripes on either side of a white stripe run along each side from the top of the shoulder to near the top of the hip. The back is grizzled grey. The belly and feet are pinkish buff to creamy white. The top of the tail is blackish, bordered with cinnamon buff. The lower surface of

RANGE: This rock-dwelling squirrel's range is restricted to the Rocky Mountains and the Southern Cascade–Sierra axis.

Total Length: 27–32 cm

Tail Length: 9.4–12 cm

Weight: 170–350 g

the tail is cinnamon buff in the centre. There is a black subterminal band and a cinnamon buff fringe on the hairs along the edge of the tail.

HABITAT: This squirrel inhabits the montane and subalpine forests of the Columbian Mountains and the Rocky Mountains, wherever rock outcrops or talus slopes provide adequate cover. In summer, if not permanently, low numbers reside in or along the alpine tundra zone. Some populations may be found at elevations as low as 300 m.

FOOD: Green vegetation forms a large part of the early summer diet. Later, more seeds, fruits, insects and carrion are eaten. Conifer seeds are a major component of the autumn diet. Fungi are another common food.

DEN: The burrow typically begins beneath a log or rock. The entrance is 8 cm in diameter and lacks an earth mound at the entrance. The tunnel soon constricts to 5 cm, and though most

burrows are about 1 m long, others may extend to 5 m. Two or more entrances are common. The nest burrow ends in a chamber that is 15 cm in diameter and has a mat of vegetation on the floor. Nearby blind tunnels serve as either latrine or food storage sites. Like many ground squirrels, this species closes the burrow with an earth plug upon entering hibernation and sometimes when it retires for the night.

REPRODUCTION: Breeding follows soon after the female emerges from hibernation in spring. After a gestation of 27 to 28 days, four to six naked, blind pups are born between mid-May and early July. At birth, the young weigh about 3.5 g. The eyes open and the upper incisor teeth erupt at 27 to 31 days. The young are weaned when they are 40 days old. They enter hibernation between August and October and are sexually mature when they emerge in spring.

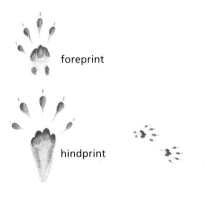

foreprint

hindprint

running trail

Cascade Golden-mantled Ground Squirrel

SIMILAR SPECIES: The **Cascade Golden-mantled Ground Squirrel** (p. 341) has more muted colours and a different range. **Chipmunks** (pp. 316–22) are much smaller, and their stripes extend through the face.

Arctic Ground Squirrel
Spermophilus parryii

HABITAT: Ideal habitat includes sandy banks, lakeshores and meadows with good drainage. This squirrel prefers den sites with a south or southwestern exposure and avoids permafrost areas.

FOOD: Grasslike plants, berries, inner bark, roots, willows and mushrooms are eaten where they are found or are carried to the den. Carrion, eggs and fledgling ground-nesting birds are avidly consumed.

Not surprisingly, the Arctic Ground Squirrel is the only member of its genus that lives in the northern tundra and alpine zone. It survives in these severe habitats by hibernating for more than six months of the year.

These squirrels are sometimes found individually, but they usually form colonies with a single dominant male and many females. They may forage as far as 1 km from their burrows, scurrying from cover to cover to avoid detection by a predator.

ALSO CALLED: Parka Squirrel.

DESCRIPTION: This colourful squirrel is cinnamon-buff over its head, underparts and legs, and greyish or russet with whitish dapples over its back. The eye ring is whitish. The brownish tail is edged with black-tipped hairs and is bright tawny underneath.

DEN: The intertwined burrows of a colony may amount to almost 25 m of tunnels on multiple levels and with many openings. Each nest, inside a 25-cm-wide chamber, is composed of dry grasses, lichens, hair and leaves. At least one tunnel is used as a latrine.

REPRODUCTION: Mating occurs when these ground squirrels emerge from hibernation in late April or early May. Following a 25-day gestation, a litter of 1 to 10 altricial young is born. At three weeks, the eyes open and the young emerge to feed on vegetation. By September, they must either find an empty burrow or dig one of their own for hibernation.

SIMILAR SPECIES: The **Columbian Ground Squirrel** (p. 332) is also richly coloured but has a different range.

RANGE: The Arctic Ground Squirrel occurs from the western coast of Hudson Bay east to Alaska, dipping southward into northern British Columbia.

Total Length: 30–47 cm

Tail Length: 10–12 cm

Weight: 600–900 g

Richardson's Ground Squirrel
Spermophilus richardsonii

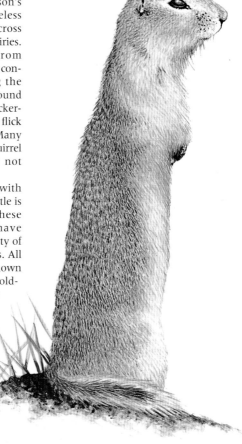

Maligned and vilified over much of its range, the Richardson's Ground Squirrel nonetheless continues to be a common sight across much of the southern half of the Prairies. Although it often hibernates from September to February or March, it is conspicuous, vocal and active during the warmer months of the year. This ground squirrel is frequently nicknamed "Flickertail," because of the conspicuous tail flick that accompanies its shrill whistle. Many people mistakenly call this ground squirrel a "gopher," but true gophers are not members of the squirrel family.

Beyond the habits that conflict with agricultural practices, surprisingly little is known about the behaviour of these prairie diggers. Only recently have researchers unravelled the complexity of their hibernation and activity cycles. All Richardson's Ground Squirrels are known to hibernate through much of the coldest winter weather, but the males begin to emerge regularly in spring, often before the last snows have retreated. Courtship and reproductive activities are condensed between late spring and early summer, so that many adult ground squirrels have already begun to enter hibernation by the end of July. The first hibernators

RANGE: This species is associated with the Prairies and the Great Plains from southern Alberta east to southwestern Manitoba and south to northeastern Montana and extreme western Minnesota.

Total Length: 28–32 cm

Tail Length: 6–8.3 cm

Weight: 260–700 g

are typically mature males, followed by adult females and then juveniles in late August. When hibernation begins in the heat of the summer months, it is called aestivation. Sometimes ground squirrels will wake from aestivation and venture aboveground in late August and September, but often the aestivation continues right into winter hibernation. The waxing and waning of these ground squirrels' waistlines is impressive—their weight can double or even triple from the time they emerge in spring to when they are ready to hibernate.

ALSO CALLED: Flickertail, Picket Pin, Gopher, Wyoming Ground Squirrel.

DESCRIPTION: The upperparts are a pinkish buffy grey to cinnamon buff and are indistinctly mottled. The underparts are pale yellowish, pinkish or grey. The buffy brown tail is about one-third the length of the body and is fringed with short, black, white-tipped hairs. One of this ground squirrel's distinguishing characteristics is its habit of standing erect on its hindlegs to survey its surroundings, a trait that earned it the name "Picket Pin."

HABITAT: An open country specialist, the Richardson's Ground Squirrel is common on prairies and in meadows and pastures.

FOOD: Like others of its kind, this ground squirrel eats flowers, fruits, seeds, grasses, green vegetation, insects and even some animal protein, mostly as carrion. When foraging, it stuffs its cheek pouches with seeds, which it carries back to its burrow for storage.

DEN: Richardson's Ground Squirrels often live in loose colonies, particularly in favourable habitats. Families live in intricate burrows with the entrances marked at the side by large mounds of excavated dirt. The main burrow is about 4 to 10 m in length and ends in a grass-lined nest chamber. There are usually many secondary entrances and plunge holes.

REPRODUCTION: Mating occurs after the females emerge from hibernation in spring. The gestation period is about 22 days, and the litter of 3 to 11 (usually seven or eight) young is born in May. The newborns are helpless, but they grow quickly and appear aboveground after about three weeks. They are weaned when they are about one month old, and their growth continues throughout summer as they prepare for their first hibernation.

SIMILAR SPECIES: The **Columbian Ground Squirrel** (p. 332) has a rich tawny nose. The **Franklin's Ground Squirrel** (p. 334) has a long, bushy tail. The **Thirteen-lined Ground Squirrel** (p. 342) has 13 alternately broken and solid buffy lines running down its back.

Columbian Ground Squirrel

Cascade Golden-mantled Ground Squirrel

Spermophilus saturatus

In the Cascade Mountains of southern British Columbia, this ground squirrel shares its environs with the Yellow-pine Chipmunk and the Townsend's Chipmunk. Look for them in Manning and Cathedral Lakes provincial parks. These three squirrels manage to avoid direct competition for food and den sites because of slight differences in their specific niches.

Because it is an important prey species, the maximum life expectancy for this ground-dweller is only four years.

DESCRIPTION: The body is mainly dull pinkish or greyish brown with pale buff underparts. The head, neck and shoulders are tinged with russet. A whitish stripe runs down each side of the back and is bordered on either side by a faint dark stripe. Overall, the colour pattern is indistinct. The upper side of the tail is greyish, and the underside is yellowish brown. There is a white eye ring.

HABITAT: Talus slopes and rocky outcroppings are preferred, at elevations of no more than 2300 m.

FOOD: Vegetation, fruits, seeds and fungi make up the bulk of the diet, but insects and some vertebrates are eaten.

DEN: The burrow is short and lacks a large spoil pile at the entrance. The hibernation nest is a mat of vegetation on the floor of a cavity 15 cm in diameter. Short tunnels off the sleeping chambers serve as either storage areas or latrines.

REPRODUCTION: Most males emerge from hibernation in April. Mating occurs when females emerge a week later, and following a 28-day gestation, two to eight altricial young are born. When they are a month old and weigh about 40 g, they are weaned.

SIMILAR SPECIES: The **Golden-mantled Ground Squirrel** (p. 336) has a different range. The **Yellow-pine Chipmunk** (p. 316) and the **Townsend's Chipmunk** (p. 321) are smaller and have stripes on the face.

RANGE: This ground squirrel is found only in the Cascade Mountains of southwestern British Columbia and Washington.

Total Length: 25–32 cm

Tail Length: 9–12 cm

Weight: 215–290 g

Thirteen-lined Ground Squirrel
Spermophilus tridecemlineatus

Where one Thirteen-lined Ground Squirrel is found, others are likely present, but this species rarely exhibits the same degree of colonialism as the Richardson's Ground Squirrel.

Although the Thirteen-lined Ground Squirrel occurs over much of the Prairies, this species is one that can easily be missed by all but the most observant of people because the squirrel's striped back blends perfectly with the alternating pattern of sun and shade created by the tall grass. A Thirteen-lined Ground Squirrel seems easy to distinguish from the far more common Richardson's Ground Squirrel, but a quick glance may be insufficient for such a determination.

The Thirteen-lined Ground Squirrel tends to favour areas with taller grass, through which it cuts paths reminiscent of voles' runways. It usually moves in a series of rushes interspersed with stops of irregular length. When alarmed, this rodent utters a shrill *seek-seek* or a high-pitched trill. It will often stop short of entering its burrow, posing near the entrance instead and allowing an observer to approach closely before disappearing from sight.

RANGE: This prairie species ranges over much of central North America, from southeastern Alberta to southern Manitoba south to New Mexico and Texas and southeast to Ohio.

Total Length: 21–30 cm

Tail Length: 6.8–11 cm

Weight: 83–230 g

The Thirteen-lined Ground Squirrel is one of the most predacious ground squirrels in Canada. Its diet is primarily vegetarian upon emergence from hibernation but shifts markedly during late May and June, when insects, bird nestlings and mice can make up almost half of the daily intake. This ground squirrel may even climb up into small trees or shrubs in search of promising bird nests. Perhaps for this reason, its burrow is often located near trees and shrubs.

ALSO CALLED: Striped Gopher.

DESCRIPTION: The back bears 13 alternately dotted and entire buffy stripes separated by brown. The top of the head is buff sprinkled with brown. The eye ring, nose, cheeks, feet and underparts are buffy. The sides are grey. The central colour of the thin, cylindrical tail is tawny. The longest tail hairs have a blackish subterminal band and buffy tips. The head appears long and narrow, with large eyes and small ears.

HABITAT: This squirrel favours the brushy edges of tallgrass prairie with herbaceous vegetation nearby. Although it is most common at lower altitudes, it is found in certain localities in the mountains.

FOOD: Seeds seem to be the staple component of the diet, but this ground squirrel

DID YOU KNOW?

True to its name, there actually are 13 light-coloured lines on this animal's back, though seven of the "lines" are rows of spots.

eats more animals than other species. Insects, slugs, other invertebrates, young birds, mice and carrion are sought out and devoured. This squirrel also eats berries and native fruits, and it is sometimes a garden pest that consumes peas, beans, strawberries and melons.

DEN: A burrow's entrances, of which there are seldom more than two, are almost never marked by earth mounds. The tunnels are 5 to 6.5 cm in diameter, and they descend steeply for 10 to 100 cm before typically making a right-angle bend and levelling out. The nest chamber, about 23 cm in diameter, is up to 2 m from the entrance in a passageway off the main burrow. It is typically filled with fine grasses and dried roots. The maximum reported diameter of a burrow system is 9 m.

REPRODUCTION: These ground squirrels mate soon after the females emerge from hibernation. The gestation period is about 27 to 28 days, and a litter usually contains 8 to 10 naked, blind, helpless young. The eyes open at 26 to 28 days, soon after which the young emerge from the burrow, switch to a diet of vegetation and meat and are weaned. They are sexually mature following their first hibernation.

SIMILAR SPECIES: No other ground squirrel has a similar pattern of alternating broken and complete buffy stripes.

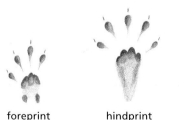

foreprint   hindprint

running trail

Black-tailed Prairie Dog
Cynomys ludovicianus

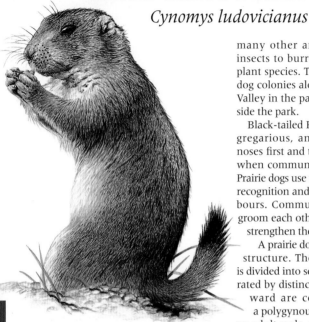

many other animals, from toads and insects to burrowing owls, and several plant species. There are about 15 prairie dog colonies along the Frenchman River Valley in the park and nine colonies outside the park.

Black-tailed Prairie Dogs are extremely gregarious, and kissing—by touching noses first and then incisors—is common when community members cross paths. Prairie dogs use this greeting as a gesture of recognition and welcome between neighbours. Community members routinely groom each other, a behaviour that helps strengthen the social bonds in a colony.

A prairie dog colony has a hierarchical structure. The largest unit, the town, is divided into several wards that are separated by distinct landmarks. Within each ward are coteries; each coterie is a polygynous family unit consisting of an adult male, two to three females and many juveniles and nonbreeding yearlings.

At any one time, several colony members will sit at burrow entrances scanning for predators. If an alarm call is given, all the colony members retreat into the burrows while a few remain at special listening posts just inside the tunnel. When all is clear, the prairie dogs leap from their burrows and run around in a playful, noisy display. They throw their heads back and leap into the air while wheezing high-pitched squeaks. This behaviour is contagious, and soon

Black-tailed Prairie Dogs are one of the curiosities of prairie regions—living in colonies or "towns" that can number several thousand individuals, these squirrels are among the most gregarious of all North American mammals. Their colonies in southern Saskatchewan do not reach the immensity of those of their kin in the southern U.S., but the importance of their colonies is still impressive. Found almost exclusively within Grasslands National Park, prairie dog towns provide important habitat for

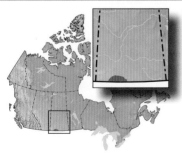

RANGE: The Black-tailed Prairie Dog is found from extreme southern Saskatchewan, through the central U.S. and just into Mexico.

Total Length: 34–42 cm

Tail Length: 7–12 cm

Weight: 680–1490 g

all the prairie dogs have emerged to join in the "leap-squeak" display.

Prairie dogs numbers today are only a tiny fraction of what they were scarcely 100 years ago, when they were actively exterminated from rangelands as "vermin." Today, these creatures inhabit only about two percent of their original range.

DESCRIPTION: The body is mainly yellowish buff with paler underparts. There are prominent brownish patches behind each eye and on each cheek. The ears are pale cinnamon. Black-tipped guard hairs on the back wear unevenly, creating a banded, mottled or speckled appearance in different individuals. The claws are black with light tips. The end of the tail is black or blackish. Sometimes the entire animal is stained the colour of the soil where it lives.

HABITAT: The Black-tailed Prairie Dog typically occupies open grasslands and some rangelands.

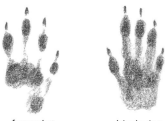

foreprint   hindprint

DID YOU KNOW?

The largest known Black-tailed Prairie Dog town was recorded in Texas. It covered 64,000 km^2 and had some 400 million individuals.

FOOD: Leaves, stems and a few roots make up most of the diet. Insects and carrion are sometimes consumed, but sedges and grasses are of paramount importance.

DEN: The complex, intertwined burrows of the Black-tailed Prairie Dog are a remarkable feat of engineering. A prairie dog town has thousands of branches and rooms, including sleeping chambers, excrement chambers and listening posts. Unlike most ground-dwelling squirrels, this species is active most of the year, but in regions with harsh winters, they may remain in their burrows in a kind of torpor for about four months.

REPRODUCTION: Females mate in March or April. Following a gestation period of 33 to 38 days, up to six (typically three) naked, blind, helpless pups are born. The pups stay below ground and nurse until they are five to seven weeks old. When they emerge above ground, they are fully weaned and begin eating vegetation. Although pups are sexually mature at about 12 months of age, most do not mate until they are at least 21 months old.

running trail

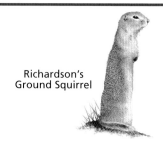

Richardson's
Ground Squirrel

SIMILAR SPECIES: The **Richardson's Ground Squirrel** (p. 339) is smaller and has faint dappling over its back.

Eastern Grey Squirrel
Sciurus carolinensis

black form

grey form

T he Eastern Grey
Squirrel is the most frequently
encountered large squirrel in eastern North America and can be locally abundant in both cities and natural areas. It is active throughout the entire year, sometimes even digging through snow to retrieve its buried nuts.

Many stories are told of the great migrations of Eastern Grey Squirrels, but squirrels do not actually migrate. In autumn, young animals disperse, and in times of food shortage, adults may also disperse to find better homes. Technically, a migration is a movement of animals to and from specific regions in response to changing seasons, whereas these squirrels "reshuffle" in response to food and population stresses over longer periods of time. When large nut crops and high reproduction rates among the squirrels are followed by a year of little food, hundreds or even thousands of Eastern Grey Squirrels may move to find new food sources. When these squirrels travel, they can cover large tracts of forests without ever touching the ground.

The mainstay diet is nuts and seeds, and these squirrels will have dozens of nut caches buried just under the surface of the soil. The caches of most other squirrels germinate if left for too long, but Eastern Grey Squirrels determinedly nip off the germinating end of the nuts before burying them. These squirrels routinely travel throughout their home range, keeping apprised of the fresh food sources available.

The classic "park and garden squirrel" of much of the western world, this animal entertains millions of urban residents who restrict their wild adventures to places in

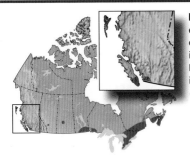

RANGE: This squirrel's native range encompasses all of the eastern U.S. and parts of Canada to southern Manitoba and Ontario in the north and eastern Texas in the south. Introduced populations have been established in Calgary, Vancouver, Victoria, Seattle and other western cities.

Total Length: 43–50 cm

Tail Length: 21–24 cm

Weight: 400–710 g

cities. This species has even been introduced to Great Britain, where it inhabits city parks and natural areas and is known as the American Grey Squirrel. Introductions of animals to new areas are never without complications—this squirrel has spread across Britain, displacing the native Red Squirrel (*S. vulgaris*), and may have a negative effect on native songbirds by feeding on eggs and hatchlings in spring.

DESCRIPTION: The Eastern Grey Squirrel has two distinct colour forms. Some adults are dusty grey with pale underparts and a silvery, flattened tail. The long tail hairs are white-tipped. There may be cinnamon highlights on the head, back and tail. The other form is solid black. Albinos also occur, but they are rare.

HABITAT: These squirrels prefer mature deciduous or mixed forests, with lots of nut-bearing trees. Older forests with larger trees support higher populations of squirrels—bigger trees provide more food and numerous suitable nesting sites.

DID YOU KNOW?

Eastern Grey Squirrels, as well as several other squirrel species, "live and learn," becoming wiser as they age. Although many die before they are a year old, some may live quite a long time, and a few individuals make it to 10 years of age or more.

FOOD: These nut lovers feed mainly on the seeds of oak, maple, ash and elm. In spring and summer, they also eat buds, flowers, leaves and occasionally animal matter such as eggs or nestling birds.

DEN: Eastern Grey Squirrels den in trees year-round. They build nests lined with dry vegetation in natural tree cavities or in refurbished woodpecker holes. Where cavities are not available, they build dreys, which are spherical leaf-and-twig nests in tree branches. In cold regions, they have been known to make ground nests, but this behaviour is not common.

REPRODUCTION: Breeding occurs from December to February and rarely in July or August. Most females have only one litter per year. Gestation is 40 to 45 days, after which a litter of one to eight (usually two to four) helpless young is born. The eyes open at 32 to 40 days, and weaning occurs about three weeks later.

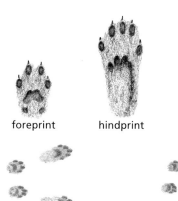

foreprint   hindprint

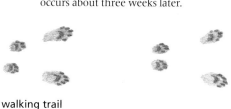

walking trail

Eastern Fox Squirrel

SIMILAR SPECIES: The slightly larger **Eastern Fox Squirrel** (p. 348) usually has more red in its fur (especially on the belly) and yellow-tipped hairs on the tail. The **Red Squirrel** (p. 350) is smaller and reddish brown. The **Northern Flying Squirrel** (p. 352) and the **Southern Flying Squirrel** (p. 354) are smaller, nocturnal and sooty grey.

Eastern Fox Squirrel
Sciurus niger

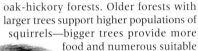

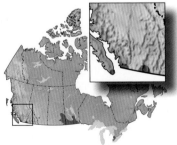

This large, colourful tree squirrel is rare in Canada, and like the Eastern Grey Squirrel, it has been introduced to some city parks. The Eastern Fox Squirrel usually leads a solitary life. Its most gregarious behaviour occurs in winter when several adults, who are often related and have overlapping territories, share food caches and tree cavities. This squirrel remains active in winter, and the individuals sharing a tree cavity come and go regardless of the cold.

Eastern Fox Squirrels are not as social as other squirrel species. They do not groom each other, "kiss" or nuzzle to maintain friendly ties.

DESCRIPTION: This squirrel has three different colour phases, but only one is common in Canada—rusty orange or greyish above with cinnamon-coloured underparts. The long tail hairs are reddish, as are the ears.

HABITAT: These squirrels prefer mature mixed or deciduous forests, particularly oak-hickory forests. Older forests with larger trees support higher populations of squirrels—bigger trees provide more food and numerous suitable nesting sites.

FOOD: Fox squirrels avidly feed on seeds, fruits, fungi, green pine cones and corn kernels. Any nonperishables such as seeds and nuts are buried in caches just under the ground surface. Like other squirrels, they may consume animal matter.

DEN: Fox squirrels den in trees year-round. They either build dreys or use natural tree cavities or woodpecker holes.

REPRODUCTION: Breeding occurs in midwinter, and a litter of two to four is born in February or March. Females may have a second litter in August or September. The young open their eyes at 32 to 40 days, and weaning occurs about three weeks later.

SIMILAR SPECIES: The slightly smaller **Eastern Grey Squirrel** (p. 346) is not as red and has white-tipped hairs on the tail. The **Red Squirrel** (p. 350) is smaller and reddish brown. The **Northern Flying Squirrel** (p. 352) and the **Southern Flying Squirrel** (p. 354) are much smaller, nocturnal and sooty grey.

RANGE: This squirrel's native range encompasses all of the eastern U.S. and parts of Canada to southern Manitoba and Ontario in the north and eastern Texas in the south. Introduced populations now exist in Calgary, Vancouver, Victoria, Seattle and other western cities.

Total Length: 43–50 cm

Tail Length: 21–24 cm

Weight: 400–710 g

Douglas's Squirrel
Tamiasciurus douglasii

Famous for their large repertoire of calls, Douglas's Squirrels are frequently heard denizens of the West Coast. Look for them in Garibaldi, Golden Ears and Manning provincial parks. These squirrels live almost exclusively in coniferous forests, and they leap easily from limb to limb as they search for food.

Douglas's Squirrels are renowned for their appetite for conifer cones. Running along conifer branches, they nip the cones free and let them fall to the ground. On a busy day, these squirrels bombard the forest floor with cones for much of the morning. When enough have been cut, the squirrels eagerly transfer them to large cone caches beside tree stumps or under fallen logs.

ALSO CALLED: Chickaree.

DESCRIPTION: In summer, the eye ring, feet and underparts are pumpkin orange, and the back and head are grizzled olive brown. The ear tufts are black, as are the flank stripes separating the brown back and orange underside. The bushy tail is dark reddish. The winter coat is more grizzled and grey.

HABITAT: Coniferous coastal rainforests are home to this squirrel, though it sometimes ventures into logged areas.

FOOD: This squirrel feeds mainly on fir, pine, spruce and hemlock seeds and green cones, but it also consumes maple samaras, alder catkins, other seeds and nuts, berries and mushrooms.

DEN: The nest is typically in a hollow tree, but this squirrel may construct a drey high in a conifer, often using an abandoned hawk or crow nest as a foundation.

REPRODUCTION: Following mating in early April, two to eight (usually four) young are born after a gestation of 35 days. The young are weaned at two months, and then have to establish their own territories. Juveniles are sexually mature following their first winter.

SIMILAR SPECIES: The **Red Squirrel** (p. 350) has a white or silvery grey belly and a redder back.

RANGE: The Douglas's Squirrel is found from southwestern British Columbia south through Washington and Oregon to the Sierra Nevada of southern California.

Total Length: 27–35 cm

Tail Length: 10–16 cm

Weight: 150–300 g

Red Squirrel
Tamiasciurus hudsonicus

gathering and storage. It urgently collects conifer cones, mushrooms, fruits and seeds in preparation for the winter months. This squirrel remains active throughout winter, except in severely cold weather. At temperatures below −25°C, it stays warm but awake in its nest.

Because the Red Squirrel does not hibernate, it needs to store massive amounts of food in winter caches. Most food caches are smaller than a cubic metre, but in extreme cases, they can reach the size of a garage. Much of this animal's efforts throughout the year are concentrated on filling these larders, and biologists speculate that its characteristically antagonistic disposition is a result of having to continually protect its food stores.

Few squirrels have earned such a reputation for playfulness and agility as the Red Squirrel. It is a well-known backyard and ravine inhabitant with a saucy disregard for its human neighbours. Like a one-man band, the Red Squirrel firmly scolds all intruders with shrill chatters, clucks and sputters, tail flicking and feet stamping. Even when it is undisturbed, this loquacious squirrel often chirps as it goes about its daily routine.

The industrious Red Squirrel devotes the daytime hours almost entirely to food

By the end of winter, Red Squirrels are ready to mate. Their courtship involves daredevil leaps through the trees and high-speed chases over the forest floor. Later, the youngsters are playful and frequently challenge nuts or mushrooms to bouts of mock combat.

DESCRIPTION: The shiny, clove brown summer coat sometimes has a reddish

RANGE: The Red Squirrel occupies coniferous forests across most of Alaska and Canada. In the West, its range extends south through the Rocky Mountains to southern New Mexico. In the East, this squirrel occurs south to Iowa and Virginia and through the Allegheny Mountains.

Total Length: 28–35 cm

Tail Length: 11–15 cm

Weight: 170–310 g

wash along the centre of the back. A black longitudinal line on each side separates the dorsal colour from the greyish to white underparts. There is a white eye ring. The backs of the ears and the legs are rufous to yellowish. The longest tail hairs have a black subterminal band and a buffy tip, which gives the tail a light fringe. The longer, softer winter fur tends to be bright to dusky rufous on the upperparts, with fewer buffy areas, and the head and belly tend to be greyer. The whiskers are black.

HABITAT: Boreal coniferous forests and mixed forests make up the major habitat, but towns with trees more than 40 years old also support Red Squirrel populations.

FOOD: Most of the diet consists of seeds extracted from conifer cones. A midden is formed where discarded cone scales and cores pile up below a favoured feeding perch. Flowers, berries, mushrooms, eggs,

DID YOU KNOW?

In a race against time, a Red Squirrel works to store spruce cones—as many as 14,000—in damp caches that prevent the cones from opening. If the cones open naturally on the tree, the valuable, fat-rich seeds are lost to the wind.

birds, mice, insects and even baby Snowshoe Hares or chipmunks may be eaten.

DEN: Tree cavities, witch's broom (deformities created in trees in response to mistletoe or fungal infections), logs and burrows may serve as den sites. The burrows and entrances are about 15 cm in diameter, with an expanded cavity housing a nest ball that is 40 cm across.

REPRODUCTION: Northern populations bear just one litter per year. Peak breeding, in April and May, is associated with frenetic chases and multiple copulations lasting up to seven minutes each. After a 35- to 38-day gestation, a litter of two to seven (usually four or five) pink, blind, helpless young is born. The eyes open at four to five weeks, and the young are weaned when they are seven to eight weeks old. Red Squirrels are sexually mature by the following spring.

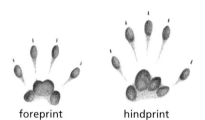

foreprint   hindprint

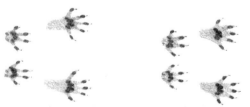

running trail

Douglas's Squirrel

SIMILAR SPECIES: The **Douglas's Squirrel** (p. 349) has orange underparts. The **Northern Flying Squirrel** (p. 352) and the **Southern Flying Squirrel** (p. 354) are similar in size, but they are a sooty pewter colour and are nocturnal. The **Eastern Grey Squirrel** (p. 346) and the **Eastern Fox Squirrel** (p. 348) are both much larger.

Northern Flying Squirrel
Glaucomys sabrinus

Like drifting leaves, Northern Flying Squirrels seem to float from tree to tree in forests throughout much of Canada. Although a flying squirrel is not capable of true flapping flight—bats are the only mammals to have mastered that feat—its aerial travels are no less impressive, with extreme glides of up to 100 m. Enabling this squirrel to "fly" are its glide membranes, called the patagium—capelike, furred skin that extends down the length of the body from the forelegs to the hindlegs.

Before a glide, a flying squirrel identifies a target and manoeuvres into the launch position—a head-down, tail-up orientation in the tree. Then, using its strong hindlegs, the squirrel propels itself into the air with its legs extended. Once airborne, it can make rapid side-to-side manoeuvres and tight downward spirals. Such control is accomplished by making minor adjustments to the orientation of the wrists and forelegs. On the ground and in trees, a flying squirrel can hop or leap, but the skin folds prevent it from running. It is not able to swim, either.

The call of the Northern Flying Squirrel is a loud *chuck chuck chuck*, which increases in pitch to a shrill falsetto when the animal is disturbed. Like other tree squirrels, the Northern Flying Squirrel does not hibernate. On severely cold days, however, groups of 5 to 10 individuals can be found huddled in a nest to keep warm.

DESCRIPTION: A unique web or fold of skin (patagium) extends laterally to the level of the ankles and wrists to become the abbreviated "wings" with which the animal glides. The eyes are

RANGE: This flying squirrel occurs in eastern Alaska and across most of Canada in appropriate habitats. Its range extends south through the western mountains to Utah and California, around the Great Lakes and through the Appalachians.

Total Length: 25–37 cm

Tail Length: 11–18 cm

Weight: 75–185 g

large and dark. The back is light brown, with hints of grey from the lead-coloured hair bases. The belly hairs are white tipped but are grey or lead grey at the base. The underparts are light grey to cinnamon precisely to the edge of the gliding membrane and the edge of the tail. The tail is noticeably flattened top to bottom and functions much as the rudder and elevators do on an airplane.

HABITAT: Coniferous and mixed forests are prime flying squirrel habitat.

FOOD: The bulk of the diet consists of lichens and fungi, but flying squirrels also eat buds, berries, some seeds, a few arthropods, bird eggs and nestlings and the protein-rich, pollen-filled male cones of conifers. They cache cones and nuts.

DEN: Nests in tree cavities, which are the most common, are lined with lichens and grasses. Leaf nests, called dreys, are located in tree forks close to the trunk. Twigs and strips of bark are used on the outside, with progressively finer materials used inside to the centre. If the drey is for winter use, it is additionally insulated to a diameter of 40 cm.

> **DID YOU KNOW?**
>
> Northern Flying Squirrels are often just as common in an area as Red Squirrels, but they are nocturnal and are therefore rarely seen. Flying squirrels routinely visit bird feeders at night; they value the seeds as much as sparrows and finches do.

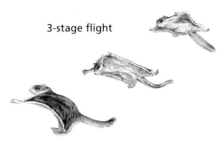

3-stage flight

REPRODUCTION: Mating takes place between late March and the end of May. After a six-week gestation, typically two to four young are born. They weigh about 5 g at birth. The eyes open after about 52 days. Ten days later, the young first leave the nest, and they are weaned when they are about 65 days old. Young squirrels first glide at three months of age, and it takes them about a month to become skilled. Flying squirrels are not sexually mature until after their second winter.

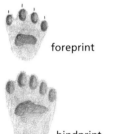

foreprint

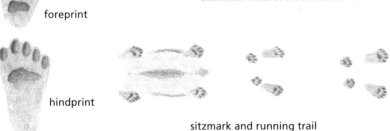

hindprint

sitzmark and running trail

Southern Flying Squirrel

SIMILAR SPECIES: The **Southern Flying Squirrel** (p. 354) is smaller, often more pewter-coloured, and the belly hairs are white to the base. The **Red Squirrel** (p. 350) is reddish brown overall. The **Eastern Grey Squirrel** (p. 346) and the **Eastern Fox Squirrel** (p. 348) are much larger. All nonflying tree squirrels lack the glide membranes on their sides.

Southern Flying Squirrel
Glaucomys volans

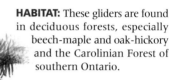

HABITAT: These gliders are found in deciduous forests, especially beech-maple and oak-hickory and the Carolinian Forest of southern Ontario.

FOOD: These squirrels collect and eat nuts, seeds, fruits, mushrooms and insects. They regularly eat small vertebrates and carrion.

DEN: Flying squirrels may refurbish old corvid nests, but most often occupy tree cavities. They sometimes make spherical dreys that are occupied by several individuals over many generations.

Small and steel grey, the Southern Flying Squirrel is a common resident of southeastern forests. It is active after sunset and just before dawn, so very few people ever see it.

Although these squirrels are primarily nonhibernating, severe winter conditions may induce a metabolic torpor. This torpid state outwardly resembles hibernation, but it is short-lived and not as deep. After the cold snap has passed, the squirrel rouses and resumes its activity. Huddling is a more common response to cold weather—in winter, 5 to 50 flying squirrels are often found together in a tree cavity, relying on their body temperature to warm the den space.

DESCRIPTION: The fine, cool-grey coat usually has cinnamon highlights. The underparts are nearly or completely white, right to the base of the hairs. The glide membranes are black-edged. The tail is broadly plumed and flattened.

REPRODUCTION: Mating occurs in early spring, and females bear a litter of two to seven young after 41 days of gestation. The young are altricial and are not weaned until they are about seven weeks old. The young stay with the mother for the summer unless another litter is born.

SIMILAR SPECIES: The **Northern Flying Squirrel** (p. 352) is larger and often has more brown in its coat. The belly fur is white tipped, but greyish at the base. The **Red Squirrel** (p. 350) is reddish brown overall. The **Eastern Grey Squirrel** (p. 346) and **Eastern Fox Squirrel** (p. 348) are much larger. All nonflying tree squirrels lack the glide membranes on their sides.

RANGE: The Southern Flying Squirrel is found in most of the eastern U.S. and north into southern Ontario, Québec, New Brunswick and Nova Scotia.

Total Length: 20–25.5 cm

Tail Length: 8–12 cm

Weight: 45–100 g

Mountain Beaver

Aplodontia rufa

The Mountain Beaver is considered the most "primitive" living rodent. Unlike other rodents, it depends on the availability of ferns in its environment. Although ferns are toxic to most other rodents, they are the primary food for this creature, hinting at its ancient origin.

The Mountain Beaver's cheek teeth are also unlike those of other rodents—they have a single central lobe of dentine surrounded by a ridge of enamel. Other rodents have teeth that show complex folding with a proliferation of enamel. When digging a burrow, a Mountain Beaver may come across stones and lumps of clay that it keeps in its burrow and occasionally gnaws upon to sharpen its teeth. These "Mountain Beaver baseballs" are also used to block the entrances of vacated burrows.

The Mountain Beaver can climb a sapling tree to a height of 3 to 4 m, allowing it to eat the tender shoots. It can also swim for short distances. When foraging, it collects leafy branches and other vegetation and carries the spoils to its burrow, where it sits on its short tail and grasps the vegetation in its forepaws using its semi-opposable thumbs.

Like a lagomorph, the Mountain Beaver reingests its soft fecal pellets, allowing better absorption of nutrients the second time through the digestive system. Hard fecal pellets, the result of the reingested soft pellets, are seized with the incisors and thrown into a burrow latrine.

ALSO CALLED: Sewellel, Ground Bear, Boomer, Aplodontia.

DESCRIPTION: At first glance, the Mountain Beaver looks similar to a Muskrat or giant pocket gopher, except that it has a very short, well-furred tail. The stocky body is covered with coarse, reddish brown or greyish brown fur. The underparts are

RANGE: The Mountain Beaver is a western North American species that ranges from the Nicola Valley in British Columbia to southeast of San Francisco near the Nevada-California border in the Sierra Nevada.

Total Length: 30–47 cm

Tail Length: 2–5 cm

Weight: 0.3–1.4 kg

light greyish brown or tawny. The short, strong limbs each have five toes, and the forelimbs are equipped with long, laterally compressed, cream-coloured claws. The soles of the feet are naked to the heel. There is a light spot below each of the short, round ears, and the white whiskers are abundant.

HABITAT: Within its limited range, the Mountain Beaver can be found in wooded areas from near sea level to treeline. It favours early seral vegetative stages with an abundance of shrubs, forbs and young trees. The highest densities of this species appear to be in deciduous forests of mountain parks; few occupy dense, old, coniferous forests.

FOOD: This animal consumes a wide variety of plants, but sword and bracken ferns form the bulk of the diet during all seasons. Bracken fern is poisonous to most mammals, but the Mountain Beaver is unharmed by it. In October, when the protein content of red alder leaves is highest, up to three-quarters of a male's diet may be composed of these leaves. Pregnant and lactating females eat a high volume of new, succulent growth.

> **DID YOU KNOW?**
>
> Many biologists consider the Mountain Beaver to be a keystone species. More than 20 different species of mammals and many kinds of insects and amphibians are known to inhabit its burrows.

DEN: The Mountain Beaver constructs an extensive burrow system consisting of tunnels radiating out from a nest chamber. Numerous burrows penetrate to the surface, but only a few have dirt piles around the opening. The nest chamber, about 30 cm in diameter, contains dried leaves and grass trampled into a flat pad. Generally, the nest is 30 cm to 1.5 m beneath the surface, but tunnels may penetrate up to 3 m underground. Pockets in the walls of some larger tunnels may contain roots, stems and leaves. Sometimes, tentlike structures of sticks, leaves and succulent vegetation are found over burrow entrances.

REPRODUCTION: After a gestation of 28 to 30 days, the young are born in March or April in the subterranean nest. Each of the one to four altricial young weighs about 25 g. Their eyes open at 45 to 54 days, and they then begin to eat vegetation and grow rapidly. Neither sex appears to be sexually mature until the second winter.

foreprint

hindprint

walking trail

SIMILAR SPECIES: All **marmots** (pp. 324–30) have longer, bushy tails. The **Muskrat** (p. 295) and introduced **Nutria** (p. 254) have nearly naked tails and are associated with water.

Muskrat

HARES & PIKAS

These rodentlike mammals are often called "lagomorphs" after the scientific name of the order, Lagomorpha, which means "hare-shaped." Rabbits, hares and pikas share the rodents' trademark chisel-like upper incisors, and taxonomists once grouped the two orders together. Unlike rodents, however, lagomorphs have a second pair of upper incisors. Casual observers never see these peglike teeth, which lie immediately behind the first pair of upper incisors.

Lagomorphs are herbivores, but they have relatively inefficient, nonruminant stomachs that have trouble digesting such a diet. To make the most of their meals, they defecate pellets of soft, green, partially digested material that they then reingest to obtain maximum nutrition. Bacteria that enter the food in the intestines contribute to better digestion and absorption of nutrients the second time around. The pellets excreted after the second digestive process are brown and fibrous.

Hare Family (Leporidae)

Rabbits and hares are characterized by their long, upright ears, long hindlegs built for jumping and short, cottony tails. These timid animals are primarily nocturnal, and many spend the day resting in shallow depressions called "forms." Rabbits and hares both belong to the same family, but they have distinct differences. Luckily, differentiating between the two is easy, with the right information. Native hares turn white in winter, are larger overall and tend to have longer ears and hindlegs than rabbits. The chief difference between the two involves reproduction. Rabbits build a maternity nest for their young, and when the young are born, they are altricial, meaning that they are helpless, blind and naked at birth. Hares, on the other hand, do not make a maternity nest and their young are precocial—they are born well developed, fully furred and with open eyes. Soon after birth, the leverets (young hares) begin to feed on vegetation, and the mothers nurse them only once per day.

Pika Family (Ochotonidae)

Pikas are the most rodentlike lagomorphs, and their short, rounded ears and squat bodies give them the appearance of small guinea pigs. Their front and rear limbs are about the same length, so pikas scurry, rather than hop, through the rocky outcrops and talus of their home territories. Pikas are most active during the day, so they are often seen by hikers in mountain parks.

Eastern Cottontail
Sylvilagus floridanus

The Eastern Cottontail is native only to the eastern half of North America. It has been widely introduced throughout many of the western states, and it is now the most widespread cottontail in North America. Its success is partly a result of its ability to adapt to a wide variety of habitats. The main requirement is tall grasses or shrubs that provide adequate cover for protection from predators. This cottontail has high fecundity, and its numbers can increase quickly in favourable conditions.

During twilight, Eastern Cottontails emerge from their daytime hideouts to graze on succulent vegetation. If you are a patient observer, you may see them as they daintily nip at grasses, always just a short leap from dense bushes or a rocky shelter. Sometimes cottontails can be seen during the day, especially young ones or those in urban areas.

The prime habitat for an Eastern Cottontail is neither fully wooded nor completely open—this rabbit requires good protective cover, but in areas where

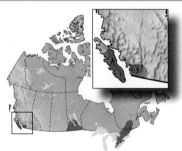

RANGE: This rabbit is found in the U.S. and along the southern borders of several eastern Canadian provinces. Introduced populations live in British Columbia, California, Oregon and Washington.

Total Length: 40–45 cm

Tail Length: 4–7 cm

Weight: 0.8–1.6 kg

foliage is too dense, it is handicapped in its ability to detect an approaching predator. Coyotes, foxes, owls and hawks prey heavily on this species.

Eastern Cottontails spend most of their days sitting quietly in forms (shallow depressions) beneath impenetrable vegetation or under boards, rocks, abandoned machinery or buildings. These midsized herbivores have small home ranges of about 1 to 2 ha. Heavy rains greatly diminish cottontail activity, restricting them to their hideouts for the duration of the storm. Eastern Cottontails do not hibernate during winter, but they limit their movements to traditional trails that they can easily locate after a snowfall. During very cold winters, they may take shelter in Woodchuck burrows.

DESCRIPTION: This rabbit is pale buffy grey above, with somewhat paler sides. The back may be grizzled with blackish hairs. The nape of the neck is orangey, and the legs are cinnamon. The underparts are whitish. The tail is brown above and cottony white below, but the white only shows when the animal is running.

HABITAT: The only major habitat requirement is cover, whether it is brush, rocky outcroppings or buildings. This rabbit likes edge situations where trees meet meadows or where brushy areas meet agricultural land. It also inhabits brushy riparian areas and is

> **DID YOU KNOW?**
>
> No other rabbit or hare is as widespread as the Eastern Cottontail. In fact, this species' range overlaps that of seven other species of cottontails and six species of hares.

occasionally seen on lawns, city parks or golf courses with nearby shrubby cover.

FOOD: This species favours clover, grasses and herbaceous vegetation in summer and woody plants in winter. It may also eat the bark of young trees.

DEN: Brush piles, holes and leaf litter are used for escape cover, but there is no true den other than for a pregnant female. Before giving birth, the female digs a nest about 25 cm long and 15 cm wide and lines it with grasses and her own fur. She then covers the depression so it is nearly impossible to see.

REPRODUCTION: These rabbits are among the most fecund of all lagomorphs and may breed at anytime of the year in suitable climates. Following mating and a gestation of 18 to 30 days, 1 to 9 bunnies are delivered in the nest. Within hours of giving birth, the mother is in estrus again and can breed. At one month of age, the young are independent, and young females can breed when they are just four months old.

SIMILAR SPECIES: Both the **Snowshoe Hare** (p. 364) and the **White-tailed Jackrabbit** (p. 370) are much larger and become white in winter. The **European Hare** (p. 368) is larger and has longer ears.

Snowshoe Hare

Mountain Cottontail
Sylvilagus nuttallii

This cottontail fits the image of the classic cute bunny—the coal black eyes, rounded ears and soft features endear it to almost every wildlife watcher who sees it. The plush-toy caricature, however, masks a tough animal that is quite capable of surviving in a harsh, unforgiving landscape full of predators.

Mountain Cottontails spend most of their days sitting quietly in forms (shallow depressions) beneath clumps of impenetrable vegetation or under rocks, abandoned machinery or buildings. These midsized herbivores have small home ranges that rarely exceed the size of a baseball field. During heavy rains, cottontails remain in their hideouts for the duration of the storm. Although these rabbits do not hibernate during winter, they confine their movements to well-used trails that they can easily locate after a snowfall.

When the sun lowers to meet the horizon, flooding mid-summer evenings in golden light, Mountain Cottontails emerge from their daytime hideouts to graze on succulent vegetation. This species has only a small range in Canada—and ironically, it does not include very much of the mountains—but if you are a patient observer, you may see this rabbit as it daintily dines on grasses, always just a short leap from protective cover.

Thomas Nuttall (1786–1859) is a fine choice to be immortalized in the name of this inquisitive and endearing animal. Although Nuttall was primarily a botanist, he made significant contributions to all fields of natural history. He was also renowned for his absent-mindedness and misadventures. Many of his most famous gaffes occurred during his journey across

RANGE: The western limit of this rabbit's range parallels the eastern border of California, running north into the Okanagan region of British Columbia. The eastern edge of the range runs south from southern Saskatchewan along the Montana-Dakotas border to northern New Mexico.

Total Length: 33–41 cm

Tail Length: 3–6 cm

Weight: 680–1020 g

the continent to the Pacific Ocean on the Wyeth expedition in 1834. He became lost on several occasions, and should any animal provide a sense of comfort to one so error-prone, it would surely be the cottontail that now bears his name.

ALSO CALLED: Nuttall's Cottontail.

DESCRIPTION: This rabbit has dark, grizzled, yellowish grey upperparts and whitish underparts year-round. The tail is blackish above and white below. There is a rusty orange patch on the nape of the neck. The front and back edges of the ears are white, and the ears are usually held erect when the rabbit runs.

HABITAT: A major habitat requirement is cover, whether it is brush, fractured rock outcrops or buildings. These rabbits like edge situations where trees meet meadows or where brushy areas meet agricultural land.

FOOD: Grasses and forbs are the primary foods, but in many areas, these cottontails feed heavily on sagebrush and juniper berries.

DID YOU KNOW?

Rabbits and hares depend on intestinal bacteria to break down the cellulose in their diets. Because the bacterial products reenter the gut beyond the site of absorption, rabbits eat their fecal pellets to process the material through the digestive tract a second time.

DEN: There is no true den, but a Mountain Cottontail will shelter in a form dug out beneath and among rocks, boards, a building and the like. The young are born in a nest that is dug out by the female and lined with grass and fur. The doe lies over the top of the nest while the young nurse. The nest is essentially invisible, and a casual observer would never suspect that the female was nursing, or that a nest of babies lay beneath her.

REPRODUCTION: Breeding begins in April, and after a 28- to 30-day gestation, a litter of one to eight (usually four or five) young is born. The female enters estrus and can breed within hours of giving birth, so there may be two litters per season. The young are born blind, hairless and with their eyes closed. They grow quickly and are weaned just before the birth of the subsequent litter.

foreprint

hindprint

hopping trail

Snowshoe Hare

SIMILAR SPECIES: Both the **Snowshoe Hare** (p. 364) and the **White-tailed Jackrabbit** (p. 370) are much larger and usually become white in winter.

European Rabbit
Oryctolagus cuniculus

Symbolically, rabbits are associated with reproduction—hence the phrase "breeds like a rabbit." For Easter celebrations, the rabbit is a popular image that represents new life, renewal and fertility. The European Rabbit is the species that particularly inspired this imagery—this creature raises many large litters in each season and multiplies at an astonishing rate.

This species has a high reproductive rate and is very adaptable, thriving almost everywhere it has been introduced. The Hudson's Bay Company, as well as many settlers, raised European Rabbits and introduced them to the wild in the Pacific Northwest. The idea was to populate common hunting areas with a rapidly reproducing game rabbit that was larger than the native cottontails. As with most species introductions, the people responsible for it did not foresee the potentially serious consequences. In the San Juan Islands of Washington, for example, the introduced European Rabbits caused drastic problems. These prolific burrowers honeycombed so much of the island that the local lighthouse nearly fell over and parts of the shoreline crumbled into the ocean.

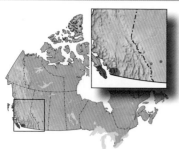

RANGE: These rabbits are well naturalized on southeastern Vancouver Island and the Lower Mainland. Intentional releases in Washington, Pennsylvania, Illinois, Indiana, New Jersey, Wisconsin and Maryland have resulted in large populations in those states. Scattered populations may exist elsewhere in Canada and the U.S. as a result of escaped pets.

Total Length: 45–60 cm

Tail Length: 7–9 cm

Weight: 1.4–2.3 kg

The European Rabbit's tendency to dig extensive burrow systems is reflected in the species name *cuniculus*, which is Latin for "underground passage." Hundreds of rabbits may live and breed together in these networked tunnels, forming colonial groups called "warrens."

The distribution and abundance of European Rabbits in Canada is not well documented, and so far they have only been recorded in British Columbia. On Vancouver Island, especially around Victoria, and on Triangle Island, the European Rabbit appears to be a common naturalized resident. It probably also occurs on many islands in the Strait of Georgia.

ALSO CALLED: Belgian Hare, Domestic Rabbit, San Juan Rabbit.

DESCRIPTION: This rabbit's colouration is highly variable, and it appears in an array of greys and browns, as well as black and white. Some individuals have multicoloured coats, owing to their ancestry as a favourite pet animal that was bred for colour variation. The ears are medium-sized, about 6 to 10 cm long, and the bicoloured tail is dark above and white below.

HABITAT: This Old World rabbit generally avoids heavily wooded areas in favour of open fields, brushy areas and parkland. It does not have strict dietary requirements and can quickly populate most areas where it is introduced or has escaped.

DID YOU KNOW?

This rabbit originated in the western Mediterranean and is the forerunner of all domestic rabbits. As with the domestic dog, there are hundreds of rabbit breeds and varieties.

FOOD: European Rabbits feed on short grasses and leafy herbaceous plants. In exceptional circumstances, when such plants are not available, they will consume any available vegetation. In croplands, they often overbrowse and become serious, costly pests. In some areas, these rabbits seriously threaten both native plant species and the animals that depend on them.

DEN: Although these rabbits may be found singly, they often live in large colonies of dozens or even hundreds of individuals and dig extensive burrows. The area around a warren is typically denuded of vegetation. Nests are made of grass and other soft fibres and are built in special chambers of the burrow system.

REPRODUCTION: A female European Rabbit can have as many as six litters per year, with up to 12 young in each litter. The gestation period is 30 days. The young are altricial but grow rapidly. Predation on juvenile rabbits is high.

SIMILAR SPECIES: The **Eastern Cottontail** (p. 358) and **Mountain Cottontail** (p. 360) are usually smaller. **Hares** (pp. 364–70) have distinctly longer ears. Any rabbit with unusual colouration is a European Rabbit.

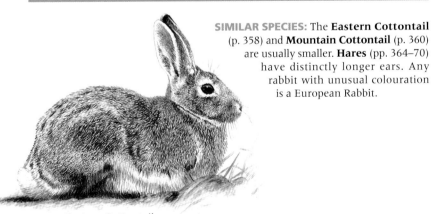

Eastern Cottontail

Snowshoe Hare
Lepus americanus

summer
coat

It is well known that lynx and hare populations fluctuate in close correlation with one another, but few people realize that other species are involved in the cycle. Recent studies have shown that as hares increase in number, they overgraze willows and alders in their habitat. In response to overgrazing, the trees produce a distasteful, toxic substance in their shoots. This substance protects the plants but initiates starvation in the hares. The change in food availability impacts the hare population at a time when the animals are already stressed by predation. The 10-year cycle between lynx, Snowshoe Hares and plants is complex and only partially understood, but it is one of the most dramatic documented cycles within the boreal forest ecosystem.

The Snowshoe Hare is able to withstand the most unforgiving aspects of northern wilderness because it possesses several fascinating adaptations for winter. As its name implies, the Snowshoe Hare has very large hindfeet and can easily walk on areas of soft snow, whereas other animals sink into the powder. This is a tremendous advantage for an animal that is preyed upon by so many different carnivores. Unfortunately, it is of minimal help against the equally big-footed Canada Lynx, a specialized hunter of the Snowshoe Hare.

In response to shorter periods of daylight at the onset of winter, Snowshoe Hares start moulting into their white winter camouflage whether snow falls or not. If the year's first snowfall is late, some individuals will lose their usual concealment and become visible from great distances—to naturalists and predators alike—as bright white balls in a brown world. The hares seem to be aware of this predicament, and they often seek out any small patch of snow on which to squat.

RANGE: The range of the Snowshoe Hare is associated with the boreal coniferous forest and mountain forests from northern Alaska to Labrador, south to California and New Mexico.

Total Length: 38–53 cm

Tail Length: 4.8–5.4 cm

Weight: 1–1.5 kg

DESCRIPTION: The summer coat is rusty brown above, with the crown of the head darker and less reddish than the back. The nape of the neck is greyish brown, and the ear tips are black. The chin, belly and lower surface of the tail are white. Adults have white feet, whereas juveniles have dark feet. In winter, the terminal portion of nearly all the body hair becomes white, but the hair bases and underfur are lead grey to brownish. The ear tips remain black.

HABITAT: Snowshoe Hares may be found in most of Canada, primarily where there is forest or dense shrubs.

FOOD: In summer, a wide variety of grasses, forbs and brush may be consumed. In winter, mostly the buds, twigs and bark of willows and alders are eaten. These hares occasionally eat carrion.

DEN: This hare does not keep a customary den but sometimes enters a hollow log or the burrow of another animal. In areas of human activity, it may shelter beneath a building.

DID YOU KNOW?

Between their highs and lows, Snowshoe Hare densities can vary by a factor of 100. During highs, there may be 12 to 15 hares per hectare; after a population crashes, hares may be relatively uncommon over huge geographical areas for years.

REPRODUCTION: Breeding activity begins in March and continues through August. After a gestation of 35 to 37 days, one to seven (usually three or four) young are born under cover, but often not in an established form or nest. The female breeds again within hours of the birth, and she may have as many as three litters in a season. The young hares are precocial and can hop within a day; they feed on grassy vegetation within 10 days. In five months, they are fully grown.

winter coat

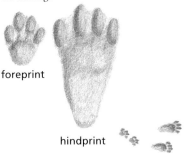
foreprint

hindprint

hopping trail

White-tailed Jackrabbit

SIMILAR SPECIES: The **White-tailed Jackrabbit** (p. 370) has longer ears and a slightly longer tail, and its winter underfur is creamy white. The **Eastern Cottontail** (p. 358) and **Mountain Cottontail** (p. 360), which do not turn white in winter, are generally smaller. The **European Hare** (p. 368) is larger and also does not turn white in winter.

Arctic Hare
Lepus arcticus

winter
coat

Arctic Hares, though sometimes found singly or in pairs, can be seen in groups of hundreds and sometimes even thousands on the open tundra. These hares do not hibernate during the long Arctic winter. Their coat becomes nearly all white, and they have other physiological and behavioural adaptations that allow them to survive. Long claws on their forefeet help them dig into the snow to find the plants they need to survive.

One of the more famous behavioural traits of hares is their "boxing" matches. These were originally thought to be males fighting for territory or to determine hierarchies, but studies of Arctic Hares have

shown that the matches occur between a male and female, and that they happen during the breeding season. The boxing match begins when a female turns to face a male that is following her and rises up on her hindlegs. The male rises up as well, but it is the female who throws the first punch. She bats at him with her forefeet, claws extended. If the male does not back down right away, the two may continue batting at each other for several minutes. It is not uncommon for individuals to be injured, and many bear the scars of previous battles.

This hare's main predator is the Arctic Wolf, but adults are usually fast enough

RANGE: The Arctic Hare is generally found north of treeline in the Arctic tundra of Alaska, Canada and Greenland.

Total Length: 55–65 cm

Tail Length: 4–8 cm

Weight: 4–6 kg

and agile enough to avoid being caught. Weasels, Arctic Foxes, owls, falcons and hawks prey on the leverets.

ALSO CALLED: Labrador Hare, Polar Hare, Greenland Hare.

DESCRIPTION: This large hare has very large feet that help it run over the snow. The heavy winter coat is all white, except for black tips on the ears. In summer, the coat can be grey, bluish grey, brown and sometimes cinnamon brown. Even though most of the coat changes colour, the legs are always white.

HABITAT: The Arctic Hare lives on the Arctic tundra. It prefers areas with shrubby growth, such as willows, for food and cover. During winter, it will move to more exposed slopes where the wind helps to scour the snow from the much-needed vegetation.

FOOD: Commonly eaten foods include the leaves, roots and bark of willows as well as herbaceous plants and grasses. This hare will even venture out onto sea ice to find seaweed. When available, the Arctic Hare happily feeds on berries, flowers and buds. It is also known to eat carrion and even garbage scraps near areas of human habitation. This hare engages in coprophagy to maximize nutrition. Hares feed quickly in the open to limit their exposure to predators, and then return to a safe place to reingest

DID YOU KNOW?

Arctic Hares are good swimmers and readily swim across streams in the tundra.

their fecal pellets and digest the vegetation properly.

DEN: This hare scrapes forms in the snow, where it huddles in a spherical shape with its ears pressed against its back and its feet tucked under its body to keep warm. It may also seek shelter under rocks or in crevices in snow and ice. Unlike other hares, the Arctic Hare may tunnel into a snowbank for shelter. The tunnels can be over 1 m in length and may end in a slightly enlarged chamber. In coastal areas, this hare may move out onto sea ice in winter, where it can eat seaweed and find good shelter among ice blocks.

REPRODUCTION: Groups of hares disperse as males and females pair up and define a breeding territory. A female gives birth to two to eight precocial leverets in summer, after a gestation of about 50 days. She does not leave the young until they are two to four days old and have learned to stay still. The leverets are fuzzy and grey at birth but turn white at two to three weeks of age. They are weaned at eight or nine weeks and are sexually mature after their first winter. Females have only one litter per year.

Snowshoe Hare

SIMILAR SPECIES: The **Snowshoe Hare** (p. 364) is almost as large, but there is very little range overlap.

European Hare
Lepus europaeus

Originally brought in as private game animals, some hares soon escaped and began to breed freely. By 1919, they had established themselves in wild populations throughout the farmlands of southern Ontario.

European Hares that were introduced to other countries easily adapted to different habitats. The hares released in Australia, for example, quickly adapted to the arid conditions, and their population exploded to pest proportions. Australians have tried many expensive but ineffective ways to eradicate this introduced mammal, which overgrazes and outcompetes native marsupials.

Have you ever wondered why the Easter Bunny—clearly a mammal—lays eggs? This story is not just a popular gimmick that sells Easter paraphernalia in spring, it is actually rooted in Germanic legend, with the European Hare at its heart. According to the myth, Eostre, the goddess of spring, created the hare by transforming a bird. Ever since this unusual conception, it is said that all hares lay eggs during the week of Easter in gratitude to Eostre and in celebration of their ancestry.

From Germany, European Hares have been widely introduced. The first introduction in Canada was to Ontario in 1912.

This athletic hare is an agile jumper and a swift runner. When chased, it runs wildly in circles, zigzagging and backtracking to confuse and elude its pursuer. When this hare is exerting itself to its limit, it can reach speeds of up to 75 km/h. Its ability to jump is remarkable—it is able to clear obstacles up to 1.5 m high and can bound a distance of up to 3.7 m. Few animals can keep up with the European Hare, though foxes and Coyotes are agile and fast enough to be its major predators.

RANGE: The European Hare is found in southern Ontario and Québec and east to New York and New England.

Total Length: 38–53 cm

Tail Length: 4.8–5.4 cm

Weight: 1–1.5 kg

DESCRIPTION: This large hare is tawny to grey brown, sometimes with flecks of black, and the underparts are white. Unlike the Snowshoe Hare, it does not turn white in winter. The winter coat is mainly grey with white below. The fur has a unique kinky or rough appearance and is very thick. The short tail is black on top and white below. The long ears are the same colour as the back.

HABITAT: These adaptable hares live in a variety of habitats. They prefer open meadows, golf courses or agricultural land that is close to woodlands and hedgerows.

FOOD: Strictly vegetarian, these hares feed on clover, grass, grain, fruits, vegetables

DID YOU KNOW?

When a European Hare makes a form in which to rest, it must have a good view and receive a gentle breeze. The breeze carries sound and odour, so the hare can detect the approach of a predator.

and green vegetation. They exhibit coprophagy as a means of maximizing nutrition.

DEN: Like other hares, European Hares sleep at night in forms. A female with young makes several separate forms hidden from view. She distributes her litter among the forms, with one leveret in each. She visits each form in rotation at night to feed her young.

REPRODUCTION: Mating usually peaks in spring but can occur at any time of year. After a gestation of 30 to 40 days, the female gives birth to two to four young. The young are born with fur and with their eyes open. Although the young are quite developed, they nurse for up to three weeks. The mother weans them, and they disperse by the fourth week. A female may have up to four litters per year.

foreprint

hindprint

hopping trail

Snowshoe Hare

SIMILAR SPECIES: The **Snowshoe Hare** (p. 364) is smaller and turns almost completely white in winter. The **White-tailed Jackrabbit** (p. 370) has larger ears and also turns white in winter. The **Eastern Cottontail** (p. 358) does not turn white in winter, is smaller and has much smaller ears.

White-tailed Jackrabbit
Lepus townsendii

summer coat

This need for salt, together with their preference for travelling on solid surfaces, may partly explain the large numbers of road-killed jackrabbits encountered on some highways.

Their protruding eyes and straight limbs are adaptations for detecting and avoiding predators. Known for their raw speed, long-legged jackrabbits can outdistance most land-based predators in an all-out run. Ambush appears to be the most effective method of catching these hares, but the open country in which these animals live gives little opportunity for cover-seeking predators.

The lean White-tailed Jackrabbit is the largest and the most commonly encountered hare through much of the Prairies. A creature of open country, this jackrabbit is most frequently seen either by day as it bursts from a hiding place with its ears erect and its tail extended, bounding out of danger with ease, or at night in the flash of a car's headlights. It is customarily solitary when foraging, but in winter, up to 50 individuals may gather in one place, often where food is abundant. These aggregations usually occur at night, when the jackrabbit is most active.

Like most herbivores, White-tailed Jackrabbits are drawn to salt, which, unfortunately, they often lick from roads.

White-tailed Jackrabbits can sometimes be found using the same rest areas day after day, so if you spook one from its hideout during a walk, you can return the next day to look for it in the same area. Quite often, these large hares are found using exactly the same forms, and they can be surprisingly approachable if your movements are slow and unthreatening.

DESCRIPTION: In summer, the upperparts are light greyish brown, and the belly is nearly white. By mid-November, the entire coat is white, except for the

RANGE: This hare is currently found from eastern Washington east to southern Manitoba, north to central Alberta and Saskatchewan and south to central California and eastern Kansas.

Total Length: 53–64 cm

Tail Length: 7–11 cm

Weight: 3–5 kg

greyish forehead and the black ear tips. The fairly long, white tail sometimes bears a greyish band on the upper surface and is held rigidly behind the animal as it runs.

HABITAT: This hare is a creature of open areas. It will enter open woodlands to seek shelter in winter but avoids dense, timbered stands.

FOOD: Grasses and forbs are the most commonly eaten plants, but jackrabbits also enjoy alfalfa and clover in agricultural regions. More shrubs and weedy plants appear in the winter diet. Like all hares, they eat their fecal pellets to run the bacteria and bacterial breakdown products from their cecum through the digestive system a second time, allowing the hare to better absorb the nutrients.

DEN: There is no den, but a shallow hollow beside a rock or beneath sagebrush serves as a daytime shelter. In winter, jackrabbits may dig depressions or short burrows in snowdrifts as shelters.

REPRODUCTION: After a 40-day gestation, the female gives birth to one to nine (usually three or four) leverets in a shallow depression. The fully furred newborns have open eyes. They soon disperse, meeting their mother to nurse only once or twice a day. By two weeks of age, they are eating some green vegetation, and they are weaned at five to six weeks, often just before the birth of the next litter. Jackrabbits reach adult weight in three to four months. Females may have up to four litters per year, though only one or two are likely in Canada.

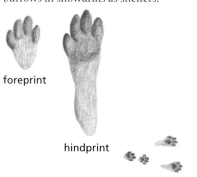

foreprint

hindprint

winter coat

hopping trail

SIMILAR SPECIES: In winter, the **Snowshoe Hare** (p. 364) has lead grey, not creamy white, hair bases. It is nearly always associated with treed or brushy areas, not the jackrabbit's prairie or alpine tundra habitats.

Snowshoe Hare

Collared Pika
Ochotona collaris

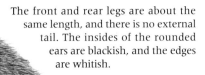

The front and rear legs are about the same length, and there is no external tail. The insides of the rounded ears are blackish, and the edges are whitish.

HABITAT: This pika prefers high-elevation talus slopes consisting of rocks more than 30 cm in diameter. There must be abundant grasses and forbs nearby.

The Collared Pika is similar to the more common American Pika, but its range in Canada is confined to the Yukon and parts of the Northwest Territories and British Columbia. Most people will never see one, but if you want to try your luck, look for it in high-elevation, boulder-strewn areas with plenty of crevices.

The presence of this round-bodied and round-headed lagomorph may be obvious in certain high-mountain habitats—it emits distinctive, high-pitched bleats when it is alarmed. The sound is so loud and far-reaching that many visitors are surprised to learn that such a grand sound can come from such a tiny animal.

DESCRIPTION: The body is grizzled grey brown above and light greyish brown to whitish below. There is a pale grey "collar" behind the ears on each side of the neck.

FOOD: A wide variety of forbs, grasses and sedges are clipped, gathered into a mouthful and carried back to the boulder-strewn domain. Some clippings maybe consumed immediately, but most are left between the rocks to be eaten during winter.

DEN: A grass-lined nest is hidden deep between rocks in talus areas.

REPRODUCTION: A litter of three or four young is born during summer. The babies each weigh 8 to 9 g and have closed eyes, but they are able to crawl in just a few days. The eyes open at 10 days, and at 12 days, the young begin to feed on vegetation and gain weight rapidly.

SIMILAR SPECIES: The **American Pika** (p. 373) lacks the indistinct greyish "collar" and has a more southern distribution.

RANGE: The Collared Pika is found in Alaska, the Yukon, the western Northwest Territories and extreme north-western British Columbia.

Total Length: 18–20 cm
Weight: 180–230 g

American Pika
Ochotona princeps

nhabiting an intricate landscape of boulders high in the mountains, the American Pika is often regarded as one of the cutest animals in the alpine wilderness. This relative of the rabbit scurries among the rocks of a talus slope as it makes its way between feeding areas and shelter. When it gathers food, the American Pika carries vegetation clippings crossways in its mouth—a bundle sometimes half as large as the animal itself. Large piles of clippings accumulate on or under the rocks in the pika's territory, and although some clippings are consumed immediately, most are left between the rocks to be eaten during winter.

Pikas are extremely vocal animals that are often heard before they are seen. The proper pronunciation of their name is *pee-ka*, which mimics their high-pitched voices—they emit bleating calls whenever they see something out of the ordinary. These sounds are often the best clues of their activity because pikas are difficult to pick out in their boulder-strewn habitat—when a pika is momentarily glimpsed from afar, one is never quite sure whether it is a genuine sighting or just a pika-sized rock.

To the patient naturalist intent on pika observation, viewing can be intimate and

RANGE: American Pikas occur from the mountains of west-central Alberta and southern British Columbia south to California, Utah and New Mexico.

Total Length: 16–21 cm

Weight: 150–300 g

rewarding because this animal often permits a close approach. When you see a pika escape into a crevice beneath the rocks, sit quietly and wait—soon it will come out again, seemingly oblivious to your unobtrusive presence.

In winter, pikas dig snow tunnels as far as 90 m out from their rock shelters to allow them to collect and eat plants. The talus slopes that are their homes often receive great quantities of snow, which helps insulate the animals from the mountain winters. Rarely venturing into the chill of the open air, pikas tend to remain beneath the snow, feeding upon the grass they so meticulously gathered and dried earlier in the year.

DESCRIPTION: This grey to tawny grey, chunky, soft-looking mammal has large, rounded ears and beady black eyes. The whiskers are long. There is no external tail. The front and rear legs are nearly equal in length, so the pika runs instead of hopping.

HABITAT: Pikas generally occupy talus slopes in the mountains, though they

> **DID YOU KNOW?**
>
> Pikas seem to be able to throw their voices, so although they are often heard, they can be difficult to locate. This ventriloquist-like ability is a great advantage to an animal that wants to warn fellow pikas without revealing itself to a potential predator.

occasionally live among jumbled logs swept down by avalanches. Surprisingly, they may live at sea level in British Columbia where there is suitable habitat.

FOOD: The pika's diet includes a wide variety of vegetation that is found in the vicinity of this animal's rocky shelter. Broad-leaved plants, grasses and sedges are all clipped and consumed.

DEN: The American Pika builds a grass-lined nest, in which the young are born, beneath the rocks of its home.

REPRODUCTION: Mating occurs in spring, and after a 30-day gestation, a litter of two to five (usually three) young is born. The newborns are furry, weigh 7 to 9 g and have closed eyes. The eyes open after 10 days. The young are weaned when they are 30 days old and two-thirds grown. Pikas are sexually mature after their first winter. There is sometimes a second summer breeding period.

foreprint	hindprint

walking trail

SIMILAR SPECIES: The **Bushy-tailed Woodrat** (p. 270) is the only other grey mammal of comparable size that might occupy the same rocky slopes as the American Pika, but it has a long, bushy tail.

Bushy-tailed Woodrat

BATS

In an evolutionary sense, bats are a very successful group of mammals. Worldwide, nearly a quarter of all mammalian species are bats, and they are second only to rodents in both diversity of species and number of individuals. Unfortunately, across North America, populations of several bat species appear to be declining.

A bat's wing consists of a double layer of skin stretched across the modified bones of the fingers and down to the legs. A small bone, the calcar, juts backward from the foot to help support the tail membrane, which stretches between the tail and each leg. The calcar is said to be keeled when there is a small projection of skin from its side.

Bats generate lift by pushing their wings against the air's resistance, so they tend to have a large wing surface area relative to their body size. This method of flight is less efficient than the airfoil lift provided by bird or airplane wings, but it allows bats to fly slower and gives them more manoeuverability. Slower flight is a real advantage when trying to catch insects.

Although bats have good vision, their nocturnal habits have led to an increased dependence on their sense of hearing—most people are acquainted with the ability of many bat species to navigate or capture prey in the dark using echolocation. The tragus—a slender lobe that projects from the inner base of many bats' ears—is thought to help in determining an echo's direction.

Unlike most other types of small mammals, bats have small litters. The high energy requirements for flight limit most female bats to having only one offspring per year. Bat populations can reach high numbers, however, because these mammals live for many years.

Evening Bat Family (Vespertilionidae)

All species of bats that occur in Canada belong to this family. True to their name, most members of this family are active in the evening and often again before dawn. A few species migrate to warmer regions for winter, but most hibernate in caves, abandoned mines or even buildings.

Key to the Bats

1a. Ear length >28 mm **2**
1b. Ear length <28 mm **4**

2a. Fur on back black; 3 white spots on rump and shoulders...**Spotted Bat** (p. 396)
2b. Fur on back not black; lacks white spots... **3**

3a. Nose with 2 prominent bumps; forearm length 40–45 mm ....**Townsend's Big-eared Bat** (p. 397)
3b. Nose lacking 2 prominent bumps; forearm length 48–57 mm.......................................**Pallid Bat** (p. 399)

4a. Fur orange or rusty red **5**
4b. Fur not orange or rusty red... **6**

5a. Reddish fur appears frosted or white-tipped...**Eastern Red Bat** (p. 389)
5b. Reddish fur is not or only slightly frosted; specimen found in southwestern BC ....**Western Red Bat** (p. 388)

6a. Fur on back with frosted or silver-tipped hairs **7**
6b. Fur on back without frosted or silver-tipped hairs.. **8**

7a. Upper surface of tail membrane covered with fur; forearm length 50–57 mm.............................**Hoary Bat** (p. 390)
7b. Upper surface of tail membrane furred at base only; forearm length 39–44 mm ....**Silver-haired Bat** (p. 392)

8a. Visible fringe of hairs on outer edge of tail membrane... **Fringed Myotis** (p. 385)
8b. No visible fringe of hairs on outer edge of tail membrane **9**

9a. Calcar with prominent keel... **10**
9b. Calcar without prominent keel **14**

10a. Forearm length >35 mm; hindfoot >9 mm long.. **11**
10b. Forearm length <35 mm; hindfoot <9 mm long.. **12**

11a. Underwing furred outward to a line extending from knee to elbow; forearm length 34–44 mm ....
....**Long-legged Myotis** (p. 386)
11b. Underwing not furred outward to a line extending from knee to elbow; forearm length 43–52 mm...............
....**Big Brown Bat** (p. 395)

12a. Fur on back pale blond to orange-yellow, contrasts sharply with blackish ears, face and wings; length of bare area on snout
about 1.5 times width across nostrils **13**
12b. Fur on back chestnut to brown, not in sharp contrast to colour of ears, face and wings; length of bare area on snout
about equal to width across nostrils...**California Myotis** (p. 377)

13a. Specimen is found in western Canada ....**Western Small-footed Myotis** (p. 378)
13b. Specimen is found in eastern Canada..**Eastern Small-footed Myotis** (p. 381)

14a. Little to no difference of fur colour between dorsum and venter; hairs are tricoloured ....**Eastern Pipistrelle** (p. 393)
14b. Fur colour lighter on venter than dorsum; hairs not tricoloured... **15**

15a. Ears long (14–22 mm), extending well beyond tip of nose when pushed forward................................ **16**
15b. Ears short (10–16 mm), not extending well beyond tip of nose when pushed forward........................... **18**

16a. Ears black, extending >5 mm beyond tip of nose when pushed forward ....**Long-eared Myotis** (p. 379)
16b. Ears dark but not black, extending <5 mm beyond nose when pushed forward.................................. **17**

17a. Poorly defined dark spot on shoulders; minute hairs on edge of tail membrane ....**Keen's Myotis** (p. 380)
17b. No dark spot on shoulders; few or no hairs on edge of tail membrane ....**Northern Myotis** (p. 384)

18a. Fur on back long, sleek and glossy; forearm length usually >36 mm ....**Little Brown Myotis** (p. 382)
18b. Fur on back short and dull; forearm length usually <36 mm ....**Yuma Myotis** (p. 387)

California Myotis
Myotis californicus

California Myotis emerge shortly after nightfall, and for a few minutes in the dying daylight, they can be followed as they fly erratically through the sky. As if surfing on invisible waves in the air, these fluttering bats rise and dive at variable speeds in the pursuit of unseen prey. These activities appear disorganized and random, but they are actually deliberate and calculated.

DESCRIPTION: This bat is small and yellowish brown to rusty or dark brown. Its wingspan is about 23 cm. The feet are tiny, and the calcar is keeled.

HABITAT: The California Myotis can be found in arid grasslands, humid coastal forests and montane forests. It roosts in rock crevices, mines and buildings and under bridges and loose tree bark. It forages mainly in or over forested areas, lakes and grasslands of the West.

FOOD: During the night, this bat forages opportunistically in areas that concentrate night-flying insects, such as cliffs or poplar groves, or over water for emerging adult caddisflies and mayflies. Additionally, it can be observed in tree canopies feeding on moths, beetles and flies. It generally flies 2 to 3 m above the ground when foraging.

DEN: The California Myotis is not too selective in its night roosts and can be found in both buildings and natural structures. Its day roosts are typically in rock crevices, but it may also roost in mineshafts, tree cavities and buildings and under bridges.

REPRODUCTION: A single young is born in late June to early July. Gestation is believed to be 50 to 60 days, but delayed implantation can extend this considerably.

SIMILAR SPECIES: All the **Mouse-eared Bats** (*Myotis* spp., pp. 377–87) are essentially impossible to identify in flight. Even with a specimen in hand, one needs a technical key.

RANGE: The California Myotis, truly a western bat, is found in coastal regions from southern Alaska south to California and southern Mexico. In the U.S., it ranges east into Montana, Colorado, New Mexico and Texas.

Total Length: 6.6–9.5 cm

Tail Length: 2.6–4.1 cm

Forearm: 2.9–3.5 cm

Weight: 3.3–5.4 g

Western Small-footed Myotis
Myotis ciliolabrum

True to its name, this bat has noticeably small feet. The calcar is strongly keeled.

T he Western Small-footed Myotis is one of the "rock bats" and occupies daytime roosts in such rocky habitats as badlands, cliffs and talus slopes. Contrary to popular belief, this bat is not a creature of arid environments—though it may live among dry rocks, it lacks the physiological adaptations that prevent water loss. Unlike a true desert animal, it is never far from water.

DESCRIPTION: The glossy fur of this attractive bat is yellowish brown to grey or even coppery brown above, and the underparts are a paler tan. The flight membranes and ears are black, and the tail membrane is dark brown. The wingspan is 20 to 25 cm, and some fur may be found on both the undersurface of the wing and the upper surface of the tail membrane. A dark brown or black mask stretches across the face from ear to ear.

HABITAT: This species prefers arid rocky or grassland regions, especially riverbanks, ridges and outcroppings with abundant rocks for roosting.

FOOD: Like most bats, the Western Small-footed Myotis eats primarily flying insects, including moths, flies, bugs and beetles.

DEN: In summer, this bat roosts in trees, buildings or rock crevices. It hibernates in caves or mines in winter. Nursery colonies are located in bank crevices, under bridges or under the shingles of old buildings.

REPRODUCTION: The females gather in small nursery colonies in late spring. One young per female is born from late May to early June.

SIMILAR SPECIES: All the **Mouse-eared Bats** (*Myotis* spp., pp. 377–87) are essentially impossible to identify in flight. Even with a specimen in hand, one needs a technical key.

RANGE: The Western Small-footed Myotis is found from southern British Columbia east to southwestern Saskatchewan and south through most of the western U.S.

Total Length: 7.6–8.9 cm

Tail Length: 3–4.5 cm

Forearm: 2.9–3.5 cm

Weight: 3.5–7.1 g

Long-eared Myotis
Myotis evotis

The dramatic nightly bat sagas that take place in summer skies go largely unnoticed by humans. In apparent silence, bats navigate and locate prey by producing ultrasonic pulses (up to five times higher in pitch than our ears can detect) and listening for the echoes of these sounds as they bounce off objects. The Long-eared Myotis uses very short-duration echolocation clicks as an adaptation for foraging in heavy vegetation. Bats with large ears tend to be insectivorous, and the aptly named Long-eared Myotis appears to be well equipped for insect hunting.

DESCRIPTION: The wingspan is about 28 cm. The upperparts can vary in colour from light yellowish brown to dark brown. The underparts are lighter. The black, naked ears are 1.9 to 2.2 cm long, with a long, narrow tragus. The wings are mainly naked, and only the lower fifth of the tail membrane is furred. The calcar is keeled.

HABITAT: This bat occurs in a variety of habitats, wherever there are suitable roosting sites. It is commonly found in forests, as well as shrublands, some grasslands and even agricultural areas.

FOOD: Feeding peaks at about 30 minutes after full darkness, somewhat later than most other bats. Moths, flies, beetles and spiders are the primary prey.

DEN: Both sexes of this mainly solitary species hibernate in caves and mines in winter. In spring, groups of up to 30 females gather in nursery colonies in tree cavities, under loose bark, in old buildings, under bridges or in loose roof shingles. Males typically roost in caves and mines in summer.

REPRODUCTION: Mating takes place just before hibernation begins, but fertilization is delayed until spring. In June or early July, after a gestation of about 40 days, each female bears a single young. The young mature quickly and are able to fly on their own in four weeks.

SIMILAR SPECIES: All the **Mouse-eared Bats** (*Myotis* spp., pp. 377–87) are essentially impossible to identify in flight. Even with a specimen in hand, one needs a technical key.

RANGE: The Long-eared Myotis is found from southern British Columbia east to southern Saskatchewan and south to northwestern New Mexico and Baja California.

Total Length: 8.3–11 cm

Tail Length: 3.5–4.8 cm

Forearm: 3.8–4.1 cm

Weight: 3.5–8.9 g

Keen's Myotis
Myotis keenii

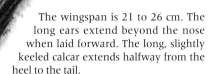

The wingspan is 21 to 26 cm. The long ears extend beyond the nose when laid forward. The long, slightly keeled calcar extends halfway from the heel to the tail.

HABITAT: This species seems to be restricted to temperate coastal rainforests.

This uncommon bat was once considered a subspecies of the Northern Myotis. The Keen's Myotis has not been well studied, but it is thought to have habits similar to the other Mouse-eared Bats (*Myotis* spp.). It is solitary and roosts in tree cavities or rock crevices during the day, and then flies out over waterbodies to feed at night. There is one known nursery colony, near a hot spring on Haida Gwaii (Queen Charlotte Islands), where the bats roost under naturally heated rocks. Otherwise, little is known about the colony formation of this species.

FOOD: The Keen's Myotis flies quite slowly but directly, taking high-flying insects along forest edges and over ponds and clearings.

DEN: Adults roost in tree cavities, rock crevices and caves, and under bark.

REPRODUCTION: In early June or July, each female in a nursery colony has one young. Little else is known about this species.

DESCRIPTION: The fur is glossy, dark brown with light underparts. Dark shoulder spots are usually visible. The ears and flight membranes are dark brown.

SIMILAR SPECIES: All the **Mouse-eared Bats** (*Myotis* spp., pp. 377–87) are essentially impossible to identify in flight. Even with a specimen in hand, one needs a technical key.

RANGE: This species has one of the smallest ranges of any North American bat. It is found west of the Coast Mountains and the Cascades from southeastern Alaska to northwestern Washington.

Total Length: 6.3–9.3 cm

Tail Length: 3.5–4.4 cm

Forearm: 3.5–3.8 cm

Weight: 4.3–5.6 g

Eastern Small-footed Myotis
Myotis leibii

The Eastern Small-footed Myotis is a very small and understudied member of this genus. It appears to be extremely rare, and throughout its range, its distribution is thin and spotty. Wherever this bat is found, however, it is always near water such as streams and ponds. When it emerges at dusk to forage, this bat flies above the nearby waterbodies to capitalize on the abundance of night-flying insects over the water. This sensitive species may show decline when forests are removed from around its hibernation sites or foraging areas.

DESCRIPTION: The glossy fur of this attractive bat is yellowish brown to grey or even coppery brown above, and its underparts are buffy or whitish. The flight membranes and tail membrane are dark brown, and the ears are black. Its wingspan is 21 to 25 cm, and some fur may be found on both the undersurface of the wing and the upper surface of the tail membrane. Across its face, from ear to ear, is a dark brown or black mask. True to its name, this bat has noticeably small feet, 6 to 8 mm in length. The calcar is strongly keeled, and the tragus is long and pointed.

HABITAT: The Eastern Small-footed Myotis prefers hilly regions with deciduous or coniferous forests. It sometimes occurs in somewhat open farmland and grassy areas.

FOOD: Like most bats, the Eastern Small-footed Myotis eats primarily small, flying insects.

DEN: In summer, this bat roosts under rock ledges and bridges (in expansion joints) and occasionally in or under the eaves of buildings. In winter, it hibernates either singly in cracks and crevices or in a group in caves and mines. Nursery colonies occur in bank crevices, under bridges or under the shingles of old buildings.

REPRODUCTION: In small nursery colonies, one young per female is born in June or July.

SIMILAR SPECIES: All the **Mouse-eared Bats** (*Myotis* spp., pp. 377–87) are essentially impossible to identify in flight. Even with a specimen in hand, one needs a technical key. The similar-sized **Eastern Pipistrelle** (p. 393) lacks the black mask.

RANGE: The Eastern Small-footed Myotis has a spotty distribution from southern Ontario, Québec and the northeastern U.S. through the Appalachian Mountains to Kentucky and Missouri.

Total Length: 7.2–8.4 cm

Tail Length: 3.0–3.9 cm

Forearm: 3.0–3.4 cm

Weight: 3–7 g

Little Brown Myotis
Myotis lucifugus

On nearly every warm, calm summer night, the skies are filled with marvellously complex screams and shrill chirps. Unfortunately for people interested in the world of bats, these magnificent vocalizations occur at frequencies higher than our ears can detect. The most common of these nighttime screamers, and quite likely the first bat most people will encounter, is the Little Brown Myotis.

Once the cool days of late August and September arrive, Little Brown Myotis begin to migrate to wintering areas. While it is not known where all of these

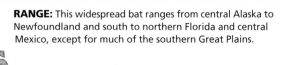

RANGE: This widespread bat ranges from central Alaska to Newfoundland and south to northern Florida and central Mexico, except for much of the southern Great Plains.

Total Length: 7–10 cm

Tail Length: 2.5–5.4 cm

Forearm: 3.5–4.1 cm

Weight: 5.3–8.9 g

bats spend the winter, thousands of them travel to natural caves and abandoned mineshafts. Large wintering populations are known to occur in certain caves, and cave adventurers are advised to take special care not to disturb these hibernating animals. All bats may rouse on occasion during hibernation, and some may even fly out on warm nights, but if a bat is roused too often by disturbance or intruders, it risks speeding up its metabolism too much and running out of reserves before the arrival of insects in spring.

DESCRIPTION: As its name suggests, this bat is little and brown. Its colouration ranges from light to dark brown on the back, with somewhat paler underparts. The tips of the hairs are glossy, giving this bat a coppery appearance. The wing and tail membranes are mainly unfurred, though fur may appear around the edges. The calcar is long and unkeeled. The tragus, which is bent, is half the length of the ear. The wingspan is 22 to 27 cm.

HABITAT: The Little Brown Myotis is the most frequently encountered bat in much of North America. At home almost anywhere, it can be found in buildings, attics or roof crevices and under bridges or loose bark on trees. Wherever colonies of this bat roost, water is sure to be nearby because the bats need a place to drink and a large supply of insects for their nightly forays.

FOOD: This bat feeds exclusively on night-flying insects, especially new

> **DID YOU KNOW?**
>
> An individual Little Brown Myotis can consume 900 insects per hour during its nighttime forays. A typical colony may eat 45 kg of insects a year.

aquatic emergents. In the evening, it leaves its day roost and swoops down to the nearest water source to snatch a drink on the wing. Foraging for insects can last for up to five hours. Later, the bat rests in a night roost, which is in a different place from its day roost. Another short feeding period occurs just prior to dawn before the bat returns to its day roost.

DEN: These bats may roost alone or in groups. Females in maternity colonies can number from just a few to 1000, and males roost singly or in small groups. A loose shingle, an open attic or a hollow tree are all suitable roost for these bats. In winter, they hibernate in large numbers in caves and old mines.

REPRODUCTION: Mating occurs either in late autumn or in the hibernation colonies. Fertilization of the egg is delayed until the female ovulates in spring, and, by June, pregnant females form nursery colonies in a protected location. In late June or early July, one young is born to each female after about 50 to 60 days of gestation. The young are blind and hairless, but their development is rapid and their eyes open in about three days. After one month, the young are on their own.

SIMILAR SPECIES: All the **Mouse-eared Bats** (*Myotis* spp., pp. 377–87) are essentially impossible to identify in flight. Even with a specimen in hand, one needs a technical key. The **Eastern Pipistrelle** (p. 393) has tricoloured hairs on its back.

Eastern Pipistrelle

Northern Myotis
Myotis septentrionalis

HABITAT: The Northern Myotis occurs primarily in forested and sometimes brushy areas. It prefers to be close to water.

FOOD: This bat hunts at dusk and again just before dawn. It feeds on small insects, especially flies and mosquitoes.

The Northern Myotis tends to roost in natural cavities and under peeling or lifting bark on old or dead trees. There is concern that this makes it vulnerable to forestry operations, which often select older trees for harvesting. This bat is considered a "gleaner," because when feeding, it grabs insects off branches, leaves and other surfaces instead of catching them in flight.

ALSO CALLED: Northern Long-eared Myotis.

DESCRIPTION: This bat is medium to dark brown with somewhat lighter underparts. The face has a distinct, dark mask. The tips of the hairs are lighter than the bases, giving a sheen to the fur. The wingspan is 23 to 25 cm. The tragus is long and splinter-shaped. The ears extend past the tip of the nose if pushed forward. The calcar is not keeled.

DEN: In September and October, this mainly solitary bat seeks out caves and mines in which to hibernate. In spring, females form nursery colonies in tree cavities, under loose bark on trees or loose shingles on rooftops, under bridges or occasionally in old buildings. These colonies may contain up to 30 females, but smaller groups are more common.

REPRODUCTION: These bats mate in autumn, but fertilization is delayed until spring. The single young is born in June or early July, after a gestation of about 40 days. The young are able to fly in about four weeks.

SIMILAR SPECIES: All the **Mouse-eared Bats** (*Myotis* spp., pp. 377–87) are essentially impossible to identify in flight. Even with a specimen in hand, one needs a technical key. The slightly smaller **Eastern Pipistrelle** (p. 393) has shorter, paler ears.

RANGE: The Northern Myotis is found from eastern British Columbia to Newfoundland and south to Nebraska, Arkansas, western Georgia and Virginia.

Total Length: 8.3–10.0 cm

Tail Length: 2.9–4.5 cm

Forearm: 3.2–4.1 cm

Weight: 3.5–8.9 g

Fringed Myotis
Myotis thysanodes

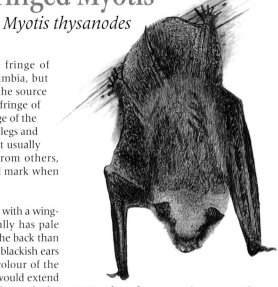

This bat exists on the fringe of southern British Columbia, but its distribution is not the source of its name. A conspicuous fringe of stiff hairs along the outer edge of the membrane between the hindlegs and the tail is a characteristic that usually distinguishes this species from others, though it is useless as a field mark when the bat is in flight.

DESCRIPTION: This large bat, with a wingspan of up to 30 cm, typically has pale brown fur that is darker on the back than on the underparts. The dark, blackish ears contrast strongly with the colour of the back. The ears are long and would extend well past the nose if pushed forward. The most unique characteristic is the fringe of small, stiff hairs on the outer edge of the tail membrane. The calcar is keeled.

HABITAT: The Fringed Myotis is most frequently encountered in grasslands near water sources and pine forests. It occurs mainly at low elevations in British Columbia.

FOOD: Foraging typically occurs soon after sunset. Moths, beetles, flies, lacewings and crickets are commonly eaten. The presence of flightless insects in the diet has led to the speculation that these bats may glean some insects from foliage.

DEN: These bats roost in caves, mines and buildings. Up to several hundred individuals will cluster in maternal roosts in summer.

REPRODUCTION: One, or uncommonly two, young are born in June or early July. They reach adult size by three weeks of age, at which time they are capable of limited flight. Maternal colonies contain only females and the young of the year.

SIMILAR SPECIES: All the **Mouse-eared Bats** (*Myotis* spp., pp. 377–87) are essentially impossible to identify in flight. Even with a specimen in hand, one needs a technical key.

RANGE: The Fringed Myotis is found from extreme southern British Columbia south through the western U.S. and into Mexico.

Total Length: 8.6–9.5 cm

Tail Length: 3.8–4.5 cm

Forearm: 4.1–4.5 cm

Weight: 5.3–8.9 g

Long-legged Myotis
Myotis volans

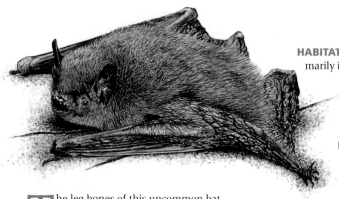

HABITAT: This bat lives primarily in coniferous forests near waterbodies. It may forage along the edges of mountain lakes.

FOOD: The diet is composed primarily of moths, flies, bugs and beetles.

The leg bones of this uncommon bat are more than 1.6 cm long, a characteristic responsible for its common name. Unfortunately, the leg bones are only fractionally longer than those of other *Myotis* species, so they are not a good field mark. Noticeable differences do occur, however, in habitat selection. The Long-legged Myotis prefers mountainous and rugged areas near waterbodies for its foraging opportunities.

DESCRIPTION: This bat is the heaviest of the "little brown bats," but weighs only an imperceptible amount more than its kin. The wingspan is 25 to 28 cm. The fur can be light brown to reddish to dark chocolate brown but is mainly dark brown. The calcar has a well-defined keel. The underwing is usually furred out to a line connecting the elbow and knee.

DEN: The Long-legged Myotis spends winter hibernating in caves or mines. In summer, it roosts in trees, buildings or rock crevices. Nursery colonies are located in bank crevices, under bridges or under south-facing shingles on old buildings.

REPRODUCTION: Mating occurs in autumn, but fertilization is delayed until spring. One young per female is born in July or August, in large nursery colonies. The young mature quickly, flying on their own in about four weeks. The longest recorded lifespan for this species is 21 years.

SIMILAR SPECIES: All the **Mouse-eared Bats** (*Myotis* spp., pp. 377–87) are essentially impossible to identify in flight. Even with a specimen in hand, one needs a technical key.

RANGE: The Long-legged Myotis ranges from northwestern British Columbia southeast to western North Dakota and south through most of the western U.S.

Total Length: 8.6–10.0 cm

Tail Length: 3.5–5.4 cm

Forearm: 3.5–4.5 cm

Weight: 5–11 g

Yuma Myotis
Myotis yumanensis

L ike most of their kin, Yuma Myotis spend much of the summer hanging comfortably in hot roosts, shifting slightly as the temperature rises and falls. They rest, relax and snuggle against one another until twilight draws them outside. With nightfall, Yuma Myotis fly out over the nearest wetland, snapping up rising insects in the cool, calm night air. Often, their stomachs are full within 15 minutes, and their foraging is finished for the night. The bats then retreat to their night roost, where they digest their meal before foraging again just before dawn.

DESCRIPTION: This medium-sized bat has brown to black fur on the back, with paler underparts. The ears are long enough to extend to the nose when pushed forward. The tragus is blunt and only about half the length of the ear. The wingspan is about 23 cm. The calcar is not keeled.

HABITAT: The Yuma Myotis tends to occur in open or shrubby areas and near water. It usually forages over lakes and streams.

FOOD: Much of the diet consists of water-hatching insects such as caddisflies, mayflies and midges.

DEN: These bats typically roost and form their maternal colonies in buildings, trees and under south-facing siding and shingles. These structures must be within foraging distance of a source of water. Yuma Myotis are common in southwestern British Columbia, and nursery colonies found around Victoria and Vancouver are of this bat, not the Little Brown Myotis, as many people assume.

REPRODUCTION: As with many bat species, mating occurs during autumn, but the sperm is stored within and fertilization occurs in spring. A single young is usually born in June or July.

SIMILAR SPECIES: All the **Mouse-eared Bats** (*Myotis* spp., pp. 377–87) are essentially impossible to identify in flight. Even with a specimen in hand, one needs a technical key.

RANGE: The Yuma Myotis is found from west-central British Columbia south to California and Mexico and east to Colorado and western Texas.

Total Length: 7.6–9.2 cm

Tail Length: 3.2–4.5 cm

Forearm: 3.2–3.8 cm

Weight: 3.5–5.3 g

Western Red Bat
Lasiurus borealis

may have a slightly frosted appearance. The wingspan is 28 to 33 cm. The top of the membrane between the hindlimbs is densely furred on the anterior portion. The short, broad, rounded, pale ears are almost hairless inside.

HABITAT: The Western Red Bat favours low-elevation forest edges alongside rivers.

FOOD: When it forages near farmlands, this bat may feed heavily on agricultural pests. It primarily takes moths, planthoppers, flies and beetles, and it may sometimes alight on vegetation to pick off insects. The peak feeding period is well after dusk.

Considering that most bats have brownish fur, the Western Red Bat stands out and should be easy to recognize. Unfortunately, this bat is not common, and the chances of seeing one are slim. It is virtually an unknown animal in Canada, with only a smattering of records reported from southern British Columbia. The Skagit Valley and the southern Okanagan appear to be the extreme northern limit for this species, and the individuals recorded there may be wanderers from farther south.

DESCRIPTION: This colourful bat has orange or reddish fur, 12 to 18 mm long, over the back. The male tends to be brighter than the female, and some individuals

DEN: In summer, this solitary bat roosts in foliage, which provides shade. The space beneath the roost must be free of obstacles to allow the bat to drop into flight.

REPRODUCTION: Mating takes place in August and September, but ovulation and fertilization are delayed until spring. Gestation appears to be 80 to 90 days, and one to four young are born in June. The young are thought to be able to fly at three or four weeks of age, and they are weaned at five or six weeks. Age at sexual maturity is not known.

SIMILAR SPECIES: No other bat in its range has the colour and measurements of the Western Red Bat.

RANGE: The Western Red Bat ranges from South America north to extreme southern British Columbia.

Total Length: 8.7–12.0 cm

Tail Length: 3.6–5.5 cm

Forearm: 3.6–4.4 cm

Weight: 10–17 g

Eastern Red Bat
Lasiurus borealis

This pretty bat is a close cousin of the more common Hoary Bat. Most other bats in Canada have brownish fur, so the Eastern Red Bat stands out and should be easy to recognize. Unfortunately, it begins foraging as late as two hours after sunset, at a time when it is too dark to see its reddish hue. If you do find one, you may be able to identify whether it is male or female because the male is much brighter red. On occasion, you might see one feeding on insects attracted to a streetlight, and the flush of light should allow you to see the colour.

DESCRIPTION: This bat has mainly yellowish orange to red fur. Some may appear slightly frosted as a result of having white-tipped hairs. The wingspan is 29 to 33 cm. The ears are small and rounded, with a small tragus. The backsides of the ears and head are covered in orangey fur. The upper surface of the tail membrane is furred. *Lasiurus* bats have four mammae; all other bats found in Canada have two.

HABITAT: This bat lives in or near forests, both deciduous and coniferous, and often in range of open, grassy areas.

FOOD: When foraging near farmlands, this bat may feed heavily on agricultural pests. It primarily eats moths, planthoppers, flies and beetles and may sometimes alight on vegetation to pick off insects.

DEN: In summer, this solitary bat roosts among branches, which provide shade. The space beneath the roost must be free of obstacles to allow the bat to drop into flight. Beginning in autumn, it migrates south for winter.

REPRODUCTION: Mating occurs in August and September, but fertilization is delayed until spring. Gestation appears to be 80 to 90 days, and one to four young are born in June. They are able to fly when three or four weeks old and are weaned at five or six weeks.

SIMILAR SPECIES: The **Western Red Bat** (p. 388) has a different range. The larger **Hoary Bat** (p. 390) is frosted over its back. The **Big Brown Bat** (p. 395) is dark brown. *Myotis* **bats** (pp. 377–87) are smaller and brown.

RANGE: The Eastern Red Bat is found in the southern Prairie provinces and from southern Ontario to the Maritimes south through most of the eastern U.S.

Total Length: 8.7–12.0 cm

Tail Length: 4.5–6.5 cm

Forearm: 3.6–4.4 cm

Weight: 7–15 g

Hoary Bat
Lasiurus cinereus

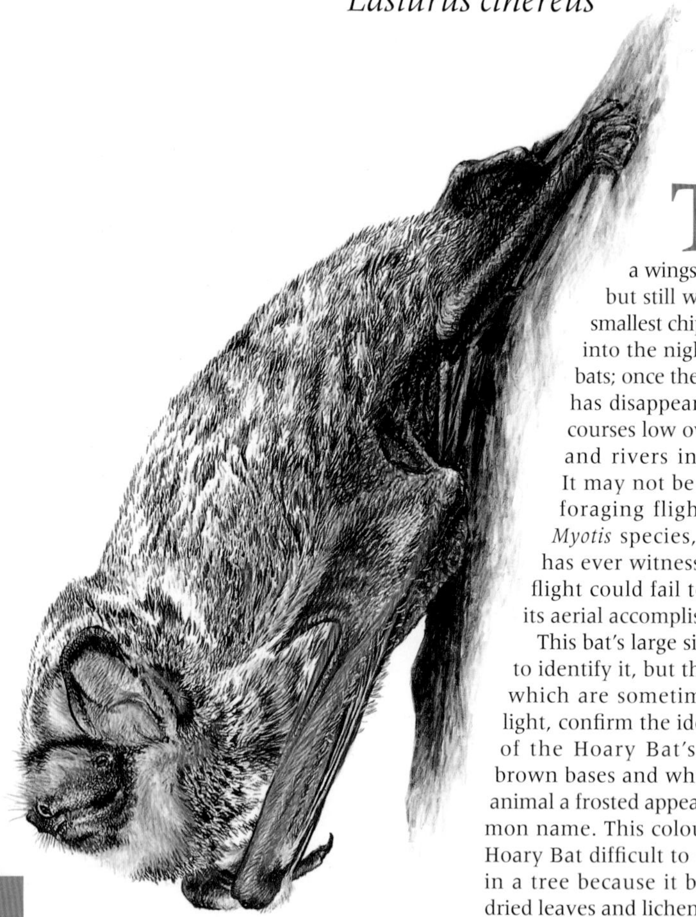

The largest bat in Canada, the Hoary Bat has a wingspan of about 40 cm but still weighs less than the smallest chipmunk. It flies later into the night than most other bats; once the last of the daylight has disappeared, the Hoary Bat courses low over wetlands, lakes and rivers in conifer country. It may not be as acrobatic in its foraging flights as the smaller *Myotis* species, but no one who has ever witnessed a Hoary Bat in flight could fail to be impressed by its aerial accomplishments.

This bat's large size is often enough to identify it, but the pale wrist spots, which are sometimes visible at twilight, confirm the identification. Many of the Hoary Bat's long hairs have brown bases and white tips, giving the animal a frosted appearance and its common name. This colouration makes the Hoary Bat difficult to see when it roosts in a tree because it blends in with the dried leaves and lichens.

Hoary Bats, as well as other tree-dwelling bats, have been the focus of recent scientific

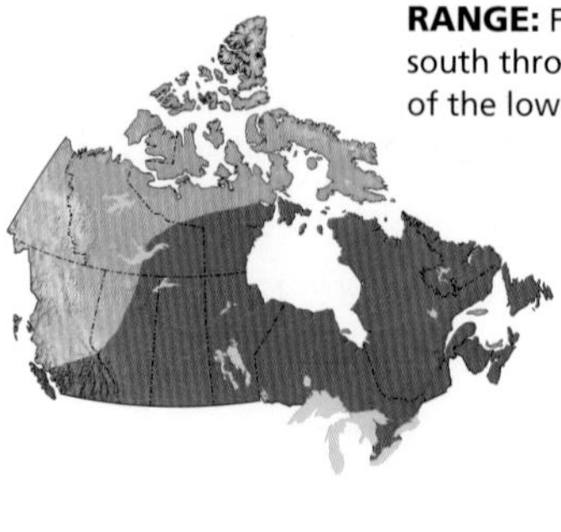

RANGE: From north-central Canada, the Hoary Bat ranges south through most of Canada and almost all of the lower 48 states.

Total Length: 11–15 cm

Tail Length: 4.1–6.7 cm

Forearm: 4.5–5.7 cm

Weight: 19–35 g

study to determine the importance of old roost trees in their habitat. These bats have complex requirements—old trees may well be important, but water quality and the availability of hatching insects in wetlands may be equally significant.

The few records from the northern parts of Canada suggest that female Hoary Bats may move quite far north. The males, it is thought, migrate only as far as the northern U.S., where they likely court and mate with the females. The males remain at these sites for summer, but some impregnated females appear to push farther north, where the young are born.

DESCRIPTION: This large bat has greyish fur with white tips that give it a heavily frosted appearance. The throat and shoulders are buffy yellow or toffee-coloured. Its wingspan is 38 to 41 cm. The ears are short, rounded and furred, with naked, black edges. The tragus is blunt and triangular. The upper surfaces of the feet and tail membrane are completely furred. The calcar is modestly keeled. Like the Western Red Bat and the Eastern Red Bat, the Hoary Bat has four mammae.

HABITAT: The Hoary Bat is often found near open, grassy areas in coniferous and deciduous forests and over lakes. It is also common in cities and can be seen feeding around streetlights at night.

DID YOU KNOW?

The Hoary Bat is the most widespread bat species in North America, and it is the only "terrestrial" mammal native to the Hawaiian Islands.

FOOD: The diet consists mainly of moths, planthoppers, flies and beetles. When this bat forages over farmland, it consumes high numbers of agricultural pests. It sometimes alights on vegetation to pick off insects. Feeding activity does not peak until well after dusk.

DEN: This migratory bat usually returns to Canada in May from as far south as Guatemala. During summer, it roosts alone in the shade of foliage, with an open space beneath the roost to allow it to drop into flight. Beginning in August or September, it migrates south, sometimes in large flocks, or "clouds."

REPRODUCTION: Hoary Bats mate in autumn, but the young are not born until late May or June because fertilization is delayed until the female ovulates in spring. Gestation lasts about 90 days, and a female usually bears two young. She places the first young on her back while she delivers the next. Before they are able to fly, young bats roost in trees and nurse between their mother's nighttime foraging flights.

SIMILAR SPECIES: The **Western Red Bat** (p. 388) and the **Eastern Red Bat** (p. 389) are smaller and have distinctly red fur. The **Silver-haired Bat** (p. 392) is black with silver-tipped hairs and is slightly smaller. The **Big Brown Bat** (p. 395) is almost as large but does not have a frosted appearance.

Western Red Bat

Silver-haired Bat
Lasionycteris noctivagans

DESCRIPTION: The fur is nearly black, with long, white-tipped hairs on the back giving the bat a frosted appearance. The naked ears and tragus are short, rounded and black. The wingspan is 28 to 30 cm. A light covering of fur may be seen over the entire surface of the tail membrane.

HABITAT: Forests are the primary habitat, but this bat can easily adapt to parks, cities and farmlands.

FOOD: The Silver-haired Bat feeds mainly on moths. It forages over standing water or in open areas near water.

DEN: Summer roosts are usually in tree cavities, under loose bark or in old buildings. In winter, these bats migrate south of the border. Females form nursery colonies in protected areas such as tree cavities, narrow crevices or old buildings.

The handsome Silver-haired Bat flies slowly and leisurely during twilight hours throughout much of southern Canada. Twilight actually happens twice in a 24-hour period—once after sunset (vesperal twilight) and again before sunrise (auroral twilight).

The feeding forays of this bat also happen twice a day. It usually flies fairly low to the ground, and if you happen to see one, either at night or in the very early morning, you may be able to watch it for some time as it dips and flops about the twilight sky catching insects. Listen closely as well and you may hear the subtle *wicka-wicka-wicka* of its leathery wings.

REPRODUCTION: Breeding takes place in autumn, but fertilization is delayed until the female ovulates in spring. In early summer, after a gestation of about two months, each female gives birth to one or two young.

SIMILAR SPECIES: This bat's white-tipped, black hairs are unique. The **Big Brown Bat** (p. 395) has mainly brown, glossy fur. The **Hoary Bat** (p. 390) has greyish rather than black fur.

RANGE: This bat is found along the southeastern coast of Alaska, across the southern half of Canada and south through most of the U.S.

Total Length: 9–11 cm

Tail Length: 3.5–5.1 cm

Forearm: 3.8–4.5 cm

Weight: 7–18 g

Eastern Pipistrelle
Pipistrellus subflavus

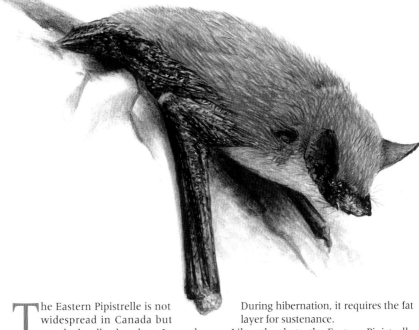

The Eastern Pipistrelle is not widespread in Canada but may be locally abundant. In southeastern regions, however, it may be the first bat many people see because it begins foraging in the evening while there is still sunlight, and in the morning, it feeds well after dawn. During most of the night, it rests, having ceased its foraging activity at about 9:30 PM. This bat forages heavily throughout summer to develop a layer of fat. In winter, the Esatern Pipistrelle moves into a mine or cave, where it hibernates. During hibernation, it requires the fat layer for sustenance.

Like other bats, the Eastern Pipistrelle is very clean. After foraging each night, it spends as much as 30 minutes grooming its fur and cleaning out debris or bugs that have accumulated. The bat uses its tongue wherever it can reach, and somewhat like a cat, it moistens its hindfeet to clean the remaining areas. Special attention is given to cleaning the ears—bats are dependent on their hearing to "see" the world with echolocation, so dirty ears

---

RANGE: This bat is found mainly from southern Ontario, Québec and the Maritimes south through the eastern U.S.

Total Length: 8–9 cm

Tail Length: 3.6–4.5 cm

Forearm: 3.1–3.5 cm

Weight: about 6 g

would be intolerable. Most bats have good eyesight, too, but as nighttime fliers, they find their hearing more useful.

The flight of the Eastern Pipistrelle is weak and erratic. Its slow flight is advantageous when enthusiastic naturalists are attempting to net it—once the bat detects the net, it has enough time to turn in the air and avoid being caught. Being a weak flier, however, means it cannot cover great distances for food, shelter or water. Another downfall of its feeble flight is that a swift breeze nearly halts its flight, and strong winds will force the bat back to its roost.

DESCRIPTION: The Eastern Pipistrelle is yellow or drab brown overall. The hairs on its back are tricoloured—dark at the base, yellowish brown in the middle and dusky grey at the tip. The wings, interfemoral membrane, ears, nose and feet are dark brown, but the leading edge of the wing is very light. This bat has a wingspan of only 19 to 22 cm. The tragus is long and slender, and the calcar is keeled. The contrast of the dark face and light fur makes the bat appear to be wearing a mask.

HABITAT: Eastern Pipistrelles are most common in shrubby areas and open woodlands close to waterbodies. Sometimes found close to cities, they are usually the first bats out in the evening and may even be seen in broad daylight. One great threat to these miniature bats is

> **DID YOU KNOW?**
>
> In German, the word for "bat" is *fledermaus*, which translates as "flittermouse," a reference to the erratic, jerky flight of the pipistrelle bats in Europe.

desiccation—they must always be near water or they will die in less than a day.

FOOD: This bat feeds on tiny flying insects such as flies, some beetles, leafhoppers and small moths. Because of its small size, it cannot eat large insects.

DEN: When roosting or hibernating, pipistrelles can be found in caves, mines, crevices and old buildings. Some bats in northern areas migrate southward. Maternity colonies are found in the crevices of rocky cliffs or in sheltered nooks of old buildings.

REPRODUCTION: In June, females give birth to two young (rarely only one), usually in protected maternity colonies. These maternity colonies do not contain more than about 12 females, and sometimes a female may roost alone to bear her young. The young require their mother's care for several weeks until they are mature. Lactating females are at great risk of dehydration, and they can be seen any hour of the night at waterbodies near their roost.

SIMILAR SPECIES: The **Eastern Small-footed Myotis** (p. 381) occupies similar habitats, but it is larger and has a much longer tragus. The **Northern Myotis** (p. 384) is slightly larger and has darker ears. The **Little Brown Myotis** (p. 382) lacks the tricoloured hairs on the back.

Eastern Small-footed Myotis

Big Brown Bat
Eptesicus fuscus

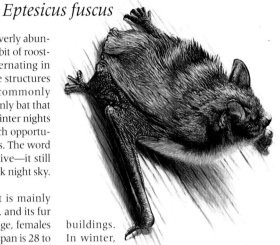

The Big Brown Bat is not overly abundant anywhere, but its habit of roosting and occasionally hibernating in houses and other human-made structures makes it one of the more commonly encountered bats. It is also the only bat that may be seen, rarely, on warm winter nights because it occasionally takes such opportunities to change hibernating sites. The word "big" in this bat's name is relative—it still looks awfully small against a dark night sky.

DESCRIPTION: This large bat is mainly brown, with lighter underparts, and its fur appears glossy or oily. On average, females are larger than males. The wingspan is 28 to 35 cm. The face, ears and flight membranes are black and mainly unfurred. The blunt tragus is about half as long as the ear. The calcar is usually keeled.

HABITAT: This species easily adapts to parks, cities, farmlands and buildings. In the wild, it typically inhabits forests.

FOOD: A fast flier, the Big Brown Bat feeds mainly on beetles and planthoppers, but rarely moths or flies. Near farmlands, it feeds heavily on agricultural pests. Foraging usually occurs at heights of no more than 9 m, and the two peak feeding periods are at dusk and just before dawn.

DEN: In summer, this bat usually roosts in tree cavities, under loose bark or in buildings. In winter, it hibernates in caves, mines or old buildings. Nursery colonies are found in protected areas such as tree cavities, large cliff crevices or old buildings.

REPRODUCTION: These bats breed in autumn or during a wakeful period in winter, but fertilization is delayed until the female ovulates in spring. A female gives birth to one or two young in early summer after about a two-month gestation.

SIMILAR SPECIES: The **Hoary Bat** (p. 390) has frosted brown or grey fur. The **Western Red Bat** (p. 388) and the **Eastern Red Bat** (p. 389) have reddish fur. The **Silver-haired Bat** (p. 392) has frosted, black fur. The *Myotis* **bats** (pp. 377–87) are all smaller.

RANGE: This bat occurs from southern Canada through most of the U.S.

Total Length: 9–14 cm

Tail Length: 2–6 cm

Forearm: 4.1–5.4 cm

Weight: 12–23 g

Spotted Bat
Euderma maculatum

sometimes light-coloured hairs on the nape of the neck. The belly is whitish. The long, pinkish to light tan ears project forward in flight but are folded back when the bat hangs up. The wingspan is about 35 cm, and the calcar is not keeled.

HABITAT: These bats are found in highland ponderosa pine regions in early summer. They descend to lower elevation deserts in August.

The Spotted Bat is an exhibitionist among a guild of committed conformists—one glance instantly reveals that it is no ordinary bat. The long, pink ears and the three huge, white spots that adorn its back are sufficiently distinctive for this bat to stand out in a crowd, but its flare is not restricted to visual appeal. While feeding, the Spotted Bat gives loud, high-pitched, metallic squeaks that are easily heard by humans. Because most bat vocalizations are beyond our hearing, it is unusually pleasing to listen to the aerial drama of the rare Spotted Bat.

ALSO CALLED: Jackass Bat, Death's Head Bat.

DESCRIPTION: The back is primarily black. There is a large white spot on each shoulder, below each ear, another on the rump and

FOOD: Spotted Bats appear to be specialized predators of noctuid moths, a large and diverse family of night-flying insects. A few beetles have also been found in the stomachs of these bats.

DEN: In summer, Spotted Bats seem to roost primarily in rock cracks and crevices on steep cliffs and in caves.

REPRODUCTION: These bats breed in autumn or during a wakeful period in winter, but fertilization is delayed until the female ovulates in spring. Usually one young is born in early summer. Even a young Spotted Bat has large ears, but the white spots on the back are absent on newborns.

SIMILAR SPECIES: The exceptionally large ears and large white spots make Spotted Bats unique among the bats of Canada.

RANGE: This bat occurs from southern British Columbia, southern Idaho and southeastern Montana south to Arizona and New Mexico.

Total Length: 11–12 cm

Tail Length: 4.5–5.1 cm

Forearm: 4.8–5.1 cm

Weight: 14 g

Townsend's Big-eared Bat
Plecotus townsendii

We all know that in *Dumbo*, Walt Disney went far beyond the realm of possibility in suggesting that this playful pachyderm could soar through the skies using its ears. Had the Disney creators searched for a more realistic character, they just might have turned to the Townsend's Big-eared Bat. The oversized ears of this bat serve no more function in flight than do the ears of any other mammal, but for sheer size and believability, few animals have ears to match.

Bats hold unquestionable supremacy in the aerial world of sound. Typically, the sounds that are produced by bats range between 20 kHz and 100 kHz. Human hearing extends only to about 20 kHz at best, so bat call frequencies begin where our hearing leaves off and can be as much as five times higher pitched than the sharpest human ears can detect. Each species of bat in Canada echolocates at a different frequency, so a person equipped with a bat detector—these things actually exist—can identify the species from its ultrasonic nighttime clicks. The Townsends's Big-eared Bat gives two different types of signals at different frequencies, believed to be used in different habitats for different styles of feeding: 12 to 25 kHz for gleaning insects and 45 to 60 kHz for in-flight meals.

As well as catching flying insects directly in its mouth, this bat also uses the membranes of its wings and tail almost like a baseball glove to deftly scoop up an insect and then pass it up to its mouth.

ALSO CALLED: Western Big-eared Bat.

DESCRIPTION: This medium-sized bat is brown overall. The belly is lighter brown than the back. Its most noticeable feature is its large ears, which are more than half

RANGE: The Townsend's Big-eared Bat ranges through all of western North America south of central British Columbia, Montana and South Dakota.

Total Length: 9–11 cm

Tail Length: 3.2–6.4 cm

Forearm: 4.1–4.8 cm

Weight: 7–11 g

the length of the forearm. The ears are jointed across the forehead at their bases. The median ear edges are double, and a prominent network of blood vessels is visible in the extended ears. At rest, the ears are curled and folded, almost resembling Bighorn Sheep horns. There is a set of conspicuous facial glands between the eye and nostril on each side of the snout. This bat's wingspan is about 30 cm, and the calcar is not keeled.

DID YOU KNOW?

Despite derogatory references to bats as "flying rats," they are actually more closely related to primates than they are to mice and other rodents.

HABITAT: This bat can be found in open areas near coniferous forests and in arid areas.

FOOD: Townsend's Big-eared Bats emerge quite late in the evening, so they are seldom observed while feeding. They forage along forest edges and can glean insects if necessary. They catch mainly small moths in the air, but also readily take beetles and flies.

DEN: In Canada, known maternity colonies are always in old buildings; in other areas, colonies are found in warm areas of caves and abandoned mines. These colonies are not as large as in other species, and clusters of more than 100 females and young are uncommon. Males tend to be solitary in summer. During winter hibernation, Townsend's Big-eared Bats tend to move deep into caves, where temperatures are constant.

REPRODUCTION: Mating occurs following ritualized courtship behaviour in October and November. The young bats are born between May and July. They are a quarter of their mother's weight at birth.

SIMILAR SPECIES: Because of its huge ears, this species can only possibly be confused with the **Pallid Bat** (p. 399), which is larger and lacks the lumps on its nose. The **Spotted Bat** (p. 396) also has huge ears, but its large, white spots make it unmistakable.

Pallid Bat

Pallid Bat
Antrozous pallidus

It might seem absurd that after millions of years of flight specialization, bats would be found foraging on the ground, but such is the case with the Pallid Bat. This bat is still a committed flier but frequently lands to take insects, other invertebrates and small vertebrates from the ground and from vegetation.

While bats have few predators in the night sky, on the ground they are vulnerable to many threats. The light dorsal colour of the Pallid Bat might be a protective adaptation that helps it blend in with the pale sands of its typically arid habitat.

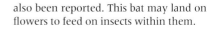

DESCRIPTION: The back is light yellow. The underparts are pale cream or almost white. Individual hairs are always darker at the tip than at the base, which is a reverse of the typical situation for bats. The broad, tan ears are extremely long, and, if pushed forward, they may extend past the muzzle. The median edge of the ears is not folded. The wingspan is about 38 cm, and the calcar is not keeled.

HABITAT: The Pallid Bat is typically associated with rocky outcrops near dry, open areas, but it is occasionally found in evergreen forests, especially within its range in British Columbia.

FOOD: Insects are the main food, but some small vertebrates such as lizards have also been reported. This bat may land on flowers to feed on insects within them.

DEN: Pallid Bats gather in night roosts following foraging. In Canada, these sites are typically in ponderosa pines. The day roost may be the same or may be in a nearby building or rock crevice.

REPRODUCTION: These bats mate from October through December and occasionally into February. The sperm is stored in the female's reproductive tract until ovulation in spring. Young are born in May and June, and twins are common.

SIMILAR SPECIES: The **Townsend's Big-eared Bat** (p. 397) is brown, somewhat smaller and has lumps on its nose.

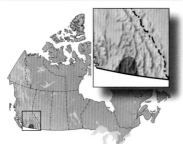

RANGE: The Pallid Bat ranges west of the Rockies from southern British Columbia to Baja California, into Utah, Colorado and western Texas, and south to central Mexico.

Total Length: 9.5–14 cm

Tail Length: 3.5–5.1 cm

Forearm: 4.8–6 cm

Weight: 16–35 g

INSECTIVORES & OPOSSUMS

T his grouping encompasses two separate orders of mammals: shrews and moles belong to the order Insectivora (insectivores), and opossums are in the order Didelphimorpha (New World opossums), which is part of the marsupials supergroup.

Mole Family (Talpidae)

Moles are among the most subterranean of all mammals, and special physical adaptations allow them to spend the majority of their lives underground. Moles look a bit like large, rotund shrews, except that their tails are proportionately shorter and their forelimbs are highly modified. The rotund appearance results from the lack of a visible neck—the head merges smoothly with the body—and this streamlined shape makes moving in underground tunnels much easier. The forelimbs have long claws and appear enormous relative to the body. They are turned outward like paddles, enabling moles to almost "swim" through soil. Moles' fur is lax, which helps them move forward or backward in tunnels easily. Their eyesight is poor, as you might guess from their tiny eyes, but their hearing is superb. Their most important sensory organ is the snout, which is flexible and usually hairless.

Moles consume up to twice their weight in food daily, and though earthworms are their primary food, they consume great numbers of pest insect larvae.

Shrew Family (Soricidae)

Shrews first appeared in the times of the Cretaceous dinosaurs, and biologists consider the modern members of this family to be most similar to the earliest placental mammals. Many people mistake shrews for very small mice, but shrews do not have a rodent's prominent incisors, and they generally have smaller ears and long, slender, pointed snouts.

Because they are so small, shrews lose heat rapidly to their surroundings, and their metabolisms surpass those of all other mammals. These tiny mammalian furnaces use energy at such a high rate that they may eat three times their own weight in invertebrate and vertebrate food each day. Some shrews have a neurotoxic venom in their saliva that enables them to subdue amphibians and mice that outweigh them. In turn, shrews are eaten by owls, hawks, foxes, Coyotes and weasels.

Shrews do not hibernate, but their periods of intense food-searching activity, which last 30 to perhaps 45 minutes, are interspersed with hour-long, energy-conserving periods of deep sleep, during which their body temperature drops.

Of the shrews of Canada, the Northern Short-tailed, Water and Arctic shrews are reasonably easy to identify visually, provided you can get a long enough look at them. The other species must be distinguished from one another on the basis of tooth and skull characteristics, distribution and, to some extent, habitat, though in many cases, the ranges overlap.

Opossum Family (Didelphidae)

The Virginia Opossum is the only marsupial in North America north of Mexico. Marsupials get their name from the marsupium, or pouch, in which newborns are typically carried. Because they lack a well-developed placenta, marsupials bear extremely premature young that range from honeybee to bumblebee size at birth. Once in the marsupium, the young attach to a nipple and continue the rest of their development outside the uterus. There is also a difference in dentition between placental mammals and marsupials: early placental mammals typically have four premolars and three molars, whereas early marsupials have three premolars and four molars. Most of the world's marsupials live in Australia and New Zealand, with a few in South and Central America.

American Shrew Mole
Neurotrichus gibbsii

If you stumble across an American Shrew Mole in southwestern British Columbia, you may think you have found a shrew, because it has small forelimbs, spends some time above ground and has a long, sparsely haired tail.

One of the least subterranean moles, it is intermittently active above ground throughout the day, pushing its way through leaf litter and decaying vegetation instead of digging tunnels. As a shrew mole forages, it moves slowly and cautiously beneath the leafy debris. Unlike a mole, it can move its forelimbs beneath its body, and it runs with an agility impossible for other moles.

DESCRIPTION: This small, shrewlike mole is nearly black. It has a relatively long, scaly tail, tiny eyes, ear pinnae that are only 2 to 6 mm long and large, scaly feet. The claws are long but not flattened or broad as in other moles.

HABITAT: Shrew moles prefer areas with abundant leaf litter, dead vegetation and rotting logs near streams, ravines or forested hillsides.

FOOD: Earthworms and sowbugs make up more than half of the diet. A wide assortment of other invertebrates and some vegetation are also eaten.

DEN: The shallow burrow is enlarged into a nest chamber 8 to 13 cm in diameter. The nest is made of dry leaves. The nest and burrow are not more than 30 cm below the surface.

REPRODUCTION: Most mating occurs from March to May, but some individuals may breed as early as February or as late as September. Litter size varies from one to six, and newborns weigh about 0.7 g. Females may have multiple litters in one season.

SIMILAR SPECIES: Other **moles** (pp. 403–07) have flattened claws and enormous forefeet. Most **shrews** (pp. 409–29) are smaller and have smaller forelimbs. The **Pacific Water Shrew** (p. 410) and **American Water Shrew** (p. 420) have hairier tails.

RANGE: The American Shrew Mole is found from the Fraser River delta in British Columbia down the coast to San Francisco. It is not found east of the Cascades.

Total Length: 10–13 cm

Tail Length: 3–4 cm

Weight: 8–11 g

Coast Mole
Scapanus orarius

The mammalian equivalents to backhoes and bulldozers, moles toil underground, bringing deep soil to the surface. Their activity aerates the soil, encourages water absorption and circulates nutrients.

A typical molehill of the Coast Mole is about a litre's worth of soil. Between October and March, when the soil is moist and the most digging occurs, one mole can push up 200 to 400 hills.

DESCRIPTION: In winter, the upperparts are dark grey, sometimes with a silvery sheen. In summer, the fur often has a brownish tinge. The tail is pinkish and sparsely haired. The nose tip is naked and pink. The tiny, functionless eyes are hidden beneath the skin. The feet are hairless, and the enlarged forefeet are turned outward.

HABITAT: This mole inhabits a variety of soil types in meadows, deciduous woodlands, brush and even some coniferous forests, if the soil is not too acidic.

FOOD: Earthworms make up more than three-quarters of the diet. Other invertebrates and some vegetation are also eaten.

DEN: Tunnels are 5 cm in diameter and may be 8 to 90 cm beneath the surface. At regular intervals, they expand into chambers 10 cm or greater in diameter. The breeding nest, about 20 cm in diameter, is located about 15 cm beneath the surface. It is lined with coarse grasses and has several connecting tunnels.

REPRODUCTION: A single litter of two to five is born each year in late March or early April, following breeding in February.

SIMILAR SPECIES: The **Townsend's Mole** (p. 404) is larger and more rotund.

RANGE: The Coast Mole is found from the lower Fraser River basin south to northern California and slightly east into southern Idaho.

Total Length: 15–18 cm

Tail Length: 3–4 cm

Weight: 60–90 g

Townsend's Mole
Scapanus townsendii

naked snout are crescent-shaped and point upward.

HABITAT: This mole prefers loose soil or cultivated fields but also occupies open brushland in valley bottoms.

FOOD: Earthworms, insects of all stages and other invertebrates make up most of the diet, but some vegetation is also consumed.

The stocky Townsend's Mole—the largest mole in North America—practically swims through the soil with its powerfully muscled forelimbs, which are equipped with shovel-like claws. Spoil from building burrows is thrust to the surface from below, forming the proverbial "molehills." During its nighttime foraging, this mole may venture into lawns, leaving a disfiguring series of soil mounds on the surface. Aesthetic impacts aside, moles are beneficial to soil aeration, water absorption and pest control.

DEN: This mole digs shallow surface tunnels, where most feeding occurs, as well as deeper tunnels with a central nest chamber about 15 to 20 cm in diameter. The two-layered nest consists of coarse, green grasses with an inner layer of fine, dry grasses, mosses and leaves. Tunnels radiate out from the nest chamber to other parts of the burrow system.

DESCRIPTION: The rotund body is covered with short, black, velvety fur. The snout is long, the neck short, the hindfeet small and the tail short and almost naked. The enormously enlarged forefeet cannot be rotated beneath the body. The foreclaws are flattened and heavy. The minute eyes lie beneath the skin and are probably useless. The nostrils at the end of the

REPRODUCTION: Mating occurs in February, with a single litter of one to four young born in late March. The hairless, altricial young weigh about 5 g at birth. At about 30 days, they are fully furred and weaned, and they leave the maternal burrow in May or June. They are sexually mature after their first winter.

SIMILAR SPECIES: The **Coast Mole** (p. 403) is smaller.

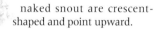

RANGE: These moles are found west of the Cascades from the U.S.-Canada border south to northern California.

Total Length: 20–24 cm

Tail Length: 3–5 cm

Weight: 65–170 g

Hairy-tailed Mole
Parascalops breweri

Hairy-tailed Moles are perfectly suited to life underground—understandably so, as they spend nearly all of their lives beneath the earth's surface. Moles are as choosy about their earthen home as we are about what house we live in. Subsurface rocks do not bother Hairy-tailed Moles—they merely dig around them—but they are very fastidious about the soil type. They like sandy loam soils because the soil structure allows for sturdy burrows to be built, but they avoid soils that have too much moisture or too much clay—burrows in this type of substrate are not durable and require too much maintenance.

DESCRIPTION: As its name suggests, this mole has a hairy tail, which is dark brown and quite short. The base of the snout and the feet are also dark brown. These brown areas may be grey or white in an older mole. The body is dark grey or nearly black above and only slightly paler below.

HABITAT: This mole favours well-drained soils in woodlands, brushy areas or meadows. It is sometimes found on golf courses and farmland.

FOOD: A voracious eater, the Hairy-tailed Mole consumes earthworms, grubs, insects and other invertebrates.

DEN: This mole lives underground in an extensive tunnel system. It is active all year, and in winter, it retreats to deeper tunnels where the temperature is constant. A nest chamber is used for sleeping and giving birth.

REPRODUCTION: Females produce one litter of four or five young per year, usually in early summer. The young are altricial and require about one month to mature enough to leave the nest.

SIMILAR SPECIES: The **Eastern Mole** (p. 406) has a short, naked tail. The **Star-nosed Mole** (p. 407) has a long, furry tail and protrusions from the nose.

RANGE: The Hairy-tailed Mole is found in southern Ontario and Québec and from New England south through the mountains to North Carolina and Tennessee.

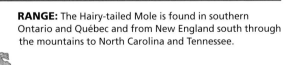

Total Length: 14–17 cm

Tail Length: 2.3–3.6 cm

Weight: 40–64 g

Eastern Mole
Scalopus aquaticus

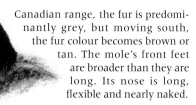

Canadian range, the fur is predominantly grey, but moving south, the fur colour becomes brown or tan. The mole's front feet are broader than they are long. Its nose is long, flexible and nearly naked.

The Eastern Mole is the most widespread mole species in North America, hence its alternate name, Common Mole. Unfortunately, its scientific name is not as accurate as its common names. The species name, *aquaticus*, was erroneously chosen to describe this mole because one of the earliest specimens was found drowned in a well, and close inspection of its forefeet revealed partial webbing between the toes. This evidence led to a belief that this mole was aquatic. Logically, of course, the webbing would help with digging through soil, and the fact that the specimen was found drowned should have been an indication that it was not aquatic.

ALSO CALLED: Common Mole.

DESCRIPTION: This rotund mole varies quite a bit in size, but its short, nearly naked tail helps distinguish it from the other two moles in its range. The colour of its fur varies regionally. In its

HABITAT: The Eastern Mole usually prefers open areas such as meadows and sparse woodlands. It favours loose, well-drained soils.

FOOD: This mole forms extensive burrows for feeding, and its primary food is earthworms, though it also consumes insects, slugs and other invertebrates.

DEN: The Eastern Mole makes a shallow underground nest, usually protected by a surface boulder or log. Some nests and burrows are used repeatedly, but others may be abandoned after one use.

REPRODUCTION: Mating occurs in February or March, and after a gestation of 30 to 42 days, the female has a litter of two to five young. The young are altricial but develop rapidly. After one month, the young are on their own.

SIMILAR SPECIES: The **Hairy-tailed Mole** (p. 405) and the **Star-nosed Mole** (p. 407) both have longer, furred tails.

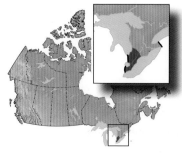

RANGE: The Eastern Mole is found in much of the eastern U.S. and just into Ontario in Essex County at the southernmost border of the province.

Total Length: 10–20 cm

Tail Length: 2–3 cm

Weight: 50–120 g

Star-nosed Mole
Condylura cristata

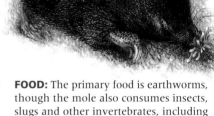

The Star-nosed Mole is a fine example of the incredible variation and diversity that exists in the animal kingdom. Unlike the rest of its kin, the Star-nosed Mole has developed a ring of "feelers" around its nose. At first glance, its nose looks gruesome and Medusa-like, but in reality, this interesting structure is an elegantly designed super-sensor. Each fleshy appendage can be collapsed or extended individually, unaffected by the movement of the ones next to it, and they move continuously in all directions to acutely and perfectly interpret the mole's surroundings.

Star-nosed Moles hunt for food either underground or in water, and their noses are perfectly suited for either habitat.

DESCRIPTION: There is no mistaking this distinctive mole. The nose is surrounded by a ring of 22 pink, fleshy, tentaclelike projections. The fur is silky and black, and the tail is long and hairy. The front feet are powerful and have long claws.

HABITAT: This mole usually prefers wet areas such as marshes, low fields or humid woodlands. It sometimes inhabits drier areas.

FOOD: The primary food is earthworms, though the mole also consumes insects, slugs and other invertebrates, including some aquatic invertebrates.

DEN: This mole digs extensive burrows for feeding, but its leaf-and-grass nest is usually made in a clump of vegetation or on a hummock on the ground surface.

REPRODUCTION: Females have one litter per year of four to seven young, usually in early spring. Gestation is about 45 days. The young are altricial but develop rapidly. By the third week, the young have left the nest.

SIMILAR SPECIES: Both the **Hairy-tailed Mole** (p. 405) and the **Eastern Mole** (p. 406) lack the distinctive nose and the long, well-haired tail.

RANGE: The Star-nosed Mole is found in southeastern Canada and the northeastern U.S. It also occurs in the Appalachian Mountains and along the Georgia coast.

Total Length: 15–21 cm

Tail Length: 5.3–8.4 cm

Weight: 30–75 g

Key to the Shrews

Arctic Shrew
Sorex arcticus

Arctic Shrews may be the handsomest of all the North American shrews. If you were able to observe one of these animals, you could even tell the season by the shrew's colour. Although many weasels and hares change colour seasonally, it is quite unusual for a shrew to do so. Not only is the Arctic Shrew's winter coat longer and denser than its summer coat, it is also more vibrant, with a black back, brown sides and a white or greyish belly. The full summer coat is less striking, with a brown back and grey underparts.

ALSO CALLED: Saddle-backed Shrew.

DESCRIPTION: The tricoloured body of this stocky shrew makes it one of the easiest shrews to recognize: the back is chocolate brown in summer and glossy black in winter, the sides are grey brown year-round, and the underparts are ashy grey in summer and silver white in winter. The tail is cinnamon-coloured year-round. Females are usually slightly larger than males.

HABITAT: This shrew typically inhabits moist areas of the boreal forest or its edges. Outside forested regions, it takes to open areas, dried-out sloughs and streamside habitats among shrubs.

FOOD: The Arctic Shrew feeds primarily on invertebrates such as insects, snails, slugs and even some carrion.

DEN: The spherical, grassy nest, 6 to 10 cm in diameter, is built in a small pocket in or under a log, under debris or in a rock crevice.

REPRODUCTION: Breeding takes place between May and August, and females generally have two litters of 4 to 10 young in per season. Females born early in the year may have their first litter in late summer of that same year, but most do not breed until the following year.

SIMILAR SPECIES: The tricoloured pelage of the Arctic Shrew best distinguishes it from other shrew species. It also tends to be heavier and stockier than most other shrews. The **American Water Shrew** (p. 420) is velvety black over its back.

RANGE: The Arctic Shrew is found from the southeastern Yukon across central Canada to eastern Québec and south to Minnesota, Wisconsin and parts of North Dakota and Michigan.

Total Length: 10–12 cm

Tail Length: 3.8–4.5 cm

Weight: 6–14 g

Pacific Water Shrew
Sorex bendirii

This fascinating shrew captures much of its food in water, and, like the American Water Shrew and a few small rodents, it can even run across the water's surface for a short distance before diving under. Beneath the water, it appears silvery because of air trapped in its fur. This trapped air makes the shrew so buoyant that, when it ceases swimming, it quickly pops to the surface like a cork.

ALSO CALLED: Marsh Shrew.

DESCRIPTION: This shrew has velvety black or blackish brown fur, with slightly paler underparts in winter. In summer, the fur is somewhat browner. The tail is dark both above and below. The nose is pointed. The hindfeet are fringed with stiff hairs that help in swimming.

HABITAT: The Pacific Water Shrew inhabits marshy areas along slow-moving streams, as well as other wetlands. During rainy winter months, it may disperse as far as 1 km from the nearest waterbody.

FOOD: This shrew avidly devours both aquatic and terrestrial invertebrates. Insects of all stages are eaten, as well as small fish. The Pacific Water Shrew immobilizes its prey with a rapid series of bites, and these paralyzed creatures may be stored for a few hours.

DEN: The nest is a ball of dry grass, often located beneath the loose bark of a fallen tree or within a rotted log or stump.

REPRODUCTION: The typical litter size is four to seven, and the young are born in a bulky nest of grass. Virtually nothing is known of gestation or time to independence. Sexual maturity probably quickly follows independence, because the maximum lifespan of this species does not exceed 1.5 years.

SIMILAR SPECIES: The **American Water Shrew** (p. 420) is about the same size but has lighter underparts and a bicoloured tail.

RANGE: From extreme southwestern British Columbia, this shrew occurs west of the Cascade Mountains as far south as San Francisco.

Total Length: 14–17 cm

Tail Length: 6–8 cm

Weight: 8–18 g

Masked Shrew

Sorex cinereus

The Masked Shrew may be the most common shrew throughout Canada. In spite of its abundance, you are most likely to see one dead in spring because starvation in late winter claims many, and their tiny bodies lie waiting after the snow melts to be recycled in the renewal of spring. To balance the high mortality rates in late winter and high year-round predation, these shrews have high fecundity. They mate from May to October, and the young that are born in autumn have a good chance of making it to spring, when they, too, will mate.

ALSO CALLED: Cinereus Shrew, Common Shrew.

DESCRIPTION: This medium-sized shrew has a dark brown back, lighter brown sides and slightly paler underparts. The winter coat is paler, and the fur is short and velvety. The shrew has a long, flexible snout, tiny eyes, small feet and a bicoloured tail, which is dark above and light below. A few may have a dark patch on the nose—the "mask" for which the species is named.

HABITAT: The Masked Shrew favours forests, either coniferous or deciduous, and sometimes tall-grass plains or brushy coulees.

FOOD: Insects account for the bulk of the diet, but this shrew also eats significant numbers of slugs, snails, young mice, carrion and even some vegetation.

DEN: The nest, located under logs, in debris, between rocks or in a burrow, is about 5 to 10 cm in diameter and looks like a woven grass ball. The nest does not have a central cavity—the shrew simply burrows inside.

REPRODUCTION: Mating occurs from April to October, and, with a gestation of about 28 days, a female may have two or three litters per year. The four to eight young are born naked, toothless and blind. Eyes and ears open in just over two weeks, and the young are weaned by three weeks of age.

SIMILAR SPECIES: Most shrews look very similar. Without a specimen and a technical key, it is almost impossible to identify a shrew reliably.

RANGE: The Masked Shrew is found across most of Alaska and Canada. Its range extends south into northern Washington, through the Rockies and across most of the northeastern U.S.

Total Length: 7–11 cm

Tail Length: 2.5–5.1 cm

Weight: 2–7 g

Long-tailed Shrew
Sorex dispar

If you are walking through high-elevation forests in southeastern New Brunswick or north-central Nova Scotia, keep an eye out for movement on talus slopes and rock slides—prime habitat for the Long-tailed Shrew. Interestingly, individuals of this species tend to have a slightly smaller body size in the northern parts of their range. A larger body has a lower surface area to volume ratio, and therefore retains heat better, so most mammals average larger body sizes in northern regions. No one knows why this shrew is the opposite.

DESCRIPTION: This medium-sized shrew is slate grey above, and the belly is sometimes lighter than the back. The long tail is quite thick around, well-furred and mainly dark grey (only faintly bicoloured). The feet are whitish.

HABITAT: The Long-tailed Shrew favours talus slopes and rock slides but is sometimes found in rock crevices or under logs in damp coniferous and deciduous forests.

FOOD: The exact diet has not been well studied, but this shrew presumably feeds on invertebrates such as earthworms, slugs, snails and both adult and larval insects.

DEN: This grassy nest is spherical and about 15 to 20 cm in diameter. Nests are found in sheltered areas such as in rock crevices or under fallen logs.

REPRODUCTION: Reproduction occurs from April to August, and females have litters of two to five young. The young are weaned and independent by the time they are four weeks old. Like other shrews, Long-tailed Shrews probably do not live for more than 14 to 17 months.

SIMILAR SPECIES: Most shrews look very similar. Without a specimen and a technical key, it is almost impossible to identify a shrew reliably.

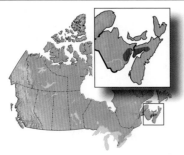

RANGE: The Long-tailed Shrew is found throughout the Appalachian Mountains. A discontiguous population is found in southeastern New Brunswick and north-central Nova Scotia.

Total Length: 10.0–13.6 cm

Tail Length: 5–6 cm

Weight: 3–8 g

Smoky Shrew
Sorex fumeus

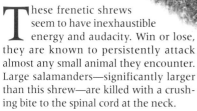

These frenetic shrews seem to have inexhaustible energy and audacity. Win or lose, they are known to persistently attack almost any small animal they encounter. Large salamanders—significantly larger than this shrew—are killed with a crushing bite to the spinal cord at the neck.

The Smoky Shrew, despite its daring, meets its match with large *Peromyscus* mice. Try as it might, this shrew seems unable to beat a large mouse and will eventually turn away defeated. The Smoky Shrew is quite talkative, and when alarmed, it utters a high-pitched, grating note. Even while it forages, it continually twitters, though the noise is almost inaudible to humans.

DESCRIPTION: The fur is nearly uniformly brown or greyish, browner in summer and greyer in winter. The long tail is brown on top and tawny or sometimes yellowish below.

HABITAT: The Smoky Shrew can be found in a variety of different moist conditions such as marshes, humid wooded areas and along streams.

FOOD: The majority of the diet consists of soft-bodied, tender invertebrates such as insects and their larvae, earthworms and slugs.

DEN: This shrew makes a small, leafy nest in a protected place such as inside a hollow log or under rocks. It also makes a burrow with an entrance about the size of a dime.

REPRODUCTION: Mating occurs in spring, especially in March. About five or six young are born in the summer months after a gestation of 20 days. The young grow quickly, and though they are sexually mature in their first year, females do not bear young until they are two years old.

SIMILAR SPECIES: Most shrews look very similar. Without a specimen and a technical key, it is almost impossible to identify a shrew reliably. The **Northern Short-tailed Shrew** (p. 428) is larger and greyer.

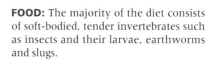

RANGE: The Smoky Shrew is found in central and southern Ontario and Québec, New Brunswick, Nova Scotia, New England and through the Appalachian Mountains to northeast Georgia.

Total Length: 7–11 cm

Tail Length: 2.5–5.1 cm

Weight: 2–7 g

Gaspé Shrew
Sorex gaspensis

As its name suggests, the Gaspé Shrew is found on the Gaspé Peninsula of Québec, but there are two other discontiguous populations, one in New Brunswick and the other in Nova Scotia. Although some researchers believe the Gaspé Shrew may be a subspecies of the Long-tailed Shrew, it currently has its own species status. Like the Long-tailed Shrew, its tail is nearly as long as its body. Observations of its grooming behaviour show that it is meticulous about keeping clean. It cleans its face with moistened forefeet, much like a cat.

DESCRIPTION: This small shrew is predominantly slate grey, with underparts the same colour or just slightly lighter than the back. The tail is the same colour as the body and almost as long.

HABITAT: The Gaspé Shrew primarily inhabits rocky areas near streams and steep, rocky slopes at high elevations. Yellow birch and sugar maple are the common overstorey trees of this shrew's preferred habitat. It is sometimes found in mixedwood areas as well.

FOOD: Like other shrews, this species feeds on a variety of invertebrates such as adult insects and larvae, slugs, centipedes, snails and earthworms, but spiders appear to be the most significant prey item.

DEN: This shrew probably builds a spherical nest like other *Sorex* species. The nest is made in a sheltered area such as in a crevice under rocks or in a decayed log. The spherical nest is a simple bundle of grasses without a central cavity.

REPRODUCTION: The Gaspé Shrew probably has a gestation similar to other shrews—less than 20 days—and it likely mates and bears young in spring and summer. Litter size ranges from two to six young, and females may have one or two litters per year.

SIMILAR SPECIES: Most shrews look very similar. Without a specimen and a technical key, it is almost impossible to identify a shrew reliably.

RANGE: This species is one of only two North American shrews whose range lies entirely within Canada. There are three main populations: one on the Gaspé Peninsula, one in western New Brunswick and the other on Cape Breton Island in Nova Scotia.

Total Length: 9.5–12.7 cm

Tail Length: 4.5–5.5 cm

Weight: 2.2–4.3 g

Hayden's Shrew
Sorex haydeni

Originally thought to be the same species as the Masked Shrew, the now-distinct Hayden's Shrew inhabits southern parts of the Prairies. Its habitat preference reflects its alternate common name, Prairie Shrew. Recent studies have found that this shrew is not restricted to native prairie, however, and it occurs with equal frequency in fields and pastures and sometimes even open parkland areas.

ALSO CALLED: Prairie Shrew.

DESCRIPTION: This medium-sized shrew has a brown or cinnamon-coloured back, lighter brown sides and pale underparts. It is paler in winter, and the fur is shorter and more velvety. It has a long, flexible snout, tiny eyes, small feet and a bicoloured tail, which is dark above and light below.

HABITAT: The Hayden's Shrew likes open areas and meadows, typically preferring short-grass prairies.

FOOD: This shrew eats primarily insects but will also consume significant numbers of slugs, snails, young mice, carrion and even some vegetation.

DEN: The nest is located under logs, in debris, between rocks or in a burrow. A nest measures about 6 to 10 cm in diameter and looks like a woven grass ball. It does not have a central cavity—the shrew simply burrows inside.

REPRODUCTION: Mating occurs from April to October, and a female may have two or three litters per year. The gestation period is about 18 days, after which four to eight young are born. At birth, they are naked, toothless and blind. Their growth is rapid—eyes and ears open in just over two weeks, and they are weaned at day 20.

SIMILAR SPECIES: Most shrews look very similar. Without a specimen and a technical key, it is almost impossible to identify a shrew reliably.

RANGE: The Hayden's Shrew inhabits prairie regions from southeastern Alberta south and east into Missouri.

Total Length: 7.4–8.8 cm

Tail Length: 2.5–3.3 cm

Weight: 2.5–5.5 g

Pygmy Shrew
Sorex hoyi

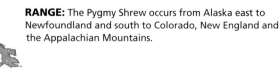

HABITAT: The Pygmy Shrew lives in a variety of different habitats, moist to dry and forested to open, including deep spruce woods, sphagnum bogs, grassy or brushy areas, cattails and rocky slopes.

Weighing only as much as a penny or two, the Pygmy Shrew represents the furthest degree of miniaturization in mammals and is considered to be the smallest of all North American mammals. The Dwarf Shrew (*S. nanus*), found in southern areas of the U.S., may weigh less, but it is longer than the Pygmy Shrew.

In spite of its size, the Pygmy Shrew is every bit as voracious as other shrews. One female on record ate about three times her body weight each day for 10 days. The Pygmy Shrew may also be one of the rarest shrews in North America.

DESCRIPTION: The coat is primarily reddish to greyish brown. The colour grades from darkest on the back to somewhat lighter underneath. It is usually greyer in winter. The third and fifth unicuspid teeth are so reduced in size that they may go unnoticed.

FOOD: This shrew feeds primarily on larval and adult insects, but earthworms, snails, slugs and carrion often make up a significant portion of the diet.

DEN: The spherical, grassy nest, 6 to 10 cm in diameter, may be under logs, under debris or in a rock crevice. There is no rounded cavity inside the nest—the shrew simply burrows its way in among the grasses.

REPRODUCTION: Breeding takes place from May until August, and 4 to 10 young are born in June, July or August. Females generally have only one litter per year. Young born early in the year may have a late-summer litter, but most females do not mate until the following year.

SIMILAR SPECIES: Most shrews look very similar. Without a specimen and a technical key, it is almost impossible to identify a shrew reliably.

RANGE: The Pygmy Shrew occurs from Alaska east to Newfoundland and south to Colorado, New England and the Appalachian Mountains.

Total Length: 5.5–6.1 cm

Tail Length: 2.7–3.3 cm

Weight: 2.5–6.3 g

Maritime Shrew
Sorex maritimensis

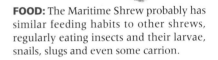

The Maritime Shrew has only recently been given species status, and it was previously thought to be a disjunct subspecies of the Arctic Shrew. Like the Arctic Shrew, it is a large, tricoloured shrew that changes colour somewhat in different seasons. Because of its recent species status, little research has been done to determine its habits. Within its range, its preferred habitat is rare and fragmented, so this species may be vulnerable to increased disturbance and other negative environmental factors.

DESCRIPTION: The tricoloured body of this stocky shrew is similar to that of the Arctic Shrew. The back is chocolate brown in summer and glossy black in winter, the sides are grey brown year-round, and the underparts are ashy grey in summer and silver white in winter. The tail is cinnamon-coloured year-round. Females are usually slightly larger than males.

HABITAT: This shrew typically inhabits wetland areas with plenty of grasses, especially Canadian reedgrass (*Calamagrostis* spp.). It prefers open habitat without heavy tree cover.

FOOD: The Maritime Shrew probably has similar feeding habits to other shrews, regularly eating insects and their larvae, snails, slugs and even some carrion.

DEN: The spherical, grassy nest, usually 6 to 10 cm in diameter, may be built in a small pocket in or under logs, under debris or in a rock crevice.

REPRODUCTION: Breeding is probably the same as the Arctic Shrew. Mating occurs between May and August, and females generally have two litters of 4 to 10 young in a season. Females born early enough in the year may have their first litter in late summer of that same year.

SIMILAR SPECIES: The identical **Arctic Shrew** (p. 409) has a different range, but otherwise the tricoloured coat of this shrew—a dark back, lighter sides and a still lighter belly—best distinguishes it from other shrews. The **American Water Shrew** (p. 420) is velvety black over its back.

RANGE: The Maritime Shrew is found in southeastern New Brunswick and mainland Nova Scotia.

Total Length: 10–12 cm

Tail Length: 3.8–4.5 cm

Weight: 6–14 g

Merriam's Shrew
Sorex merriami

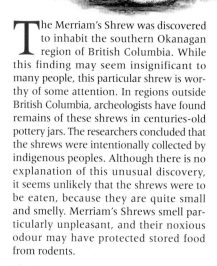

The Merriam's Shrew was discovered to inhabit the southern Okanagan region of British Columbia. While this finding may seem insignificant to many people, this particular shrew is worthy of some attention. In regions outside British Columbia, archeologists have found remains of these shrews in centuries-old pottery jars. The researchers concluded that the shrews were intentionally collected by indigenous peoples. Although there is no explanation of this unusual discovery, it seems unlikely that the shrews were to be eaten, because they are quite small and smelly. Merriam's Shrews smell particularly unpleasant, and their noxious odour may have protected stored food from rodents.

DESCRIPTION: This shrew is greyish or brownish above and whitish below. In winter, it is brighter in appearance. The bicoloured tail is sparsely furred. Males have very large flank glands. If you lift the upper lip, you can see four pointed teeth crowded behind the upper incisor.

An identifying feature of this shrew is that the second tooth (unicuspid) is the largest, and the third tooth is larger than the fourth.

HABITAT: This shrew inhabits sagebrush flats, deserts, semi-deserts and sometimes dry grasslands.

FOOD: The Merriam's Shrew is thought to eat mostly insects, including beetles, crickets, wasps and caterpillars. Spiders are likely another seasonally common food source.

DEN: This shrew makes a typical spherical, grassy nest, often under a log or in soft soil.

REPRODUCTION: Mating occurs from April through July, with females having multiple litters of typically four to seven young.

SIMILAR SPECIES: Most shrews look very similar. Without a specimen and a technical key, it is almost impossible to identify a shrew reliably.

RANGE: The Merriam's Shrew is found from Washington to North Dakota and south to New Mexico and Arizona in appropriate habitat. Recent studies show them in southern British Columbia as well.

Total Length: 8.6–11 cm

Tail Length: 3.2–4.1 cm

Weight: 3.5–7.1 g

Dusky Shrew
Sorex monticolus

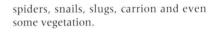

Given their tiny size, shrews can be remarkably fierce mammals. This observation may surprise people who have never experienced a shrew up close, but some biologists who have worked with both shrews and bears say they prefer to study bears. Identifying shrews almost invariably involves an examination of their teeth, and living shrews, which understandably dislike such close scrutiny, invariably try to bite the offending fingertips.

The Dusky Shrew is probably the most abundant shrew within its range.

ALSO CALLED: Montane Shrew.

DESCRIPTION: This medium-sized shrew has a pale brown back and sides in summer. Its back is slightly darker in winter. The underparts are silvery grey to buffy brown. The bicoloured tail is whitish below and the same colour as the back above.

HABITAT: The Dusky Shrew can be found in moist alpine meadows and wet sedge meadows, among willows alongside mountain brooks and in damp coniferous forests with nearby bogs.

FOOD: The diet includes a variety of adult and larval insects, earthworms, spiders, snails, slugs, carrion and even some vegetation.

DEN: This shrew usually builds a spherical nest in a decayed log. The nest is a simple bundle of grasses without a central cavity.

REPRODUCTION: Mating occurs from March to August, during which time a female likely has more than one litter of two to nine young. The young are helpless at birth and must nurse heavily to complete their rapid growth. Their eyes and ears open in about two weeks, and they are weaned soon afterward.

SIMILAR SPECIES: Most shrews look very similar. Without a specimen and a technical key, it is almost impossible to identify a shrew reliably.

RANGE: The Dusky Shrew is found from Alaska southeast to Manitoba and south along the Rocky Mountains to Mexico.

Total Length: 9–13 cm

Tail Length: 3.5–5.1 cm

Weight: 5–7 g

American Water Shrew

Sorex palustris

To the consternation of naturalists who are keen on species identification, most of the shrews in Canada have few distinguishing characteristics. Thankfully, the American Water Shrew is an exception—this finger-sized heavyweight is different enough in its appearance and habits to deserve celebrity status.

Whereas other shrews prefer to wreak terror exclusively on the small vertebrates and invertebrates roaming on land, the American Water Shrew also dives in to feed upon aquatic prey. This shrew is a particularly fierce predator, ably seizing not only insect nymphs, but even sticklebacks and other small fish. The shrew drags the catch onto land, where the prey is quickly consumed. This shrew is specially adapted for swimming—small hairs on the hindfeet widen the foot and create a flipper effect for propulsion. It is very powerful and can easily outswim most prey species. Once out of the water, its fringed feet serve as a comb with which to brush water droplets out of its fur.

An easy shrew to recognize, the American Water Shrew can be seen beneath overhangs along flowing waters, particularly small creeks and backwaters. If you are walking along these shorelines, you might see a small, black bundle rocket

RANGE: This transcontinental shrew ranges from southern Alaska east to Labrador and south along the Cascades and Sierra Nevada to California, along the Rocky Mountains to New Mexico and along the Appalachians almost to Georgia.

Total Length: 14–17 cm

Tail Length: 6–9 cm

Weight: 9–19 g

from beneath the overhang into the water. The motion at first suggests a frog, but the American Water Shrew tends to enter the water with more finesse, hardly producing a splash. Often, the shrew will first run a short distance across the surface of the water before diving in. Some voles and mice are also scared into or across water in this way, but even at a quick glance, you can distinguish this shrew from those rodents by its velvety black colour. Only the Pacific Water Shrew in British Columbia shares similar colouration and the water-loving behaviour.

ALSO CALLED: Northern Water Shrew.

DESCRIPTION: The American Water Shrew is one of the largest shrews in Canada. It has a velvety black back and contrasting light brown or silver underparts. The tail is similarly bicoloured. The third and fourth toes of the hindfeet are slightly webbed, and a stiff fringe of hairs around the hindfeet aid in swimming. Males tend to be somewhat larger than females.

HABITAT: This shrew can be found alongside flowing streams with undercut, root-entwined banks, in sphagnum mosses on the shores of lakes and occasionally in nearly dry streambeds or tundra regions.

DID YOU KNOW?

Both terrestrial and aquatic animals prey on American Water Shrews. Weasels, American Mink and otters catch them, as do large trout, bass, walleye and pike.

FOOD: Aquatic insects, spiders, snails, other invertebrates and small fish form the bulk of the diet. With true shrew frenzy, this scrappy water lover may even attack fish half as large as itself.

DEN: This shrew dens in a shallow burrow in root-entwined banks, in sphagnum moss shorelines or even in the woody debris of beaver lodges. The nest is a spherical mound about 10 cm in diameter and composed of dry vegetation such as twigs, leaves and sedges.

REPRODUCTION: Breeding occurs from February until late summer. Females have multiple litters each year. Those born early in the year usually have their first litter in that same year. Litters vary in size from five to eight young, and, as with other shrews, the young grow rapidly and are independent in a few weeks.

SIMILAR SPECIES: All other shrews lack the velvety black fur. The **Pacific Water Shrew**'s (p. 410) tail is uniformly coloured, and the shrew tends to be browner over its back with only slightly lighter undersides. The **Arctic Shrew** (p. 409) may have deep brown or nearly black fur, but only on the top of its back.

Pacific Water Shrew

Preble's Shrew
Sorex preblei

New to most people in Canada, the Preble's Shrew was only recently discovered in the southern Okanagan region of British Columbia. It is one of the smallest members of the genus *Sorex*, and it is about the same size as the Pygmy Shrew.

Preble's Shrews are much better known in the northwestern U.S., but even there, they are not well studied and information about them is limited.

DESCRIPTION: The Preble's Shrew has a brownish grey back that grades to lighter colours on the sides and underparts. If you raise the upper lip on the side of the snout, you can see four pointed teeth behind the large, lobed first incisor. A distinguishing feature of this shrew is that the third of these unicuspid teeth is not smaller than the fourth.

HABITAT: This tiny shrew seems to prefer dry sagebrush desert environments or grasslands with rocky areas. Within Canada, its only known habitat is in the southern Okanagan.

FOOD: The Preble's Shrew is thought to eat mostly invertebrates such as beetles, crickets, wasps, caterpillars and spiders.

DEN: The den is often found in soft soil, among rocks or under woody debris. The nest chamber is exceedingly small, and the entrance to the burrow is small and indistinct.

REPRODUCTION: Little is known about this shrew's reproduction, but it is probably similar to that of other shrews. Mating likely occurs from April through July, with females having multiple litters per year.

SIMILAR SPECIES: Most shrews look very similar. Without a specimen and a technical key, it is almost impossible to identify a shrew reliably. The **Vagrant Shrew** (p. 427) and the **Dusky Shrew** (p. 419) both have a fourth unicuspid tooth that is larger than the third.

RANGE: Preble's Shrews are found in British Columbia and south through Washington, Oregon, California and most of Idaho and Montana. They are known to inhabit the Nicola and Okanagan valleys, but they may be more widespread in the province.

Total Length: 7.7–9.5 cm

Tail Length: 2.8–3.8 cm

Weight: 2.1–4.1 g

Olympic Shrew
Sorex rohweri

This shrew, not described until 2007, is a recent addition to the mammal species of Canada. Although a few have been found alive in southern-most British Columbia, much of their known range here actually comes from studying mis-identified museum specimens. This shrew is closely related to, and easily confused with, both the Masked Shrew and the Vagrant Shrew. Researchers have been able to identify several Olympic Shrews amid museum specimens and assign a small Canadian range based on the collection localities. The best place in British Columbia to find an Olympic Shrew is in Burns Bog in Delta.

DESCRIPTION: This shrew is brownish grey above and lighter below. The tail is similarly bicoloured, a trait more obvious in its winter pelage. Older shrews seem to lose the fur on their tails, making their tails appear sparsely furred.

HABITAT: The few records of this shrew in Canada indicate that it prefers mixed or coniferous forests and forest edges.

FOOD: Like other shrews, the Olympic Shrew likely eats adult and larval insects, as well as earthworms, spiders, slugs, snails and carrion.

DEN: The nest is probably similar to that of other shrews—a grassy sphere measuring about 5–10 cm in diameter and located under logs, in debris, between rocks or in a burrow. Shrew nests do not have a central cavity—the shrew simply burrows inside.

REPRODUCTION: Mating probably occurs from April to October. Gestation is likely about 28 days, and females probably have two or three litters per year. Shrews are born naked, toothless and blind. Eyes and ears open in just over two weeks, and the young are weaned by three weeks of age.

SIMILAR SPECIES: Most shrews look very similar. Without a specimen and a technical key, it is almost impossible to identify a shrew reliably.

RANGE: The Olympic Shrew is found only in western Washington and extreme south-western British Columbia, from Burns Bog to Chilliwack Lake.

Total Length: 9–12 cm

Tail Length: 3.0–4.1 cm

Weight: 4–7 g

Trowbridge's Shrew
Sorex trowbridgii

Trowbridge's Shrew tends to collect and store seeds, a behaviour not reported in other North American shrews. Because its diet is more diverse than that of its kin, it has an advantage over the Vagrant Shrew and the Dusky Shrew where their ranges overlap.

Trowbridge's Shrew is active both day and night, but its periods of activity are short, followed by periods of quiescence. This shrew probably dies before it is 1.5 years old, but during late summer, the population peaks because of early summer births.

DESCRIPTION: This velvety, dark grey shrew has underparts that are nearly as dark as the back. In summer, the body colour is slightly brownish. The tail is sharply bicoloured, dark above and light below. The tail of a young animal is hairy but tends to be naked in older individuals. The ears are nearly hidden in the hair, and the vibrissae are long and abundant. The feet are whitish to light tan.

HABITAT: Throughout its range, this shrew frequents mature forests with abundant ground litter. It appears to prefer dry ground beneath Douglas-firs, but when other shrews are absent, it occupies ravines, swampy woods and areas where deep grass borders salmonberry thickets.

FOOD: The diet is primarily small insects, spiders, centipedes, snails, slugs, earthworms and flatworms, but these shrews also often eat Douglas-fir seeds and the seeds of other plants. They occasionally even eat subterranean fungi.

DEN: The nest of a Trowbridge's Shrew has not been described, but it is likely similar to that of other shrews.

REPRODUCTION: Ordinarily, three to six young are born in spring and early summer. During this time, adult females are continually pregnant.

SIMILAR SPECIES: The **Pacific Water Shrew** (p. 410), the only other shrew within the same range, has dark underparts, is much larger and heavier and has far fewer vibrissae.

RANGE: Trowbridge's Shrew is found from southwestern British Columbia along the coast into central California.

Total Length: 10–12 cm

Tail Length: 5.0–5.5 cm

Weight: 6–8 g

Tundra Shrew
Sorex tundrensis

Unlike most other shrews, the Tundra Shrew is readily identifiable in the field. Like the similar Arctic Shrew, the Tundra Shrew is distinctly tricoloured: brown on the back, light brown on the sides and light grey below. Fortunately, these two shrews inhabit different parts of Canada, so tricolouration remains a key identification feature.

For a long time, researchers were uncertain whether the Tundra Shrew and the Arctic Shrew were really separate species, but it now appears that the Tundra Shrew's closest relatives are found in Siberia.

DESCRIPTION: This medium-sized shrew is tricoloured, with a brown back, pale brown sides and pale grey underparts. The colour change on the sides is quite distinct, rather than a smooth gradient. Some individuals appear more greyish than brown, but have the same tricoloured pattern. A juvenile does not exhibit the pattern until its first moult. In winter, the fur is longer and the tricolour pattern fades into a bicolour pattern, brownish on the back and greyish underneath.

HABITAT: The Tundra Shrew lives in a variety of different habitats including tundra, areas of thick grass, dense shrubs or dwarf trees and sometimes marshy or boggy areas.

FOOD: This shrew feeds primarily on earthworms, larval and adult insects and vegetation such as floral parts and grasses.

DEN: Like other shrews, the Tundra Shrew probably makes spherical, grassy nests in sheltered areas such as under logs, under debris or in a rock crevice.

REPRODUCTION: Very little is known about this shrew, but pregnant females have been recorded in June, July and September. Litters of 8 to 12 young are typical, and females probably have more than one litter per year.

SIMILAR SPECIES: The **Arctic Shrew** (p. 409) is also tricoloured, but it is slightly larger and its range does not overlap. Other shrews are uniformly coloured.

RANGE: The Tundra Shrew occurs in Alaska, the western Yukon, the northern Northwest Territories and the northwestern corner of British Columbia.

Total Length: 8.3–12.0 cm

Tail Length: 2.2–3.6 cm

Weight: 5–10 g

Barren-ground Shrew
Sorex ugyunak

Like all shrews, the Barren-ground Shrew is strictly territorial and will resolutely defend its home range against all intruders. This tundra-loving shrew is closely related to the St. Lawrence Island Shrew (*S. jacksoni*) and the Pribilof Island Shrew (*S. pribilofensis*), both of which are isolated on islands off the coast of Alaska. At some point in history, the population was contiguous, but now the geographic separation has allowed these shrews to diverge into three separate species. The Barren-ground Shrew's species name, *ugyunak*, is Inuktitut for "shrew."

DESCRIPTION: The Barren-ground Shrew has a distinct brown stripe over its back, with light greyish brown sides and underparts. The tail is lightly bicoloured, brownish on top and paler below. Unlike other shrews, the tail has a slight tuft at the end.

HABITAT: This shrew lives in the tundra in wet areas with abundant grasses and other vegetation. Other associated vegetation includes willow thickets and some birch.

FOOD: Insects and other invertebrates are the main foods, though seeds may be eaten as well.

DEN: Very little is known about the habits of this shrew, but it likely makes a spherical, grassy nest similar to that of other shrew species. The nests usually lack a central chamber—the shrew simply burrows into the centre of the clump.

REPRODUCTION: Breeding probably occurs in late spring and summer, likely only once per year. Litter size probably numbers three to five young.

SIMILAR SPECIES: Most shrews look similar. Without a specimen and a technical key, it is almost impossible to identify a shrew reliably.

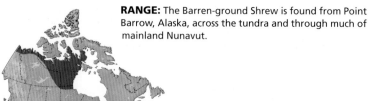

RANGE: The Barren-ground Shrew is found from Point Barrow, Alaska, across the tundra and through much of mainland Nunavut.

Total Length: 7.4–10.3 cm

Tail Length: about 3 cm

Weight: 3.0–5.2 g

Vagrant Shrew
Sorex vagrans

The Vagrant Shrew and the Dusky Shrew may be the most difficult mammals to distinguish from one another. Even experts have trouble telling whether the two tiny, medial tines on the upper incisors are located near the upper limit of the dark tooth pigment (Vagrant Shrew) or within the pigmented part of the incisor (Dusky Shrew). Naturally, live shrews would never submit to such scrutiny and so remain nearly impossible to tell apart.

Unbeknownst even to many naturalists, some shrews use a kind of echolocation. The Vagrant Shrew emits high-frequency vocalizations that it uses to orient itself.

ALSO CALLED: Wandering Shrew.

DESCRIPTION: This shrew is pale brown on the back and sides in summer. In winter, it is slightly darker over the back. The underparts vary from silvery grey to buffy brown. The tail is bicoloured, pale brown above and whitish below.

HABITAT: The Vagrant Shrew favours forested regions that have water nearby. It sometimes occurs in moister habitats such as the edges of mountain brooks with willows along the banks.

FOOD: This shrew eats a variety of adult and larval insects, earthworms, spiders, snails, slugs, carrion and even some vegetation.

DEN: The spherical, grassy nest is usually built in a decayed log. It lacks a central cavity, and the shrew simply burrows inside.

REPRODUCTION: Mating begins in March, and litters of two to nine young are born from early April to mid-August. Females likely have more than one litter per year. The young are helpless at birth and must feed heavily from their mother to complete their rapid growth. Their eyes and ears open in about two weeks, and they are weaned soon thereafter.

SIMILAR SPECIES: Most shrews look very similar. Without a specimen and a technical key, it is almost impossible to identify a shrew reliably.

RANGE: The Vagrant Shrew's range extends from southern British Columbia through Washington, Oregon, Idaho and western Montana to northern California, Nevada and northern Utah.

Total Length: 9–12 cm

Tail Length: 3.5–4.1 cm

Weight: 5–7 g

Northern Short-tailed Shrew
Blarina brevicauda

When a Northern Short-tailed Shrew digs its burrow, it digs rapidly with its front feet in a blur of motion. As the dirt accumulates under its body, the shrew kicks it out of the entranceway with its hindfeet. As it tunnels deeper and more dirt accumulates, the only way the shrew can get rid of the loose dirt is to turn a sideways somersault in the burrow and bulldoze the earth out to the ground surface using its forehead.

The Northern Short-tailed Shrew has a toxin in its saliva to neutralize prey. This toxin has been shown to destroy cancer cells, and researchers are currently working on possible cancer treatments made from this shrew's spit.

DESCRIPTION: This shrew is one of the largest in North America. Its robust body is uniformly lead grey in colour. It has a noticeably short tail that is similarly coloured.

HABITAT: The Northern Short-tailed Shrew inhabits many regions including woodlands, fields and marshy areas. In hot climates, it remains in moist areas.

FOOD: This voracious and even vicious shrew eats not only the standard shrew fare of insects and soft invertebrates but is known to attack young rabbits and once even a 38-cm-long gartersnake!

DEN: A bulky leaf or grass nest is built in a rock crevice or hollow log. This shrew also digs burrows and runways, which it patrols in search of food.

REPRODUCTION: Mating may occur throughout the year, but the peak is in spring. Gestation varies from 17 to 21 days, and litter size is four to eight young. The young are about the size of honeybees at birth and are nearly mature at 50 days old.

SIMILAR SPECIES: This species is distinctive because of its large size and short tail.

RANGE: This shrew is found in central and southeastern Canada from Saskatchewan to the Atlantic Coast and south to Nebraska, Kentucky, Alabama, the upper mid-Atlantic states and the Appalachian Mountains.

Total Length: 9.6–14.0 cm

Tail Length: 2–3 cm

Weight: 14–29 g

Least Shrew
Cryptotis parva

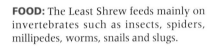

The Least Shrew is the only member of the genus *Cryptotis*, and it has an extremely short tail when compared to most other shrews. New to most people in Canada, the Least Shrew is only known from a couple of records in the southern part of Ontario. This shrew is much better known in the eastern U.S., where it can be found in a wide variety of habitats. Unfortunately, it is not well studied, and information about it is limited. The information below has been taken from populations south of Ontario.

DESCRIPTION: The Least Shrew is very small, with a brownish grey back that grades to lighter colours on the sides and underparts. In summer, this shrew is more brown; in winter, it is more grey. It has a short tail that is less than half the head and body length. If you raise the upper lip on the side of the snout, three unicuspid teeth are visible. A fourth tooth is present as well, but it is hidden behind the third.

HABITAT: This tiny shrew seems to prefer open deciduous forests, wooded ravines and some grassy or wet areas.

FOOD: The Least Shrew feeds mainly on invertebrates such as insects, spiders, millipedes, worms, snails and slugs.

DEN: The den is often found in soft soil, among rocks or under woody debris. The nest chamber is exceedingly small, and the entrance to the burrow is small and indistinct. There are some records of several Least Shrews building and sharing the same burrow.

REPRODUCTION: Little is known about this shrew's reproduction, but it is probably similar to that of other shrews. Mating occurs from March through November, with females having multiple litters per year.

SIMILAR SPECIES: Most shrews look very similar. Without a specimen and a technical key, it is almost impossible to identify a shrew reliably.

RANGE: Least Shrews are found in most of the eastern U.S., and records exist from Long Point in Ontario.

Total Length: 6.9–8.9 cm

Tail Length: 1.9–2.2 cm

Weight: 4.0–6.5 g

Virginia Opossum
Didelphis virginiana

Among the mammals of North America, the Virginia Opossum is unique because of its prehensile tail, maternal pouch, opposable "thumbs" and habit of feigning death. Made famous by its portrayals in children's literature, the opossum is widely known but poorly understood. Few people realize that this animal is a marsupial, and that it is more closely related to the kangaroos and koalas of Australia than to any other mammal native to Canada or the U.S.

Thanks to the many children's stories, we conjure up images of opossums hanging in trees by their tails, but this behaviour is not nearly as common as the literature suggests. An opossum's tail is prehensile and strong, but it is unlikely to be used in such a manner unless the animal has slipped off a branch or is reaching for something.

The phrase "playing possum" is derived from the feigned death scene that is put on by a frightened opossum. If an opossum cannot scare away an intruder with fervent hissing and screeching, it will roll over, dangle its legs, close its eyes, loll its tongue and drool. Presumably, this death pose is so startling that the predator will leave the opossum alone.

If you do much driving through opossum country,

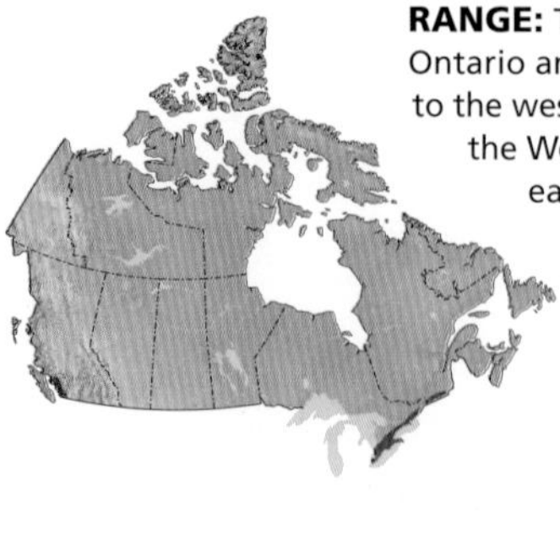

RANGE: The Virginia Opossum is found in southern Ontario and most of the eastern U.S. It was introduced to the western U.S., and its expanding range now includes the West Coast as far north as British Columbia and eastward along the Snake River into Idaho.

Total Length: 69–84 cm

Tail Length: 30–36 cm

Weight: 1.1–1.6 kg

it should not be long before you encounter one. Unfortunately, opossums are frequent victims of roadway collisions. They are slow-moving animals that forage at night and find the bounty of road-killed insects and other animals hard to resist.

With an abundance of food, opossums may become very fat. They draw upon their fat reserves in winter in colder parts of their range, but most of Canada is too cold for these creatures, with their naked ears and tails.

DESCRIPTION: The opossum is a cat-sized, grey mammal with a white face, long, pointed nose and long tail. Its ears are black, slightly rounded and nearly hairless. The tail is rounded, scaly and prehensile. The legs, the base of the tail and patches around the eyes are black. The mix of white, black and grey hairs gives the animal an overall grizzled appearance.

HABITAT: The Virginia Opossum favours moist woodlands or brushy areas near watercourses, but given a warm enough climate, it may be found almost anywhere, even in cities. Alternating warm and cold winters can cause an expansion or reduction in the northern extremes of its range.

FOOD: A full description of this animal's diet would include almost any organic matter. These omnivores eat invertebrates, insects, small mammals and birds,

> **DID YOU KNOW?**
> About the size of a honeybee at birth, an opossum begins life as one of the smallest baby mammals in North America.

grains, berries and other fruits, grasses and carrion.

DEN: By day, an opossum will hide in a burrow dug by another mammal, in a hollow tree or log, under a building or in a rock pile. In colder parts of its range, it may remain holed up in a den for days during cold weather but does not hibernate.

REPRODUCTION: Up to 25 young may be born in a litter after a gestation of 12 to 13 days. The young must crawl into the female's pouch and attach to one of the nipples (usually 13, but the number may vary); only the young attached to a nipple can survive. After about three months in the pouch, an average of eight to nine young emerge, weighing about 160 g each. Females are sexually mature at six months to one year of age.

SIMILAR SPECIES: No other mammal shares the combination of characteristics seen in the opossum. The young, newly emerged from the pouch, might be mistaken for rats, but rats do not have naked, black ears.

foreprint

walking trail

hindprint

walking trail (in snow)

Glossary

aestivation: dormancy during a hot, dry period; also known as "summer hibernation"

altricial: born in an undeveloped state, with eyes closed, unable to walk and requiring parental care

amphipod: a shrimplike crustacean of the order Amphipoda that has a laterally compressed body

anterior: near the head or the front of the body

antitragus: a lobe near the base of the outer margin of the external ear

arboreal: living in or pertaining to trees

artiodactyl: an even-toed, hoofed mammal belonging to the order Artiodactyla

auditory bulla (*pl.* bullae): in mammals, a hollow, bony structure at the back of the skull that encases the inner and middle ear

baleen: strands of keratin that hang in sheets from the upper jaws of whales in the suborder Mysticeta and are used to filter food from water

benthic: dwelling at or near the bottom of a lake or ocean

bifurcated: divided into two branches

bow riding: a behaviour seen in dolphins in which the animals swim in the bow waves of boats

breach: a whale display in which the animal rises vertically into the air, clearing the water's surface with almost all of its body before splashing back in

buff: a dull, brownish yellow

cache: a place in which food is hidden for future use; food hidden in such a place

calcar: in bats, a small projection from the inner side of each hind foot into the membrane between the hind legs

callosity: a roughened patch of skin found on the head of a right whale

cambium: a layer of tissue under the bark of a tree from which new growth develops

canid: a member of the dog family (Canidae)

carnivorous: flesh eating

Carolinian forest: the forest zone extending from southwestern Ontario to the Carolinas, dominated by deciduous or broad-leaf trees such as ash, chestnut, hickory and walnut

cecum: a pouch in the large intestine containing bacteria that digest cellulose

cervid: a hoofed mammal of the deer family (Cervidae)

cetacean: a marine mammal of the order Cetacea, which includes whales, dolphins and porpoises

colony: a group of animals living together and interacting socially

coniferous: pertaining to needle-leaved, cone-bearing trees (e.g., fir, spruce, pine)

coprophagy: the consumption of feces

corm: an underground bulb or stem

corvid: a member of the family Corvidae, which includes ravens, crows and jays

crepuscular: active at dusk or dawn

deciduous: pertaining to trees that shed their leaves in autumn (e.g., oak, maple, elm)

delphinid: a member of the ocean dolphin family (Delphinidae)

dormancy: a state of inactivity, with greatly slowed metabolism, respiration and heart rate

dorsal: pertaining to the back or spine (compare *ventral*)

dorsum: the back or dorsal region of the body

drey: a spherical tree nest made of leaves, twigs and moss

tree squirrel drey

echolocation: the ability of some animals (including bats and cetaceans among mammals) to detect an object by emitting sound waves and interpreting the returning echoes, which are changed from bouncing off the object

encephalization quotient (EQ): the ratio of an animal's brain weight to its body size; used as a measurement of the possible intelligence of an animal

endangered: facing imminent extirpation or extinction

extinct: no longer in existence anywhere

extirpated: no longer found in a given geographic area but still in existence elsewhere in the world

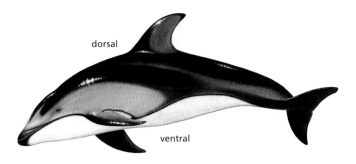

dorsal

ventral

family: a biological classification that ranks below order and designates a group of closely related genera

fecundity: the ability to produce an abundance of offspring

fluke: *n.* either of the lobes on a whale's tail; *v.* a whale behaviour in which the animal raises its tail above the water before diving

forb: a nonwoody, flowering plant other than a grass

form: a shallow depression dug by a rabbit or hare in which to rest and hide

fusiform: tapered at both ends; spindle-shaped

gastropod: an invertebrate from the Gastropoda class of molluscs, which typically have a spiralled shell, an unsegmented body and a single, muscular "foot" (e.g., snails)

genus (*pl.* genera): a group of closely related species

gestation: the time of pregnancy, from conception to birth

gregarious: preferring to living in large groups with other individuals of the same species; sociable

grizzled: having mostly dark fur that is sprinkled or streaked with grey or another light colour

guard hairs: long, coarse hairs that help protect a mammal's underfur from the weather

habitat: the environment in which an animal or plant lives

herbaceous: pertaining to plants that lack woody stems

herbivorous: plant eating

hibernaculum: the den in which an animal hibernates

hibernation: winter dormancy

hierarchy: a social order; the ranking of individuals by social status

Holarctic: a biogeographical region that circles the northern region of the globe, stretching from the North Pole to 30° to 45° N latitude

home range: the total area through which an individual animal moves during its usual activities (compare *territory*)

interbreed: for individuals of different species to mate with each other

interfemoral membrane: the fold of skin that stretches between a bat's hind legs and tail

keratin: a protein that is the prime constituent of hair and fingernails in humans, and hooves and horns in other mammals

keystone species: a species that many other organisms in an ecosystem depend on

lagomorph: a mammal from the order Lagomorpha, whose members are distinguished by double incisor teeth (e.g., rabbits, hares, pikas)

lanugo: the fine, soft hair that covers the body of newborn mammals

leveret: a young hare

lingual: the side of the tooth toward the tongue; relating to the tongue

lobtail: a display in which a whale forcefully slaps its tail flukes on the water's surface

mammae: mammary glands

marsupial: a group of mammals that bears live young that are born premature and develop in a pouch (e.g., opossums, kangaroos, wombats)

microtine: a rodent of the subfamily Microtinae (e.g., voles and lemmings)

midden: a storage pile of conifer cones and seeds or a refuse pile of seed husks and cone debris on the ground

Red Squirrel defending
its midden

migration: the journey that an animal undertakes to get from one region to another, usually in response to seasonal and reproductive cycles

mustelid: a member of the weasel family (Mustelidae)

nocturnal: active at night

order: a biological classification that designates a group of closely related families of organisms

omnivorous: feeding on both plant and animal material

palmate: branching like the fingers of a human hand

patagium: a thin membrane that extends between the limbs and body of a bat or gliding mammal

pelage: the fur or hair of a mammal

pinna (*pl.* pinnae): the outer ear

pinniped: a carnivorous, aquatic mammal with a streamlined body specialized for swimming and limbs modified into flippers (e.g., seals, sea lions, walruses)

plantigrade: walking on the soles of the feet

porpoise: *v.* a method of moving through the water that involves alternately rising above the water's surface and then diving underwater

posterior: at or toward the rear of the body

predator: an animal that kills its prey (compare *scavenger*)

precocial: young born in a well-developed state and capable of moving and feeding independently almost right away

pronking: the bouncing gait of a deer or antelope in which the animal bounds and lands with all four legs simultaneously; also called *stotting*

rookery: a colony of breeding animals

rostrum: the snout of a whale or dolphin

royal tine: on a Wapiti, the fourth tine of the antlers when numbered from the lowest ("brow") tine

runway: a beaten path made by the repeated travels of small animals

scat: fecal pellets or droppings; feces

scavenger: an animal that feeds on animals it did not kill (compare *predator*)

seral: a transitional stage of a plant community

species: a biological classification below genus that designates closely related organisms that are able to breed and produce viable offspring

spyhop: a whale behaviour in which the animal, while in an almost vertical position, raises its head out of the water just far enough to look around

stotting: the bouncing gait of a deer or antelope in which the animal bounds and lands with all four legs simultaneously; also called *pronking*

subequal: nearly equal in length

subnivean: under the snow but above the ground

subspecies: a subcategory of species that designates a geographic population that is genetically distinct from other populations of that species, but is still able to successfully breed with them

subterranean: underground

sympatric: having identical or overlapping ranges

territory: a defended area within an animal's home range

torpor: a state of dormancy in which an animal's metabolism and heart rate slow

tragus: a lobe projecting upward from inside the base of the ears, as in bats

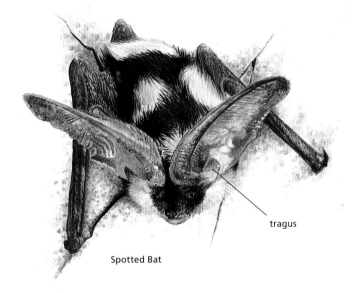

tragus

Spotted Bat

treeline: the limit of normal tree growth as determined by elevation, temperature or other environmental conditions

tubercle: a small, rounded nodule on an animal's skin

underfur: a thick, insulating undercoat of fur

ungulate: a hoofed mammal

unicuspid: in shrews, any of the small teeth between the two front teeth and the large rear teeth

uropatagium: the fold of skin that stretches from a bat's hind legs to its tail

venter: the abdominal area

ventral: pertaining to the belly (compare *dorsal*)

vibrissae: sensitive hairs or bristles ("whiskers") found on the face of a mammal and used in the sense of touch

villus (*pl.* villi): hairlike projections in the small intestine that aid in nutrient absorption

warren: an extensive burrow system created by a colony of rabbits

Selected References

Books

Chapman, Joseph A., Bruce C. Thompson, and George A. Feldhamer, eds. 2003. *Wild Mammals of North America: Biology, Management, and Conservation*. The John Hopkins University Press, Baltimore.

Eder, Tamara. 2009. *Squirrels of North America*. Lone Pine Publishing, Edmonton.

———. 2002. *Mammals of Ontario*. Lone Pine Publishing, Edmonton.

———. 2001. *Whales and other Marine Mammals of British Columbia and Alaska*. Lone Pine Publishing, Edmonton.

Eder, Tamara, and Don Pattie. 2001. *Mammals of British Columbia*. Lone Pine Publishing, Edmonton.

Fisher, Chris, and Don Pattie. 1999. *Mammals of Alberta*. Lone Pine Publishing, Edmonton.

Fisher, Chris, Don Pattie, and Tamara Hartson. 2000. *Mammals of the Rocky Mountains*. Lone Pine Publishing, Edmonton.

Kays, Roland W., and Don E. Wilson. 2009. *Mammals of North America*. Princeton University Press, Princeton, NJ.

Nagorsen, David W. 1995. *The Bats of British Columbia*. Royal British Columbia Museum and UBC Press, Vancouver.

Shackleton, David. 1999. *Hoofed Mammals of British Columbia*. Royal British Columbia Museum and UBC Press, Vancouver.

Wilson, Don E., and Sue Ruff. 1999. *The Smithsonian Book of North American Mammals*. Smithsonian Institution Press, Washington, DC.

Websites

The Canadian Atlas Online. Canadian Geographic. www.canadiangeographic.ca/atlas/intro.aspx?lang=En.

Committee on the Status of Endangered Wildlife in Canada (COSEWIC). Government of Canada. www.cosewic.gc.ca/.

North American Mammals. Smithsonian National Museum of Natural History. www.mnh.si.edu/mna/main.cfm?lang=_en.

An Online Encyclopedia of Life. NatureServe Explorer. www.natureserve.org/explorer/.

Red List of Threatened Species. International Union for Conservation of Nature (IUCN). www.iucnredlist.org/.

Index

Entries in **boldface** type refer to the primary species accounts.

Illustration
& Photography Credits

Illustrations

All prints, trails and marine mammal illustrations (pp. 90–131) are by Ian Sheldon.

All other illustrations are by Gary Ross with the exception of the following: Gary Ross/Tamara Eder, 233, 285, 287, 292, 294, 366, 368, 412, 414, 417, 423, 425, 426, 428; George Penetrante/Tamara Eder, 192–195, 198, 300–304; Kindrie Grove, 224, 269, 277, 282, 284, 399, 411.

Photographs

All photographs are copyright of the photographers and organizations listed, and all reasonable efforts have been made to obtain permission from respective copyright holders for the images used in this book. Please contact the publisher if errors have been made.

AFSC/NOAA/US Gov, 122–123; Anjela Buch/Dreamstime.com, 111; Ari S. Friedlaender, 93; Bob Balestri/iStockphoto/Thinkstock, 209; Comstock Images/Thinkstock, 21; Cory Bialecki, 27, 51, 69, 77; David B. Fleetham/Visuals Unlimited, 112–113; David Cloud/Hemera/Thinkstock, 40–41; Dohnal/Dreamstime.com, 276; Eric Stoops/Corel Corporation, 98–99, 104–105; Eyewire, 212; Getty Images, 65, 139, 140, 143; Getty Images/iStockphoto, 4; Hemera Technologies/Photos.com, 19, 21, 70–71, 73, 78, 130–131, 136, 168, 307; Hemera Technologies/Photos.com/Thinkstock, 48–49, 250–251; iStockphoto/Thinkstock, 23, 28, 52, 291, 398; J.P.A. Zoetekouw/Hemera Technologies/Thinkstock, 93; Jack Stephens, 197; John Marriott/Jem Photography/Banff Lake Louise Tourism, 218; Jupiterimages/Photos.com, 205; Jupiterimages/Photos.com/Thinkstock, 88, 171, 222; Ken Balcomb, 94–95; Kevin Walsh, 193; Leslie Degner, 154; Liquidlibrary/Photos.com, 195, 230; Mark Newman/West Stock, 206; Michael "Mike" L. Baird, 180–181; Michael Rolands/iStockphoto/Thinkstock, 241; NOAA/FLFWC, 118–119; Outdoorsman/Dreamstime.com, 226–227; Photos.com, 22, 23, 29, 60, 74, 135, 167, 176–177, 217, 245, 246, 249; Photos.com/Stockbyte, 144, 183, 221; Renee DeMartin/West Stock, 126–127, 188–189; Seattle Support Group/Photos.com, 44; Tamara Eder, 33, 179, 201, 202–203, 211, 237; Terry Parker, 66, 82, 108–109; Tom Brakefield/Photos.com, 1, 20, 24, 43, 81, 87, 175, 184, 229, 234, 257, 258, 306; Tom Tietz/iStockphoto/Thinkstock, 56, 172, 238; Wayne Lynch, 242.

ABOUT THE AUTHORS

Tamara Eder, equipped from the age of six with a canoe, a dip net and a notepad, grew up in Alberta with a fascination for nature and the diversity of life. She has a degree in environmental conservation sciences, and has photographed and written about wildlife in Bermuda, the Galapagos Islands, the Amazon Basin, Argentina, Tibet and India. With a fondness for paleontology, Tamara has studied and participated in paleo digs in Alberta and Patagonia. She has worked in both Canada and Argentina as an interpretive naturalist and guide, specializing in ecology and paleontology. An award-winning photographer, her photographs appear in numerous books, posters and online magazines. Tamara now lives in Patagonia, and she continues to write, photograph and travel.

Gregory Kennedy has been an active naturalist since he was very young. He is the author of many books on natural history, and has also produced film and television shows on environmental issues and indigenous concerns in Southeast Asia, New Guinea, South and Central America, the High Arctic and elsewhere. He has also been involved in numerous research projects around the world ranging from studies in the upper canopy of tropical and temperate rainforests to deepwater marine investigations.